Graduate Texts in Mathematics 3

Managing Editor: P. R. Halmos

Helmut H. Schaefer

Topological Vector Spaces

Third Printing Corrected

Springer-Verlag New York Heidelberg Berlin

Helmut H. Schaefer

Professor of Mathematics, University of Tübingen

AMS Subject Classifications (1970)

Primary   46-02, 46 A 05, 46 A 20, 46 A 25, 46 A 30, 46 A 40, 47 B 55
Secondary 46 F 05, 81 A 17

Third Printing Corrected 1971

ISBN 0-387-05380-8 Springer-Verlag New York Heidelberg Berlin (soft cover)
ISBN 0-387-90026-8 Springer-Verlag New York Heidelberg Berlin (hard cover)
ISBN 3-540-05380-8 Springer-Verlag Berlin Heidelberg New York (soft cover)

To my Wife

This book initially appeared in 1966. Minor errors and misprints have been corrected for this third printing. The author wishes to express his appreciation to Springer-Verlag for including this volume in the series, Graduate Texts in Mathematics.

Tübingen, December 1970   H. H. Schaefer

Preface

The present book is intended to be a systematic text on topological vector spaces and presupposes familiarity with the elements of general topology and linear algebra. The author has found it unnecessary to rederive these results, since they are equally basic for many other areas of mathematics, and every beginning graduate student is likely to have made their acquaintance. Similarly, the elementary facts on Hilbert and Banach spaces are widely known and are not discussed in detail in this book, which is mainly addressed to those readers who have attained and wish to get beyond the introductory level.

The book has its origin in courses given by the author at Washington State University, the University of Michigan, and the University of Tübingen in the years 1958–1963. At that time there existed no reasonably complete text on topological vector spaces in English, and there seemed to be a genuine need for a book on this subject. This situation changed in 1963 with the appearance of the book by Kelley, Namioka *et al.* [1] which, through its many elegant proofs, has had some influence on the final draft of this manuscript. Yet the two books appear to be sufficiently different in spirit and subject matter to justify the publication of this manuscript; in particular, the present book includes a discussion of topological tensor products, nuclear spaces, ordered topological vector spaces, and an appendix on positive operators. The author is also glad to acknowledge the strong influence of Bourbaki, whose monograph [7], [8] was (before the publication of Köthe [5]) the only modern treatment of topological vector spaces in printed form.

A few words should be said about the organization of the book. There is a preliminary chapter called "Prerequisites," which is a survey aimed at clarifying the terminology to be used and at recalling basic definitions and facts to the reader's mind. Each of the five following chapters, as well as the Appendix, is divided into sections. In each section, propositions are marked u.v, where u is the section number, v the proposition number within the section. Propositions of special importance are additionally marked "Theorem." Cross references within the chapter are (u.v), outside the chapter (r, u.v), where r (roman numeral) is the number of the chapter referred to. Each chapter is preceded by an introduction and followed by exercises. These "Exercises" (a total of 142) are devoted to further results and supplements, in particular, to examples and counter-examples. They are not meant to be worked out one after the other, but every reader should take notice of them because of their informative value. We have refrained from marking some of them as difficult, because the difficulty of a given problem is a highly subjective matter. However, hints have been given where it seemed appropriate, and occasional references indicate literature that may be needed, or at least helpful. The bibliography, far from being complete, contains (with few exceptions) only those items that are referred to in the text.

I wish to thank A. Pietsch for reading the entire manuscript, and A. L. Peressini and B. J. Walsh for reading parts of it. My special thanks are extended to H. Lotz for a close examination of the entire manuscript, and for many valuable discussions. Finally, I am indebted to H. Lotz and A. L. Peressini for reading the proofs, and to the publisher for their care and cooperation.

H. H. S.

Tübingen, Germany
December, 1964

Table of Contents

III. LINEAR MAPPINGS

IV. DUALITY

V. ORDER STRUCTURES

Appendix. SPECTRAL PROPERTIES OF POSITIVE OPERATORS

PREREQUISITES

A formal prerequisite for an intelligent reading of this book is familiarity with the most basic facts of set theory, general topology, and linear algebra. The purpose of this preliminary section is not to establish these results but to clarify terminology and notation, and to give the reader a survey of the material that will be assumed as known in the sequel. In addition, some of the literature is pointed out where adequate information and further references can be found.

Throughout the book, statements intended to represent definitions are distinguished by setting the term being defined in bold face characters.

A. SETS AND ORDER

1. *Sets and Subsets.* Let X, Y be sets. We use the standard notations $x \in X$ for "x is an element of X", $X \subset Y$ (or $Y \supset X$) for "X is a subset of Y", $X = Y$ for "$X \subset Y$ and $Y \supset X$". If (p) is a proposition in terms of given relations on X, the subset of all $x \in X$ for which (p) is true is denoted by $\{x \in X: (p)x\}$ or, if no confusion is likely to occur, by $\{x: (p)x\}$. $x \notin X$ means "x is not an element of X". The **complement** of X relative to Y is the set $\{x \in Y: x \notin X\}$, and denoted by $Y \sim X$. The empty set is denoted by $\emptyset$ and considered to be a finite set; the set (**singleton**) containing the single element x is denoted by $\{x\}$. If (p_1), (p_2) are propositions in terms of given relations on X, $(p_1) \Rightarrow (p_2)$ means "(p_1) implies (p_2)", and $(p_1) \Leftrightarrow (p_2)$ means "(p_1) is equivalent with (p_2)". The set of all subsets of X is denoted by $\mathfrak{P}(X)$.

2. *Mappings.* A mapping f of X into Y is denoted by $f: X \to Y$ or by $x \to f(x)$. X is called the **domain** of f, the image of X under f, the **range** of f; the **graph** of f is the subset $G_f = \{(x, f(x)): x \in X\}$ of $X \times Y$. The mapping of the set $\mathfrak{P}(X)$ of all subsets of X into $\mathfrak{P}(Y)$ that is associated with f, is also denoted by f; that is, for any $A \subset X$ we write $f(A)$ to denote the set

$\{f(x) : x \in A\} \subset Y$. The associated map of $\mathfrak{P}(Y)$ into $\mathfrak{P}(X)$ is denoted by f^{-1}; thus for any $B \subset Y$, $f^{-1}(B) = \{x \in X : f(x) \in B\}$. If $B = \{b\}$, we write $f^{-1}(b)$ in place of the clumsier (but more precise) notation $f^{-1}(\{b\})$. If $f: X \to Y$ and $g: Y \to Z$ are maps, the composition map $x \to g(f(x))$ is denoted by $g \circ f$.

A map $f: X \to Y$ is **biunivocal (one–to–one, injective)** if $f(x_1) = f(x_2)$ implies $x_1 = x_2$; it is **onto** Y **(surjective)** if $f(X) = Y$. A map f which is both injective and surjective is called **bijective** (or a **bijection**).

If $f: X \to Y$ is a map and $A \subset X$, the map $g : A \to Y$ defined by $g(x) = f(x)$ whenever $x \in A$ is called the **restriction** of f to A and frequently denoted by f_A. Conversely, f is called an **extension** of g (to X with values in Y).

3. *Families.* If A is a *non-empty* set and X is a set, a mapping $\alpha \to x(\alpha)$ of A into X is also called a **family** in X; in practice, the term family is used for mappings whose domain A enters only in terms of its set theoretic properties (i.e., cardinality and possibly order). One writes, in this case, x_α for $x(\alpha)$ and denotes the family by $\{x_\alpha : \alpha \in A\}$. Thus every non-empty set X can be viewed as the family (identity map) $x \to x(x \in X)$; but it is important to notice that if $\{x_\alpha : \alpha \in A\}$ is a family in X, then $\alpha \neq \beta$ does not imply $x_\alpha \neq x_\beta$. A **sequence** is a family $\{x_n : n \in N\}$, $N = \{1, 2, 3, \ldots\}$ denoting the set of natural numbers. If confusion with singletons is unlikely and the domain (index set) A is clear from the context, a family will sometimes be denoted by $\{x_\alpha\}$ (in particular, a sequence by $\{x_n\}$).

4. *Set Operations.* Let $\{X_\alpha : \alpha \in A\}$ be a family of sets. For the union of this family, we use the notations $\bigcup \{X_\alpha : \alpha \in A\}$, $\bigcup_{\alpha \in A} X_\alpha$, or briefly $\bigcup_\alpha X_\alpha$ if the index set A is clear from the context. If $\{X_n : n \in N\}$ is a sequence of sets we also write $\bigcup_1^\infty X_n$, and if $\{X_1, \ldots, X_k\}$ is a finite family of sets we write $\bigcup_1^k X_n$ or $X_1 \cup X_2 \cup \ldots \cup X_k$. Similar notations are used for intersections and Cartesian products, with $\bigcup$ replaced by $\bigcap$ and $\prod$ respectively. If $\{X_\alpha : \alpha \in A\}$ is a family such that $X_\alpha = X$ for all $\alpha \in A$, the product $\prod_\alpha X_\alpha$ is also denoted by X^A.

If R is an equivalence relation (i.e., a reflexive, symmetric, transitive binary relation) on the set X, the set of equivalence classes (the **quotient set**) by R is denoted by X/R. The map $x \to \hat{x}$ (also denoted by $x \to [x]$) which orders to each x its equivalence class $\hat{x}$ (or $[x]$), is called the **canonical** (or **quotient**) map of X onto X/R.

5. *Orderings.* An **ordering (order structure, order)** on a set X is a binary relation R, usually denoted by $\leqq$, on X which is reflexive, transitive, and anti-symmetric ($x \leqq y$ and $y \leqq x$ imply $x = y$). The set X endowed with an order $\leqq$ is called an **ordered** set. We write $y \geqq x$ to mean $x \leqq y$, and $x < y$ to mean $x \leqq y$ but $x \neq y$ (similarly for $x > y$). If R_1 and R_2 are orderings of X, we say that R_1 is **finer** than R_2 (or that R_2 is **coarser** than R_1) if $x(R_1)y$ implies $x(R_2)y$. (Note that this defines an ordering on the set of all orderings of X.)

Let $(X, \leqq)$ be an ordered set. A subset A of X is **majorized** if there exists $a_0 \in X$ such that $a \leqq a_0$ whenever $a \in A$; a_0 is a **majorant (upper bound)** of A. Dually, A is **minorized** by a_0 if $a_0 \leqq a$ whenever $a \in A$; then a_0 is a **minorant (lower bound)** of A. A subset A which is both majorized and minorized, is called **order bounded**. If A is majorized and there exists a majorant a_0 such that $a_0 \leqq b$ for any majorant b of A, then a_0 is unique and called the **supremum (least upper bound)** of A; the notation is $a_0 = \sup A$. In a dual fashion, one defines the **infimum (greatest lower bound)** of A, to be denoted by $\inf A$. For each pair $(x, y) \in X \times X$, the supremum and infimum of the set $\{x, y\}$ (whenever they exist) are denoted by $\sup(x, y)$ and $\inf(x, y)$ respectively. $(X, \leqq)$ is called a **lattice** if for each pair (x, y), $\sup(x, y)$ and $\inf(x, y)$ exist, and $(X, \leqq)$ is called a **complete lattice** if $\sup A$ and $\inf A$ exist for every non-empty subset $A \subset X$. (In general we avoid this latter terminology because of the possible confusion with uniform completeness.) $(X, \leqq)$ is **totally ordered** if for each pair (x, y), at least one of the relations $x \leqq y$ and $y \leqq x$ is true. An element $x \in X$ is **maximal** if $x \leqq y$ implies $x = y$.

Let $(X, \leqq)$ be a *non-empty* ordered set. X is called **directed** under $\leqq$ (briefly, directed $(\leqq)$) if every subset $\{x, y\}$ (hence each finite subset) possesses an upper bound. If $x_0 \in X$, the subset $\{x \in X : x_0 \leqq x\}$ is called a **section** of X (more precisely, the section of X **generated** by x_0). A family $\{y_\alpha : \alpha \in A\}$ is **directed** if A is a directed set; the **sections** of a directed family are the subfamilies $\{y_\alpha : \alpha_0 \leqq \alpha\}$, for any $\alpha_0 \in A$.

Finally, an ordered set X is **inductively ordered** if each totally ordered subset possesses an upper bound. In each inductively ordered set, there exist maximal elements (Zorn's lemma). In most applications of Zorn's lemma, the set in question is a family of subsets of a set S, ordered by set theoretical inclusion $\subset$.

6. *Filters.* Let X be a set. A set $\mathfrak{F}$ of subsets of X is called a **filter** on X if it satisfies the following axioms:

(1) $\mathfrak{F} \neq \varnothing$ and $\varnothing \notin \mathfrak{F}$.
(2) $F \in \mathfrak{F}$ and $F \subset G \subset X$ implies $G \in \mathfrak{F}$.
(3) $F \in \mathfrak{F}$ and $G \in \mathfrak{F}$ implies $F \cap G \in \mathfrak{F}$.

A set $\mathfrak{B}$ of subsets of X is a **filter base** if (1') $\mathfrak{B} \neq \varnothing$ and $\varnothing \notin \mathfrak{B}$, and (2') if $B_1 \in \mathfrak{B}$ and $B_2 \in \mathfrak{B}$ there exists $B_3 \in \mathfrak{B}$ such that $B_3 \subset B_1 \cap B_2$. Every filter base $\mathfrak{B}$ generates a unique filter $\mathfrak{F}$ on X such that $F \in \mathfrak{F}$ if and only if $B \subset F$ for at least one $B \in \mathfrak{B}$; $\mathfrak{B}$ is called a **base** of the filter $\mathfrak{F}$. The set of all filters on a non-empty set X is inductively ordered by the relation $\mathfrak{F}_1 \subset \mathfrak{F}_2$ (set theoretic inclusion of $\mathfrak{P}(X)$); $\mathfrak{F}_1 \subset \mathfrak{F}_2$ is expressed by saying that $\mathfrak{F}_1$ is **coarser** than $\mathfrak{F}_2$, or that $\mathfrak{F}_2$ is **finer** than $\mathfrak{F}_1$. Every filter on X which is maximal with respect to this ordering, is called an **ultrafilter** on X; by Zorn's lemma, for each filter $\mathfrak{F}$ on X there exists an ultrafilter finer than $\mathfrak{F}$. If $\{x_\alpha : \alpha \in A\}$ is a directed family in X, the ranges of the sections of this family form a filter base on X; the corresponding filter is called the **section filter** of the family.

An **elementary** filter is the section filter of a sequence $\{x_n : n \in N\}$ in X (N being endowed with its usual order).

Literature. Sets: Bourbaki [1], Halmos [3]. Filters: Bourbaki [4], Bushaw [1]. Order: Birkhoff [1], Bourbaki [1].

B. GENERAL TOPOLOGY

1. *Topologies.* Let X be a set, $\mathfrak{G}$ a set of subsets of X invariant under finite intersections and arbitrary unions; it follows that $X \in \mathfrak{G}$, since X is the intersection of the empty subset of $\mathfrak{G}$, and that $\varnothing \in \mathfrak{G}$, since $\varnothing$ is the union of the empty subset of $\mathfrak{G}$. We say that $\mathfrak{G}$ defines a **topology** $\mathfrak{T}$ on X; structurized in this way, X is called a **topological space** and denoted by $(X, \mathfrak{T})$ if reference to $\mathfrak{T}$ is desirable. The sets $G \in \mathfrak{G}$ are called **open**, their complements $F = X \sim G$ are called **closed** (with respect to $\mathfrak{T}$). Given $A \subset X$, the open set $\mathring{A}$ (or int A) which is the union of all open subsets of A, is called the **interior** of A; the closed set $\bar{A}$, intersection of all closed sets containing A, is called the **closure** of A. An element $x \in \mathring{A}$ is called an **interior point** of A (or interior to A), an element $x \in \bar{A}$ is called a **contact point (adherent point)** of A. If A, B are subsets of X, B is **dense** relative to A if $A \subset \bar{B}$ (**dense** *in* A if $B \subset A$ and $A \subset \bar{B}$). A topological space X is **separable** if X contains a countable dense subset; X is **connected** if X is not the union of two disjoint non-empty open subsets (otherwise, X is **disconnected**).

Let X be a topological space. A subset $U \subset X$ is a **neighborhood** of x if $x \in \mathring{U}$, and a neighborhood of A if $x \in A$ implies $x \in \mathring{U}$. The set of all neighborhoods of x (respectively, of A) is a filter on X called the **neighborhood filter** of x (respectively, of A); each base of this filter is a **neighborhood base** of x (respectively, of A). A bijection f of X onto another topological space Y such that $f(A)$ is open in Y if and only if A is open in X, is called a **homeomorphism**; X and Y are **homeomorphic** if there exists a homeomorphism of X onto Y. The **discrete topology** on X is the topology for which every subset of X is open; the **trivial** topology on X is the topology whose only open sets are $\varnothing$ and X.

2. *Continuity and Convergence.* Let X, Y be topological spaces and let $f : X \to Y$. f is **continuous at** $x \in X$ if for each neighborhood V of $y = f(x)$, $f^{-1}(V)$ is a neighborhood of x (equivalently, if the filter on Y generated by the base $f(\mathfrak{U})$ is finer than $\mathfrak{B}$, where $\mathfrak{U}$ is the neighborhood filter of x, $\mathfrak{B}$ the neighborhood filter of y). f is **continuous** on X into Y (briefly, continuous) if f is continuous at each $x \in X$ (equivalently, if $f^{-1}(G)$ is open in X for each open $G \subset Y$). If Z is also a topological space and $f : X \to Y$ and $g : Y \to Z$ are continuous, then $g \circ f : X \to Z$ is continuous.

A filter $\mathfrak{F}$ on a topological space X is said to **converge** to $x \in X$ if $\mathfrak{F}$ is finer than the neighborhood filter of x. A sequence (more generally, a directed family) in X **converges** to $x \in X$ if its section filter converges to x. If also Y

is a topological space and $\mathfrak{F}$ is a filter (or merely a filter base) on X, and if $f: X \to Y$ is a map, then f is said to **converge** to $y \in Y$ *along* $\mathfrak{F}$ if the filter generated by $f(\mathfrak{F})$ converges to y. For example, f is continuous at $x \in X$ if and only if f converges to $y = f(x)$ along the neighborhood filter of x. Given a filter $\mathfrak{F}$ on X and $x \in X$, x is a **cluster point** (**contact point, adherent point**) of $\mathfrak{F}$ if $x \in \bar{F}$ for each $F \in \mathfrak{F}$. A **cluster point** of a sequence (more generally, of a directed family) is a cluster point of the section filter of this family.

3. *Comparison of Topologies.* If X is a set and $\mathfrak{T}_1, \mathfrak{T}_2$ are topologies on X, we say that $\mathfrak{T}_2$ is **finer** than $\mathfrak{T}_1$ (or $\mathfrak{T}_1$ **coarser** than $\mathfrak{T}_2$) if every $\mathfrak{T}_1$-open set is $\mathfrak{T}_2$-open (equivalently, if every $\mathfrak{T}_1$-closed set is $\mathfrak{T}_2$-closed). (If $\mathfrak{G}_1$ and $\mathfrak{G}_2$ are the respective families of open sets in X, this amounts to the relation $\mathfrak{G}_1 \subset \mathfrak{G}_2$ in $\mathfrak{P}(\mathfrak{P}(X))$.) Let $\{\mathfrak{T}_\alpha : \alpha \in A\}$ be a family of topologies on X. There exists a finest topology $\mathfrak{T}$ on X which is coarser than each $\mathfrak{T}_\alpha (\alpha \in A)$; a set G is $\mathfrak{T}$-open if and only if G is $\mathfrak{T}_\alpha$-open for each α. Dually, there exists a coarsest topology $\mathfrak{T}_0$ which is finer than each $\mathfrak{T}_\alpha (\alpha \in A)$. If we denote by $\mathfrak{G}'_0$ the set of all finite intersections of sets open for some $\mathfrak{T}_\alpha$, the set $\mathfrak{G}_0$ of all unions of sets in $\mathfrak{G}'_0$ constitutes the $\mathfrak{T}_0$-open sets in X. Hence with respect to the relation "$\mathfrak{T}_2$ is finer than $\mathfrak{T}_1$", the set of all topologies on X is a complete lattice; the coarsest topology on X is the trivial topology, the finest topology is the discrete topology. The topology $\mathfrak{T}$ is the greatest lower bound (briefly, *the* lower bound) of the family $\{\mathfrak{T}_\alpha : \alpha \in A\}$; similarly, $\mathfrak{T}_0$ is *the* upper bound of the family $\{\mathfrak{T}_\alpha : \alpha \in A\}$.

One derives from this two general methods of defining a topology (Bourbaki [4]). Let X be a set, $\{X_\alpha : \alpha \in A\}$ a family of topological spaces. If $\{f_\alpha : \alpha \in A\}$ is a family of mappings, respectively of X into X_α, the **projective topology** (**kernel topology**) on X with respect to the family $\{(X_\alpha, f_\alpha) : \alpha \in A\}$ is the coarsest topology for which each f_α is continuous. Dually, if $\{g_\alpha : \alpha \in A\}$ is a family of mappings, respectively of X_α into X, the **inductive topology** (**hull topology**) with respect to the family $\{(X_\alpha, g_\alpha) : \alpha \in A\}$ is the finest topology on X for which each g_α is continuous. (Note that each f_α is continuous for the discrete topology on X, and that each g_α is continuous for the trivial topology on X.) If $A = \{1\}$ and $\mathfrak{T}_1$ is the topology of X_1, the projective topology on X with respect to (X_1, f_1) is called the **inverse image** of $\mathfrak{T}_1$ under f_1, and the inductive topology with respect to (X_1, g_1) is called the **direct image** of $\mathfrak{T}_1$ under g_1.

4. *Subspaces, Products, Quotients.* If $(X, \mathfrak{T})$ is a topological space, A a subset of X, f the canonical imbedding $A \to X$, then the **induced** topology on A is the inverse image of $\mathfrak{T}$ under f. (The open subsets of this topology are the intersections with A of the open subsets of X.) Under the induced topology, A is called a **topological subspace** of X (in general, we shall avoid this terminology because of possible confusion with vector subspaces). If $(X, \mathfrak{T})$ is a topological space, R an equivalence relation on X, g the canonical map $X \to X/R$, then the direct image of $\mathfrak{T}$ under g is called the **quotient** (topology) of $\mathfrak{T}$; under this topology, X/R is the **topological quotient** of X by R.

Let $\{X_\alpha: \alpha \in A\}$ be a family of topological spaces, X their Cartesian product, f_α the projection of X onto X_α. The projective topology on X with respect to the family $\{(X_\alpha, f_\alpha): \alpha \in A\}$ is called the **product topology** on X. Under this topology, X is called the **topological product** (briefly, **product**) of the family $\{X_\alpha: \alpha \in A\}$.

Let X, Y be topological spaces, f a mapping of X into Y. We say that f is **open** (or an **open map**) if for each open set $G \subset X$, $f(G)$ is open in the topological subspace $f(X)$ of Y. f is called **closed** (a **closed map**) if the graph of f is a closed subset of the topological product $X \times Y$.

5. *Separation Axioms.* Let X be a topological space. X is a **Hausdorff** (or **separated**) space if for each pair of distinct points x, y there are respective neighborhoods U_x, U_y such that $U_x \cap U_y = \varnothing$. If (and only if) X is separated, each filter $\mathfrak{F}$ that converges in X, converges to exactly one $x \in X$; x is called the **limit** of $\mathfrak{F}$. X is called **regular** if it is separated and each point possesses a base of closed neighborhoods; X is called **normal** if it is separated and for each pair A, B of disjoint closed subsets of X, there exists a neighborhood U of A and a neighborhood V of B such that $U \cap V = \varnothing$.

A Hausdorff topological space X is normal if and only if for each pair A, B of disjoint closed subsets of X, there exists a continuous function f on X into the real interval $[0,1]$ (under its usual topology) such that $f(x) = 0$ whenever $x \in A, f(x) = 1$ whenever $x \in B$ (Urysohn's theorem).

A separated space X such that for each closed subset A and each $b \notin A$, there exists a continuous function $f: X \to [0,1]$ for which $f(b) = 1$ and $f(x) = 0$ whenever $x \in A$, is called **completely regular**; clearly, every normal space is completely regular, and every completely regular space is regular.

6. *Uniform Spaces.* Let X be a set. For arbitrary subsets W, V of $X \times X$, we write $W^{-1} = \{(y, x): (x, y) \in W\}$, and $V \circ W = \{(x, z): there \; exists \; y \in X$ *such that* $(x, y) \in W, (y, z) \in V\}$. The set $\Delta = \{(x, x): x \in X\}$ is called the **diagonal** of $X \times X$. Let $\mathfrak{W}$ be a filter on $X \times X$ satisfying these axioms:

(1) *Each* $W \in \mathfrak{W}$ *contains the diagonal* Δ.
(2) $W \in \mathfrak{W}$ *implies* $W^{-1} \in \mathfrak{W}$.
(3) *For each* $W \in \mathfrak{W}$, *there exists* $V \in \mathfrak{W}$ *such that* $V \circ V \subset W$.

We say that the filter $\mathfrak{W}$ (or any one of its bases) defines a **uniformity** (or **uniform structure**) on X, each $W \in \mathfrak{W}$ being called a **vicinity** (**entourage**) of the uniformity. Let $\mathfrak{G}$ be the family of all subsets G of X such that $x \in G$ implies the existence of $W \in \mathfrak{W}$ satisfying $\{y: (x, y) \in W\} \subset G$. Then $\mathfrak{G}$ is invariant under finite intersections and arbitrary unions, and hence defines a topology $\mathfrak{T}$ on X such that for each $x \in X$, the family $W(x) = \{y: (x, y) \in W\}$, where W runs through $\mathfrak{W}$, is a neighborhood base of x. The space $(X, \mathfrak{W})$, endowed with the topology $\mathfrak{T}$ derived from the uniformity $\mathfrak{W}$, is called a **uniform space**. A topological space X is **uniformisable** if its topology can be derived from a uniformity on X; the reader should be cautioned that, in general, such a uniformity is not unique.

A uniformity is **separated** if its vicinity filter satisfies the additional axiom
(4) $\cap \{W: W \in \mathfrak{W}\} = \Delta$.

(4) is a necessary and sufficient condition for the topology derived from the uniformity to be a Hausdorff topology. A Hausdorff topological space is uniformisable if and only if it is completely regular.

Let X, Y be uniform spaces. A mapping $f: X \to Y$ is **uniformly continuous** if for each vicinity V of Y, there exists a vicinity U of X such that $(x,y) \in U$ implies $(f(x), f(y)) \in V$. Each uniformly continuous map is continuous. The uniform spaces X, Y are **isomorphic** if there exists a bijection f of X onto Y such that both f and f^{-1} are uniformly continuous; f itself is called a **uniform isomorphism**.

If $\mathfrak{W}_1$ and $\mathfrak{W}_2$ are two filters on $X \times X$, each defining a uniformity on the set X, and if $\mathfrak{W}_1 \subset \mathfrak{W}_2$, we say that the uniformity defined by $\mathfrak{W}_1$ is **coarser** than that defined by $\mathfrak{W}_2$. If X is a set, $\{X_\alpha: \alpha \in A\}$ a family of uniform spaces and $f_\alpha(\alpha \in A)$ are mappings of X into X_α, then there exists a coarsest uniformity on X for which each $f_\alpha(\alpha \in A)$ is uniformly continuous. In this way, one defines the **product uniformity** on $X = \prod_\alpha X_\alpha$ to be the coarsest uniformity for which each of the projections $X \to X_\alpha$ is uniformly continuous; similarly, if X is a uniform space and $A \subset X$, the **induced uniformity** is the coarsest uniformity on A for which the canonical imbedding $A \to X$ is uniformly continuous.

Let X be a uniform space. A filter $\mathfrak{F}$ on X is a **Cauchy filter** if, for each vicinity V, there exists $F \in \mathfrak{F}$ such that $F \times F \subset V$. If each Cauchy filter converges (to an element of X) then X is called **complete**. To each uniform space X one can construct a complete uniform space $\tilde{X}$ such that X is (uniformly) isomorphic with a dense subspace of $\tilde{X}$, and such that $\tilde{X}$ is separated if X is. If X is separated, then $\tilde{X}$ is determined by these properties to within isomorphism, and is called the **completion** of X. A base of the vicinity filter of $\tilde{X}$ can be obtained by taking the closures (in the topological product $\tilde{X} \times \tilde{X}$) of a base of vicinities of X. A **Cauchy sequence** in X is a sequence whose section filter is a Cauchy filter; if every Cauchy sequence in X converges, then X is said to be **semi-complete (sequentially complete)**.

If X is a complete uniform space and A a closed subspace, then the uniform space A is complete; if X is a separated uniform space and A a complete subspace, then A is closed in X. A product of uniform spaces is complete if and only if each factor space is complete.

If X is a uniform space, Y a complete separated space, $X_0 \subset X$ and $f: X_0 \to Y$ uniformly continuous; then f has a unique uniformly continuous extension $\bar{f}: \bar{X}_0 \to Y$.

7. *Metric and Metrizable Spaces.* If X is a set, a non-negative real function d on $X \times X$ is called a **metric** if the following axioms are satisfied:

(1) $d(x, y) = 0$ *is equivalent with* $x = y$.
(2) $d(x, y) = d(y, x)$.
(3) $d(x, z) \leqq d(x, y) + d(y, z)$ (*triangle inequality*).

Clearly, the sets $W_n = \{(x, y): d(x, y) < n^{-1}\}$, where $n \in N$, form a filter base on $X \times X$ defining a separated uniformity on X; by the **metric space** (X, d) we understand the uniform space X endowed with the metric d. Thus all uniform concepts apply to metric spaces. (It should be understood that, historically, uniform spaces are the upshot of metric spaces.) A topological space is **metrizable** if its topology can be derived from a metric in the manner indicated; a uniform space is metrizable (i.e., its uniformity can be generated by a metric) if and only if it is separated and its vicinity filter has a countable base. Clearly, a metrizable uniform space is complete if it is semi-complete.

8. *Compact and Precompact Spaces.* Let X be a Hausdorff topological space. X is called **compact** if every open cover of X has a finite subcover. For X to be compact, each of the following conditions is necessary and sufficient: (a) A family of closed subsets of X has non-empty intersection whenever each finite subfamily has non-empty intersection. (b) Each filter on X has a cluster point. (c) Each ultrafilter on X converges.

Every closed subspace of a compact space is compact. The topological product of any family of compact spaces is compact (Tychonov's theorem). If X is compact, Y a Hausdorff space, and $f: X \to Y$ continuous, then $f(X)$ is a compact subspace of Y. If f is a continuous bijection of a compact space X onto a Hausdorff space Y, then f is a homeomorphism (equivalently: If $(X, \mathfrak{T}_1)$ is compact and $\mathfrak{T}_2$ is a Hausdorff topology on X coarser than $\mathfrak{T}_1$, then $\mathfrak{T}_1 = \mathfrak{T}_2$).

There is the following important relationship between compactness and uniformities: On every compact space X, there exists a unique uniformity generating the topology of X; the vicinity filter of this uniformity is the neighborhood filter of the diagonal Δ in the topological product $X \times X$. In particular, every compact space is a complete uniform space. A separated uniform space is called **precompact** if its completion is compact. (However, note that a topological space can be precompact for several distinct uniformities yielding its topology.) X is precompact if and only if for each vicinity W, there exists a finite subset $X_0 \subset X$ such that $X \subset \bigcup \{W(x): x \in X_0\}$. A subspace of a precompact space is precompact, and the product of any family of precompact spaces is precompact.

A Hausdorff topological space is called **locally compact** if each of its points possesses a compact neighborhood.

9. *Category and Baire Spaces.* Let X be a topological space, A a subset of X. A is called **nowhere dense (rare)** in X if its closure $\bar{A}$ has empty interior; A is called **meager** (of **first category**) in X if A is the union of a countable set of rare subsets of X. A subset A which is not meager is called **non-meager** (of **second category**) in X; if every non-empty open subset is nonmeager in X, then X is called a **Baire space**. Every locally compact space and every complete

metrizable space is a Baire space (Baire's theorem). Each non-meager subset of a topological space X is non-meager in itself, but a topological subspace of X can be a Baire space while being a rare subset of X.

Literature: Berge [1]; Bourbaki [4], [5], [6]; Kelley [1]. A highly recommendable introduction to topological and uniform spaces can be found in Bushaw [1].

C. LINEAR ALGEBRA

1. *Vector Spaces.* Let L be a set, K a (not necessarily commutative) field. Suppose there are defined a mapping $(x, y) \to x + y$ of $L \times L$ into L, called **addition**, and a mapping $(\lambda, x) \to \lambda x$ of $K \times L$ into L, called **scalar multiplication**, such that the following axioms are satisfied (x, y, z denoting arbitrary elements of L, and λ, μ arbitrary elements of K):

(1) $(x + y) + z = x + (y + z)$.
(2) $x + y = y + x$.
(3) *There exists an element* $0 \in L$ *such that* $x + 0 = x$ *for all* $x \in L$.
(4) *For each* $x \in L$, *there exists* $z \in L$ *such that* $x + z = 0$.
(5) $\lambda(x + y) = \lambda x + \lambda y$.
(6) $(\lambda + \mu)x = \lambda x + \mu x$.
(7) $\lambda(\mu x) = (\lambda \mu)x$.
(8) $1x = x$.

Endowed with the structure so defined, L is called a **left vector space** over K. The element 0 postulated by (3) is unique and called the **zero element** of L. (We shall not distinguish notationally between the zero elements of L and K.) Also, for any $x \in L$ the element z postulated by (4) is unique and denoted by $-x$; moreover, one has $-x = (-1)x$, and it is customary to write $x - y$ for $x + (-y)$.

If (1)–(4) hold as before but scalar multiplication is written $(\lambda, x) \to x\lambda$ and (5)–(8) are changed accordingly, L is called a **right vector space** over K. By a **vector space** over K, we shall always understand a left vector space over K. Since there is no point in distinguishing between left and right vector spaces over K when K is commutative, there will be no need to consider right vector spaces except in C.4 below, and Chapter I, Section 4. (From Chapter II on, K is always supposed to be the real field $\boldsymbol{R}$ or the complex field $\boldsymbol{C}$.)

2. *Linear Independence.* Let L be a vector space over K. An element $\lambda_1 x_1 + \cdots + \lambda_n x_n$, where $n \in N$, is called a **linear combination** of the elements $x_i \in L (i = 1, \ldots, n)$; as usual, this is written $\sum_{i=1}^{n} \lambda_i x_i$ or $\sum_i \lambda_i x_i$. If $\{x_\alpha : \alpha \in H\}$ is a finite family, the sum of the elements x_α is denoted by $\sum_{\alpha \in H} x_\alpha$; for reasons of convenience, this is extended to the empty set by defining $\sum_{x \in \varnothing} x = 0$. (This

should not be confused with the symbol $A + B$ for subsets A, B of L, which by A.2 has the meaning $\{x + y: x \in A, y \in B\}$; thus if $A = \varnothing$, then $A + B = \varnothing$ for all subsets $B \subset L$.) A subset $A \subset L$ is called **linearly independent** if for every non-empty finite subset $\{x_i: i = 1, ..., n\}$ of A, the relation $\sum_i \lambda_i x_i = 0$ implies $\lambda_i = 0$ for $i = 1, ..., n$. Note that by this definition, the empty subset of L is linearly independent. A linearly independent subset of L which is maximal (with respect to set inclusion) is called a **basis (Hamel basis)** of L. The existence of bases in L containing a given linearly independent subset is implied by Zorn's lemma; any two bases of L have the same cardinality d, which is called the **dimension** of L (over K).

3. *Subspaces and Quotients.* Let L be a vector space over K. A **vector subspace** (briefly, **subspace**) of L is a non-empty subset M of L invariant under addition and scalar multiplication, that is, such that $M + M \subset M$ and $KM \subset M$. The set of all subspaces of L is clearly invariant under arbitrary intersections. If A is a subset of L, the **linear hull** of A is the intersection M of all subspaces of L that contain A; M is also said to be the subspace of L **generated** by A. M can also be characterized as the set of all linear combinations of elements of A (including the sum over the empty subset of A). In particular, the linear hull of $\varnothing$ is $\{0\}$.

If M is a subspace of L, the relation "$x - y \in M$" is an equivalence relation in L. The quotient set becomes a vector space over K by the definitions $\hat{x} + \hat{y} = x + y + M$, $\lambda\hat{x} = \lambda x + M$ where $\hat{x} = x + M$, $\hat{y} = y + M$, and is denoted by L/M.

4. *Linear Mappings.* Let L_1, L_2 be vector spaces over K. $f: L_1 \rightarrow L_2$ is called a **linear map** if $f(\lambda_1 x_1 + \lambda_2 x_2) = \lambda_1 f(x_1) + \lambda_2 f(x_2)$ for all $\lambda_1, \lambda_2 \in K$ and $x_1, x_2 \in L_1$. Defining addition by $(f_1 + f_2)(x) = f_1(x) + f_2(x)$ and scalar multiplication by $(f\lambda)(x) = f(\lambda x)(x \in L_1)$, the set $L(L_1, L_2)$ of all linear maps of L_1 into L_2 generates a right vector space over K. (If K is commutative, the mapping $x \rightarrow f(\lambda x)$ will be denoted by λf and $L(L_1, L_2)$ considered to be a left vector space over K.) Defining $(f\lambda)(x) = f(x)\lambda$ if L_2 is the one-dimensional vector space K_0(over K) associated with K, we obtain the **algebraic dual** L_1^* of L_1. The elements of L_1^* are called **linear forms** on L_1.

L_1 and L_2 are said to be **isomorphic** if there exists a linear bijective map $f: L_1 \rightarrow L_2$; such a map is called an **isomorphism** of L_1 onto L_2. A linear injective map $f: L_1 \rightarrow L_2$ is called an **isomorphism** of L_1 into L_2.

If $f: L_1 \rightarrow L_2$ is linear, the subspace $N = f^{-1}(0)$ of L_1 is called the **null space (kernel)** of f. f defines an isomorphism f_0 of L_1/N onto $M = f(L_1)$; f_0 is called the bijective map **associated** with f. If ϕ denotes the quotient map $L_1 \rightarrow L_1/N$ and ψ denotes the canonical imbedding $M \rightarrow L_2$, then $f = \psi \circ f_0 \circ \phi$ is called the **canonical decomposition** of f.

5. *Vector Spaces over Valuated Fields.* Let K be a field, and consider the real field R under its usual absolute value. A function $\lambda \rightarrow |\lambda|$ of K into R_+ (real numbers ≥ 0) is called an **absolute value** on K if it satisfies the following axioms:

(1) $|\lambda| = 0$ *is equivalent with* $\lambda = 0$.
(2) $|\lambda + \mu| \leq |\lambda| + |\mu|$.
(3) $|\lambda\mu| = |\lambda||\mu|$.

The function $(\lambda, \mu) \to |\lambda - \mu|$ is a metric on K; endowed with this metric and the corresponding uniformity, K is called a **valuated** field. The valuated field K is called **non-discrete** if its topology is not discrete (equivalently, if the range of $\lambda \to |\lambda|$ is distinct from $\{0,1\}$). A non-discrete valuated field is necessarily infinite.

Let L be a vector space over a non-discrete valuated field K, and let A, B be subsets of L. We say that A **absorbs** B if there exists $\lambda_0 \in K$ such that $B \subset \lambda A$ whenever $|\lambda| \geq |\lambda_0|$. A subset U of L is called **radial (absorbing)** if U absorbs every finite subset of L. A subset C of L is **circled** if $\lambda C \subset C$ whenever $|\lambda| \leq 1$.

The set of radial subsets of L is invariant under finite intersections; the set of circled subsets of L is invariant under arbitrary intersections. If $A \subset L$, the **circled hull** of A is the intersection of all circled subsets of L containing A. Let $f: L_1 \to L_2$ be linear, L_1 and L_2 being vector spaces over a non-discrete valuated field K. If $A \subset L_1$ and $B \subset L_2$ are circled, then $f(A)$ and $f^{-1}(B)$ are circled. If B is radial then $f^{-1}(B)$ is radial; if A is radial and f is surjective, then $f(A)$ is radial.

The fields R and C of real and complex numbers, respectively, are always considered to be endowed with their usual absolute value, under which they are non-discrete valuated fields. In addition, R is always considered under its usual order.

Literature: Baer [1]; Birkhoff-MacLane [1]; Bourbaki [2], [3], [7].

Chapter I

TOPOLOGICAL VECTOR SPACES

This chapter presents the most basic results on topological vector spaces. With the exception of the last section, the scalar field over which vector spaces are defined can be an arbitrary, non-discrete valued field K; K is endowed with the uniformity derived from its absolute value. The purpose of this generality is to clearly identify those properties of the commonly used real and complex number field that are essential for these basic results. Section 1 discusses the description of vector space topologies in terms of neighborhood bases of 0, and the uniformity associated with such a topology. Section 2 gives some means for constructing new topological vector spaces from given ones. The standard tools used in working with spaces of finite dimension are collected in Section 3, which is followed by a brief discussion of affine subspaces and hyperplanes (Section 4). Section 5 studies the extremely important notion of boundedness. Metrizability is treated in Section 6. This notion, although not overly important for the general theory, deserves special attention for several reasons; among them are its connection with category, its role in applications in analysis, and its role in the history of the subject (cf. Banach [1]). Restricting K to subfields of the complex numbers, Section 7 discusses the transition from real to complex fields and vice versa.

1. VECTOR SPACE TOPOLOGIES

Given a vector space L over a (not necessarily commutative) non-discrete valued field K and a topology $\mathfrak{T}$ on L, the pair $(L,\mathfrak{T})$ is called a **topological vector space** (abbreviated **t.v.s.**) over K if these two axioms are satisfied:

$(LT)_1$ $(x, y) \to x + y$ *is continuous on* $L \times L$ *into* L.
$(LT)_2$ $(\lambda, x) \to \lambda x$ *is continuous on* $K \times L$ *into* L.

Here L is endowed with $\mathfrak{T}$, K is endowed with the uniformity derived from its absolute value, and $L \times L$, $K \times L$ denote the respective topological products. Loosely speaking, these axioms require addition and scalar multiplication to be (jointly) continuous. Since, in particular, this implies the continuity of $(x, y) \to x - y$, every t.v.s. is a commutative topological group. A t.v.s. $(L, \mathfrak{T})$ will occasionally be denoted by $L(\mathfrak{T})$, or simply by L if the topology of L does not require special notation.

Two t.v.s. L_1 and L_2 over the same field K are called **isomorphic** if there exists a biunivocal linear map u of L_1 onto L_2 which is a homeomorphism; u is called an **isomorphism** of L_1 onto L_2. (Although mere algebraic isomorphisms will, in general, be designated as such, the terms "topological isomorphism" and "topologically isomorphic" will occasionally be used to avoid misunderstanding.) The following assertions are more or less immediate consequences of the definition of a t.v.s.

1.1

Let L be a t.v.s. over K.

(i) For each $x_0 \in L$ and each $\lambda_0 \in K$, $\lambda_0 \neq 0$, the mapping $x \to \lambda_0 x + x_0$ is a homeomorphism of L onto itself.

(ii) For any subset A of L and any base $\mathfrak{U}$ of the neighborhood filter of $0 \in L$, the closure $\bar{A}$ is given by $\bar{A} = \bigcap \{A + U: U \in \mathfrak{U}\}$.

(iii) If A is an open subset of L, and B is any subset of L, then $A + B$ is open.

(iv) If A, B are closed subsets of L such that every filter on A has an adherent point (in particular, such that A is compact), then $A + B$ is closed.

(v) If A is a circled subset of L, then its closure $\bar{A}$ is circled, and the interior $\mathring{A}$ of A is circled when $0 \in \mathring{A}$.

Proof. (i): Clearly, $x \to \lambda_0 x + x_0$ is onto L and, by $(LT)_1$ and $(LT)_2$, continuous with continuous inverse $x \to \lambda_0^{-1}(x - x_0)$. Note that this assertion, as well as (ii), (iii), and (v), requires only the separate continuity of addition and scalar multiplication.

(ii): Let $B = \bigcap \{A + U: U \in \mathfrak{U}\}$. By (i), $\{x - U: U \in \mathfrak{U}\}$ is a neighborhood base of x for each $x \in L$; hence $x \in B$ implies that each neighborhood of x intersects A, whence $B \subset \bar{A}$. Conversely, if $x \in \bar{A}$ then $x \in A + U$ for each 0-neighborhood U, whence $\bar{A} \subset B$.

(iii): Since $A + B = \bigcup \{A + b: b \in B\}$, $A + B$ is a union of open subsets of L if A is open, and hence an open subset of L.

(iv): We show that for each $x_0 \notin A + B$ there exists a 0-neighborhood U such that $(x_0 - U) \cap (A + B) = \varnothing$ or, equivalently, that $(B + U) \cap (x_0 - A) = \varnothing$. If this were not true, then the intersections $(B + U) \cap (x_0 - A)$ would form a filter base on $x_0 - A$ (as U runs through a 0-neighborhood base in L). By the assumption on A, this filter base would have an adherent point $z_0 \in x_0 - A$, also contained in the closure of $B + U$ and hence in $B + U + U$, for all U. Since by $(LT)_1$, $U + U$ runs through a neighborhood base of 0 as U does, (ii) implies that $z_0 \in B$, which is contradictory.

(v): Let A be circled and let $|\lambda| \leqq 1$. By $(LT)_2$, $\lambda A \subset A$ implies $\lambda \bar{A} \subset \bar{A}$; hence $\bar{A}$ is circled. Also if $\lambda \neq 0$, $\lambda \mathring{A}$ is the interior of λA by (i) and hence contained in $\mathring{A}$. The assumption $0 \in \mathring{A}$ then shows that $\lambda \mathring{A} \subset \mathring{A}$ whenever $|\lambda| \leqq 1$.

In the preceding proof we have repeatedly made use of the fact that in a t.v.s., each translation $x \rightarrow x + x_0$ is a homeomorphism (which is a special case of (i)); a topology $\mathfrak{T}$ on a vector space L is called **translation-invariant** if all translations are homeomorphisms. Such a topology is completely determined by the neighborhood filter of any point $x \in L$, in particular by the neighborhood filter of 0.

1.2

A topology $\mathfrak{T}$ on a vector space L over K satisfies the axioms $(LT)_1$ and $(LT)_2$ if and only if $\mathfrak{T}$ is translation-invariant and possesses a 0-neighborhood base $\mathfrak{B}$ with the following properties:

(a) *For each $V \in \mathfrak{B}$, there exists $U \in \mathfrak{B}$ such that $U + U \subset V$.*
(b) *Every $V \in \mathfrak{B}$ is radial and circled.*
(c) *There exists $\lambda \in K$, $0 < |\lambda| < 1$, such that $V \in \mathfrak{B}$ implies $\lambda V \in \mathfrak{B}$.*

If K is an Archimedean valuated field, condition (c) is dispensable (which is, in particular, the case if $K = R$ or $K = C$).

Proof. We first prove the existence, in every t.v.s., of a 0-neighborhood base having the listed properties. Given a 0-neighborhood W in L, there exists a 0-neighborhood U and a real number $\varepsilon > 0$ such that $\lambda U \subset W$ whenever $|\lambda| \leqq \varepsilon$, by virtue of $(LT)_2$; hence since K is non-discrete, $V = \bigcup \{\lambda U : |\lambda| \leqq \varepsilon\}$ is a 0-neighborhood which is contained in W, and obviously circled. Thus the family $\mathfrak{B}$ of all circled 0-neighborhoods in L is a base at 0. The continuity at $\lambda = 0$ of $(\lambda, x_0) \rightarrow \lambda x_0$ for each $x_0 \in L$ implies that every $V \in \mathfrak{B}$ is radial. It is obvious from $(LT)_1$ that $\mathfrak{B}$ satisfies condition (a); for (c), it suffices to observe that there exists $\lambda \in K$ such that $0 < |\lambda| < 1$, since K is non-discrete, and that λV $(V \in \mathfrak{B})$, which is a 0-neighborhood by (1.1) (i), is circled (note that if $|\mu| \leqq 1$ then $\mu = \lambda v \lambda^{-1}$ where $|v| \leqq 1$). Finally, the topology of L is translation-invariant by (1.1) (i).

Conversely, let $\mathfrak{T}$ be a translation-invariant topology on L possessing a 0-neighborhood base $\mathfrak{B}$ with properties (a), (b), and (c). We have to show that $\mathfrak{T}$ satisfies $(LT)_1$ and $(LT)_2$. It is clear that $\{x_0 + V : V \in \mathfrak{B}\}$ is a neighborhood base of $x_0 \in L$; hence if $V \in \mathfrak{B}$ is given and $U \in \mathfrak{B}$ is selected such that $U + U \subset V$, then $x - x_0 \in U$, $y - y_0 \in U$ imply that $x + y \in x_0 + y_0 + V$; so $(LT)_1$ holds. To prove the continuity of the mapping $(\lambda, x) \rightarrow \lambda x$, that is $(LT)_2$, let λ_0, x_0 be any fixed elements of K, L respectively. If $V \in \mathfrak{B}$ is given, by (a) there exists $U \in \mathfrak{B}$ such that $U + U \subset V$. Since by (b) U is radial, there exists a real number $\varepsilon > 0$ such that $(\lambda - \lambda_0) x_0 \in U$ whenever $|\lambda - \lambda_0| \leqq \varepsilon$.

Let $\mu \in K$ satisfy (c); then there exists an integer $n \in N$ such that $|\mu^{-n}| = |\mu|^{-n} \geq |\lambda_0| + \varepsilon$; let $W \in \mathfrak{B}$ be defined by $W = \mu^n U$. Now since U is circled, the relations $x - x_0 \in W$ and $|\lambda - \lambda_0| \leq \varepsilon$ imply that $\lambda(x - x_0) \in U$, and hence the identity

$$\lambda x = \lambda_0 x_0 + (\lambda - \lambda_0)x_0 + \lambda(x - x_0)$$

implies that $\lambda x \in \lambda_0 x_0 + U + U \subset \lambda_0 x_0 + V$, which proves $(LT)_2$.

Finally, if K is an Archimedean valued field, then $|2| > 1$ for $2 \in K$. Hence $|2^n| = |2|^n > |\lambda_0| + \varepsilon$ (notation of the preceding paragraph) for a suitable $n \in N$. By repeated application of (b), we can select a $W_1 \in \mathfrak{B}$ such that $2^n W_1 \subset W_1 + \cdots + W_1 \subset U$, where the sum has 2^n summands $(2 \in N)$. Since W_1 (and hence $2^n W_1$) is circled, W_1 can be substituted for W in the preceding proof of $(LT)_2$, and hence (c) is dispensable in this case. This completes the proof of (1.2).

COROLLARY. *If L is a vector space over K and $\mathfrak{B}$ is a filter base in L having the properties (a) through (c) of (1.2), then $\mathfrak{B}$ is a neighborhood base of 0 for a unique topology $\mathfrak{T}$ such that $(L, \mathfrak{T})$ is a t.v.s. over K.*

Proof. We define the topology $\mathfrak{T}$ by specifying a subset $G \subset L$ to be open whenever $x \in G$ implies $x + V \subset G$ for some $V \in \mathfrak{B}$. Clearly $\mathfrak{T}$ is the unique translation-invariant topology on L for which $\mathfrak{B}$ is a base at 0, and hence the unique topology with this property and such that $(L, \mathfrak{T})$ is a t.v.s.

Examples

In the following examples, K can be any non-discrete valued field; for instance, the field of p-adic numbers, or the field of quaternions with their usual absolute values, or any subfield of these such as the rational, real, or complex number field (with the respective induced absolute value).

1. Let A be any non-empty set, K^A the set of all mappings $\alpha \to \xi_\alpha$ of A into K; we write $x = (\xi_\alpha)$, $y = (\eta_\alpha)$ to denote elements x, y of K^A. Defining addition by $x + y = (\xi_\alpha + \eta_\alpha)$ and scalar multiplication by $\lambda x = (\lambda \xi_\alpha)$, it is immediate that K^A becomes a vector space over K. For any finite subset $H \subset A$ and any real number $\varepsilon > 0$, let $V_{H, \varepsilon}$ be the subset $\{x: |\xi_\alpha| \leq \varepsilon$ if $\alpha \in H\}$ of K^A; it is clear from (1.2) that the family of all these sets $V_{H, \varepsilon}$ is a 0-neighborhood base for a unique topology under which K^A is a t.v.s.

2. Let X be any non-empty topological space; the set of all continuous functions f on X into K such that $\sup_{t \in X} |f(t)|$ is finite is a subset of K^X, which is a vector space $\mathscr{C}_K(X)$ under the operations of addition and scalar multiplication induced by the vector space K^X (Example 1); the sets $U_n = \{f: \sup_{t \in X} |f(t)| \leq n^{-1}\}$ $(n \in N)$ form a neighborhood base of 0 for a unique topology under which $\mathscr{C}_K(X)$ is a t.v.s.

3. Let $K[t]$ be the ring of polynomials $f[t] = \sum_n \alpha_n t^n$ over K in one indeterminate t. With multiplication restricted to left multiplication by

polynomials of degree 0, $K[t]$ becomes a vector space over K. Let r be a fixed real number such that $0 < r \leqq 1$ and denote by V_ε the set of polynomials for which $\sum_n |\alpha_n|^r \leqq \varepsilon$. The family $\{V_\varepsilon : \varepsilon > 0\}$ is a 0-neighborhood base for a unique topology under which $K[t]$ is a t.v.s.

1.3

If L is a t.v.s. and $x \in L$, each neighborhood of x contains a closed neighborhood of x. In particular, the family of all closed 0-neighborhood forms a base at 0.

Proof. For any 0-neighborhood U there exists another, V, such that $V + V \subset U$. Since $y \in \bar{V}$ only if $(y - V) \cap V$ is non-empty, it follows that $\bar{V} \subset V + V \subset U$. Hence $x + U$ contains the closed neighborhood $x + \bar{V}$ of x.

Since by (1.2) any 0-neighborhood contains a circled 0-neighborhood, and hence by (1.1) (v) and (1.3) a closed, circled 0-neighborhood, we obtain the following corollary:

COROLLARY. *If L is a t.v.s. and $\mathfrak{U}$ is any neighborhood base of 0, then the closed, circled hulls of the sets $U \in \mathfrak{U}$ form again a base at 0.*

(1.3) shows that every Hausdorff t.v.s. is a regular topological space. It will be seen from the next proposition that every t.v.s. is uniformisable, hence every Hausdorff t.v.s. is completely regular. A uniformity on a vector space L is called **translation-invariant** if it has a base $\mathfrak{N}$ such that $(x, y) \in N$ is equivalent with $(x + z, y + z) \in N$ for each $z \in L$ and each $N \in \mathfrak{N}$.

1.4

The topology of any t.v.s. can be derived from a unique translation-invariant uniformity $\mathfrak{N}$. If $\mathfrak{B}$ is any neighborhood base of 0, the family $N_V = \{(x, y): x - y \in V\}$, $V \in \mathfrak{B}$ is a base for $\mathfrak{N}$.

Proof. Let $(L, \mathfrak{T})$ be a t.v.s. with 0-neighborhood base $\mathfrak{B}$. It is immediate that the sets N_V, $V \in \mathfrak{B}$ form a filter base on $L \times L$ that is a base for a translation-invariant uniformity $\mathfrak{N}$ yielding the topology $\mathfrak{T}$ of L. If $\mathfrak{N}_1$ is another uniformity with these properties, there exists a base $\mathfrak{M}$ of $\mathfrak{N}_1$, consisting of translation-invariant sets, and such that the sets

$$U_M = \{x - y \colon (x, y) \in M\} \qquad M \in \mathfrak{M}$$

form a 0-neighborhood base for $\mathfrak{T}$. Since $U_M \subset V$ implies $M \subset N_V$ and conversely, it follows that $\mathfrak{N}_1 = \mathfrak{N}$.

The fact that there is a unique translation-invariant uniformity from which the topology of a t.v.s. can be derived is of considerable importance in the theory of such spaces (as it is for topological groups), since uniformity concepts can be applied unambiguously to arbitrary subsets A of a t.v.s. L. The uniformity meant is, without exception, that induced on $A \subset L$ by the uniformity $\mathfrak{N}$ of (1.4). For example, a subset A of a t.v.s. L is complete if and only if every Cauchy filter in A converges to an element of A; A is semi-complete (or sequentially complete) if and only if every Cauchy sequence in A converges to an element of A. It follows from (1.4) that a filter $\mathfrak{F}$ in A is a Cauchy filter if and only if for each 0-neighborhood V in L, there exists $F \in \mathfrak{F}$ such that $F - F \subset V$; accordingly, a sequence $\{x_n: n \in N\}$ in A is a Cauchy sequence if and only if for each 0-neighborhood V in L there exists $n_0 \in N$ such that $x_m - x_n \in V$ whenever $m \geqq n_0$ and $n \geqq n_0$.

A t.v.s. L is a Hausdorff (or separated) topological space if and only if L is a separated uniform space; hence by (1.4), L is separated if and only if $\bigcap \{U: U \in \mathfrak{U}\} = \{0\}$, where $\mathfrak{U}$ is any neighborhood base of 0 in L. An equivalent condition is that for each non-zero $x \in L$, there exists a 0-neighborhood U such that $x \notin U$ (which is also immediate from (1.3)).

Recall that a subspace (vector subspace, linear subspace) of a vector space L over K is defined to be a subset $M \neq \varnothing$ of L such that $M + M \subset M$ and $KM \subset M$. If L is a t.v.s., by a **subspace** of L we shall understand (unless the contrary is expressly stated) a vector subspace M endowed with the topology induced by L; clearly, M is a t.v.s. which is separated if L is.

If L is a Hausdorff t.v.s., the presence of a translation-invariant separated uniformity makes it possible to imbed L as a dense subspace of a complete Hausdorff t.v.s. $\tilde{L}$ which is essentially unique, and is called the **completion** of L. (See also Exercise 2.)

1.5

Let L be a Hausdorff t.v.s. over K. There exists a complete Hausdorff t.v.s. $\tilde{L}$ over K containing L as a dense subspace; $\tilde{L}$ is unique to within isomorphism. Moreover, for any 0-neighborhood base $\mathfrak{B}$ in L, the family $\mathfrak{W} = \{\bar{V}: V \in \mathfrak{B}\}$ of closures in $\tilde{L}$ is a 0-neighborhood base in $\tilde{L}$.

Proof. We assume it known (cf. Bourbaki [4], chap. II) that there exists a separated, complete uniform space $\tilde{L}$ which contains L as a dense subspace, and which is unique up to a uniform isomorphism. By (1.4) $(x, y) \to x + y$ is uniformly continuous on $L \times L$ into $\tilde{L}$, and for each fixed $\lambda \in K$ $(\lambda, x) \to \lambda x$ is uniformly continuous on L into $\tilde{L}$; hence these mappings have unique continuous (in fact, uniformly continuous) extensions to $\tilde{L} \times \tilde{L}$ and $\tilde{L}$, respectively, with values in $\tilde{L}$. It is quickly verified (continuation of identities) that these extensions make $\tilde{L}$ into a vector space over K. Before showing that the uniform space $\tilde{L}$ is a t.v.s. over K, we prove the second assertion. Since $\{N_V: V \in \mathfrak{B}\}$ is a base of the uniformity $\mathfrak{N}$ of L (notation as in (1.4)), the closures $\bar{N}_V$ of these sets in $\tilde{L} \times \tilde{L}$ form a base of the uniformity $\tilde{\mathfrak{N}}$ of L; we assert that $N_{\bar{V}} = \bar{N}_V$ for all $V \in \mathfrak{B}$. But if $(\tilde{x}, \tilde{y}) \in \bar{N}_V$, then $\tilde{x} - \tilde{y} \in \bar{V}$, since $(\tilde{x}, \tilde{y}) \to \tilde{x} - \tilde{y}$ is continuous on $\tilde{L} \times \tilde{L}$ into $\tilde{L}$. Conversely, if $\tilde{x} - \tilde{y} \in \bar{V}$, then we have $\tilde{x} \in \tilde{y} + \bar{V}$; hence $\tilde{x}$ is in the closure (taken in $\tilde{L}$) of $\tilde{y} + V$, since translations in $\tilde{L}$ are homeomorphisms; this implies that $(\tilde{x}, \tilde{y}) \in \bar{N}_V$.

It follows that $\mathfrak{W}$ is a neighborhood base of 0 in $\tilde{L}$; we use (1.2) to show that under the topology $\tilde{\mathfrak{T}}$ defined by $\tilde{\mathfrak{N}}$, $\tilde{L}$ is a t.v.s. Clearly, $\tilde{\mathfrak{T}}$ is translation-invariant and satisfies conditions (a) and (c) of (1.2); hence it suffices to show that each $\bar{V} \in \mathfrak{W}$ contains a $\tilde{\mathfrak{T}}$-neighborhood of 0 that is radial and circled. Given $V \in \mathfrak{B}$, there exists a circled 0-neighborhood U in L such that $U + U \subset V$. The closure $(U + U)^-$ in $\tilde{L}$ is a 0-neighborhood by the preceding, is circled and clearly contained in $\bar{V}$. Let us show that it is radial. Given $\tilde{x} \in \tilde{L}$, there exists a Cauchy filter $\mathfrak{F}$ in L convergent to $\tilde{x}$, and an $F \in \mathfrak{F}$ such that $F - F \subset U$. Let x_0 be any element of F; since U is radial there exists $\lambda \in K$ such that $x_0 \in \lambda U$, and since U is circled we can assume that $|\lambda| \geqq 1$. Now $F - x_0 \subset U$; hence $F \subset x_0 + U$ and $\tilde{x} \in \bar{F} \subset \lambda(U + U)^-$, which proves the assertion.

Finally, the uniqueness of $(\tilde{L}, \tilde{\mathfrak{T}})$ (to within isomorphism) follows, by virtue of (1.4), from the uniqueness of the completion $\tilde{L}$ of the uniform space L.

REMARK. The completeness of the valuated field K is not required for the preceding construction. On the other hand, if L is a complete Hausdorff t.v.s. over K, it is not difficult to see that scalar multiplication has a unique continuous extension to $\tilde{K} \times L$, where $\tilde{K}$ is the completion of K. Thus it follows from (1.5) that for every Hausdorff t.v.s. over K there exists a (essentially unique) complete Hausdorff t.v.s. L_1 over $\tilde{K}$ such that the topological group L is isomorphic with a dense subgroup of the topological group L_1.

We conclude this section with a completeness criterion for a t.v.s. $(L, \mathfrak{T}_1)$ in terms of a coarser topology $\mathfrak{T}_2$ on L.

1.6

Let L be a vector space over K and let $\mathfrak{T}_1$, $\mathfrak{T}_2$ be Hausdorff topologies under each of which L is a t.v.s., and such that $\mathfrak{T}_1$ is finer than $\mathfrak{T}_2$. If $(L, \mathfrak{T}_1)$ has a neighborhood base of 0 consisting of sets complete in $(L, \mathfrak{T}_2)$, then $(L, \mathfrak{T}_1)$ is complete.

Proof. Let $\mathfrak{B}_1$ be a $\mathfrak{T}_1$-neighborhood base of 0 in L consisting of sets complete in $(L, \mathfrak{T}_2)$. Given a Cauchy filter $\mathfrak{F}$ in $(L, \mathfrak{T}_1)$ and $V_1 \in \mathfrak{B}_1$, there exists a set $F_0 \in \mathfrak{F}$ such that $F_0 - F_0 \subset V_1$. If y is any fixed element of F_0, the family $\{y - F: F \in \mathfrak{F}\}$ is a Cauchy filter base for the uniformity associated with $\mathfrak{T}_2$, for which V_1 is complete; since $y - F_0 \subset V_1$, this filter base has a unique $\mathfrak{T}_2$-limit $y - x_0$. It is now clear that $x_0 \in L$ is the $\mathfrak{T}_2$-limit of $\mathfrak{F}$. Since V_1 is $\mathfrak{T}_2$-closed, we have $F_0 - x_0 \subset V_1$ or $F_0 \subset x_0 + V_1$; V_1 being arbitrary, this shows $\mathfrak{F}$ to be finer than the $\mathfrak{T}_1$-neighborhood filter of x_0 and thus proves $(L, \mathfrak{T}_1)$ to be complete.

For the reader familiar with normed spaces, we point out this example for (1.6): Every reflexive normed space is complete and hence is a Banach space. For in such a space the positive multiples of the closed unit ball, which form a 0-neighborhood base for the norm topology, are weakly compact and hence weakly complete.

2. PRODUCT SPACES, SUBSPACES, DIRECT SUMS, QUOTIENT SPACES

Let $\{L_\alpha: \alpha \in A\}$ denote a family of vector spaces over the same scalar field K; the Cartesian product $L = \prod_\alpha L_\alpha$ is a vector space over K if for $x = (x_\alpha)$, $y = (y_\alpha) \in L$ and $\lambda \in K$, addition and scalar multiplication are defined by $x + y = (x_\alpha + y_\alpha)$, $\lambda x = (\lambda x_\alpha)$. If $(L_\alpha, \mathfrak{T}_\alpha)$ $(\alpha \in A)$ are t.v.s. over K, then L is a t.v.s. under the product topology $\mathfrak{T} = \prod_\alpha \mathfrak{T}_\alpha$; the simple verification of $(LT)_1$ and $(LT)_2$ is left to the reader. Moreover, it is known from general topology that $L(\mathfrak{T})$ is a Hausdorff space and a complete uniform space, respectively, if and only if each factor is. $(L, \mathfrak{T})$ will be called the **product** of the family $\{L_\alpha(\mathfrak{T}_\alpha): \alpha \in A\}$.

As has been pointed out before, by a subspace M of a vector space L over K we understand a subset $M \neq \varnothing$ invariant under addition and scalar multiplication; we record the following simple consequence of the axioms $(LT)_1$ and $(LT)_2$:

2.1

If $(L, \mathfrak{T})$ is a t.v.s. and M is a subspace of L, the closure $\overline{M}$ in $(L, \mathfrak{T})$ is again a subspace of L.

Proof. In fact, it follows from $(LT)_1$ that $\overline{M} + \overline{M} \subset \overline{M}$, and from $(LT)_2$ that $K\overline{M} \subset \overline{M}$.

We recall the following facts from linear algebra. If L is a vector space, M_i $(i = 1, \ldots, n)$ subspaces of L whose linear hull is L and such that $M_i \cap (\sum_{j \neq i} M_j) = \{0\}$ for each i, then L is called the **algebraic direct sum** of the subspaces $M_i (i = 1, \ldots, n)$. It follows that each $x \in L$ has a unique representation $x = \sum_i x_i$, where $x_i \in M_i$, and the mapping $(x_1, \ldots, x_n) \to \sum_i x_i$ is an algebraic isomorphism of $\prod_i M_i$ onto L. The mapping $u_i: x \to x_i$ is called the **projection** of L onto M_i associated with this decomposition. If each u_i is viewed as an endomorphism of L, one has the relations $u_i u_j = \delta_{ij} u_i (i, j = 1, \ldots, n)$ and $\sum_i u_i = e$, e denoting the identity map.

If $(L, \mathfrak{T})$ is a t.v.s. and L is algebraically decomposed as above, each of the projections u_i is an open map of L onto the t.v.s. M_i. In fact, if G is an open subset of L and N_i denotes the null space of u_i, then $G + N_i$ is open in L by (1.1) and $u_i(G) = u_i(G + N_i) = (G + N_i) \cap M_i$. From $(LT)_1$ it is also clear that the mapping $\psi: (x_1, \ldots, x_n) \to \sum_i x_i$ of $\prod_i M_i$ onto L is continuous; if ψ is an isomorphism, L is called the **direct sum** (or topological direct sum if this distinction is desirable) of the subspaces $M_i (i = 1, \ldots, n)$; we write $L = M_1 \oplus \cdots \oplus M_n$.

2.2

Let a t.v.s. L be the algebraic direct sum of n subspaces M_i ($i = 1, ..., n$). Then $L = M_1 \oplus \cdots \oplus M_n$ if and only if the associated projections u_i are continuous ($i = 1, ..., n$).

Proof. By definition of the product topology, the mapping $\psi^{-1} \colon x \to (u_1 x, ..., u_n x)$ of L onto $\prod_i M_i$ is continuous if and only if each u_i is.

REMARK. Since the identity map e is continuous on L, the continuity of $n - 1$ of these projections implies the continuity of the remaining one.

A subspace N of a t.v.s. L such that $L = M \oplus N$ is called a subspace **complementary** (or **supplementary**) to M; such complementary subspaces do not necessarily exist, even if M is of finite dimension (Exercise 8); cf. also Chapter IV, Exercise 12.

Let $(L, \mathfrak{T})$ be a t.v.s. over K, let M be a subspace of L, and let ϕ be the **natural** (canonical, quotient) map of L onto L/M—that is, the mapping which orders to each $x \in L$ its equivalence class $\hat{x} = x + M$. The **quotient topology** $\hat{\mathfrak{T}}$ is defined to be the finest topology on L/M for which ϕ is continuous. Thus the open sets in L/M are the sets $\phi(H)$ such that $H + M$ is open in L; since $G + M$ is open in L whenever G is, $\phi(G)$ is open in L/M for every open $G \subset L$; hence ϕ is an open map. It follows that $\phi(\mathfrak{B})$ is a 0-neighborhood base in L/M for every 0-neighborhood base $\mathfrak{B}$ in L; since ϕ is linear, $\hat{\mathfrak{T}}$ is translation-invariant and $\phi(\mathfrak{B})$ satisfies conditions (a), (b), and (c) of (1.2) if these are satisfied by $\mathfrak{B}$. Hence $(L/M, \hat{\mathfrak{T}})$ is a t.v.s. over K, called the **quotient space** of $(L, \mathfrak{T})$ over M.

2.3

If L is a t.v.s. and if M is a subspace of L, then L/M is a Hausdorff space if and only if M is closed in L.

Proof. If L/M is Hausdorff, the set $\{0\} \subset L/M$ is closed; by the continuity of ϕ, $M = \phi^{-1}(0)$ is closed. Conversely, if $\hat{x} \neq 0$ in L/M, then $\hat{x} = \phi(x)$, where $x \notin M$; if M is closed, the complement U of M in L is a neighborhood of x; hence $\phi(U)$ is a neighborhood of $\hat{x}$ not containing 0. Since $\phi(U)$ contains a closed neighborhood of $\hat{x}$ by (1.3), L/M is a Hausdorff space.

By (2.3), a Hausdorff t.v.s. L/M can be associated with every t.v.s. L by taking for M the closure in L of the subspace $\{0\}$; M is a subspace by (2.1). This space L/M is called the Hausdorff t.v.s. associated with L.

There is the following noteworthy relation between quotients and direct sums:

2.4

Let L be a t.v.s. and let L be the algebraic direct sum of the subspaces M, N. Then L is the topological direct sum of M and N: $L = M \oplus N$, if and only if the mapping v which orders to each equivalence class mod M its unique representative in N is an isomorphism of the t.v.s. L/M onto the t.v.s. N.

Proof. Denote by u the projection of L onto N vanishing on M, and by ϕ the natural map of L onto L/M. Then $u = v \circ \phi$. Let $L = M \oplus N$. Since ϕ is open and u is continuous, v is continuous; since ϕ is continuous and u is open, v is open. Conversely, if v is an isomorphism then v is continuous; hence u is continuous which implies $L = M \oplus N$.

3. TOPOLOGICAL VECTOR SPACES OF FINITE DIMENSION

By the **dimension** of a t.v.s. L over K, we understand the algebraic dimension of L over K, that is, the cardinality of any maximal linearly independent subset of L; such a set is called a **basis** (or **Hamel basis**) of L. Let K_0 denote the one-dimensional t.v.s. obtained by considering K as a vector space over itself.

3.1

Every one-dimensional Hausdorff t.v.s. L over K is isomorphic with K_0; more precisely, $\lambda \to \lambda x_0$ is an isomorphism of K_0 onto L for each $x_0 \in L$, $x_0 \neq 0$, and every isomorphism of K_0 onto L is of this form.

Proof. It follows from $(LT)_2$ that $\lambda \to \lambda x_0$ is continuous; moreover, this mapping is an algebraic isomorphism of K_0 onto L. To see that $\lambda x_0 \to \lambda$ is continuous, it is sufficient to show the continuity of this map at $0 \in L$. Let $\varepsilon < 1$ be a positive real number. Since K is non-discrete, there exists $\lambda_0 \in K$ such that $0 < |\lambda_0| < \varepsilon$, and since L is assumed to be Hausdorff, there exists a circled 0-neighborhood $V \subset L$ such that $\lambda_0 x_0 \notin V$. Hence $\lambda x_0 \in V$ implies $|\lambda| < \varepsilon$; for $|\lambda| \geq \varepsilon$ would imply $\lambda_0 x_0 \in V$, since V is circled, which is contradictory.

Finally, if u is an isomorphism of K_0 onto L such that $u(1) = x_0$, then u is clearly of the form $\lambda \to \lambda x_0$.

3.2

Theorem. *Every Hausdorff t.v.s. L of finite dimension n over a complete valuated field K is isomorphic with K_0^n. More precisely, $(\lambda_1, ..., \lambda_n) \to \lambda_1 x_1 + \cdots + \lambda_n x_n$ is an isomorphism of K_0^n onto L for each basis $\{x_1, ..., x_n\}$ of L, and every isomorphism of K_0^n onto L is of this form.*

Proof. The proof is conducted by induction. (3.1) implies the assertion to be valid for $n = 1$. Assume it to be correct for $k = n - 1$. If $\{x_1, ..., x_n\}$ is any basis of L, L is the algebraic direct sum of the subspaces M and N with bases $\{x_1, ..., x_{n-1}\}$ and $\{x_n\}$, respectively. By assumption, M is isomorphic with K_0^{n-1}; since K_0 is complete, M is complete and since L is Hausdorff, M is closed in L. By (2.3), L/M is Hausdorff and clearly of dimension 1; hence the map v, ordering to each equivalence class mod M its unique representative in N, is an isomorphism by (3.1). It follows from (2.4) that $L = M \oplus N$, which means that $(\lambda_1, ..., \lambda_n) \to \lambda_1 x_1 + \cdots + \lambda_n x_n$ is an isomorphism of $K_0^{n-1} \times K_0 = K_0^n$ onto L. Finally, it is obvious that every isomorphism of K_0^n onto L is of this form.

It is worth remarking that while (3.1) (and a fortiori (3.2)) obviously fails for non-Hausdorff spaces L, (3.2) may fail for $n > 1$ when K is not complete (Exercise 4).

Theorem (3.2) can be restated by saying that if K is a complete valuated field, then the product topology on K_0^n is the only Hausdorff topology satisfying $(LT)_1$ and $(LT)_2$ (Tychonoff [1]). This has a number of important consequences.

3.3

Let L be a t.v.s. over K and let K be complete. If M is a closed subspace of L and N is a finite dimensional subspace of L, then $M + N$ is closed in L.

Proof. Let ϕ denote the natural map of L onto L/M; L/M is Hausdorff by (2.3). Since $\phi(N)$ is a finite-dimensional subspace of L/M, it is complete by (3.2), hence closed in L/M. This implies that $M + N = \phi^{-1}(\phi(N))$ is closed, since ϕ is continuous.

3.4

Let K be complete, let N be a finite dimensional Hausdorff t.v.s. over K, and let L be any t.v.s. over K. Every linear map of N into L is continuous.

Proof. The result is trivial if N has dimension 0. If N has positive dimension n, it is isomorphic with K_0^n by (3.2). But every linear map on K_0^n into L is necessarily of the form $(\lambda_1, ..., \lambda_n) \to \lambda_1 y_1 + \cdots + \lambda_n y_n$, where $y_i \in L$, and hence continuous by $(LT)_1$ and $(LT)_2$.

We recall that the codimension of a subspace M of a vector space L is the dimension of L/M; N is an algebraic complementary subspace of M if $L = M + N$ is an algebraic direct sum.

3.5

Let L be a t.v.s. over the complete field K and let M be a closed subspace of finite codimension. Then $L = M \oplus N$ for every algebraic complementary subspace N of M.

Proof. L/M is a finite dimensional t.v.s., which is Hausdorff by (2.3); hence by (3.4), the mapping v of L/M onto N, which orders to each element of L/M its unique representative in N, is continuous. By (2.2), this implies $L = M \oplus N$, since the projection $u = v \circ \phi$ is continuous.

REMARK. It follows from (2.4) that in the circumstances of (3.5), N is necessarily a Hausdorff subspace of L. It is not difficult to verify this directly.

We now turn to the second important theorem concerning t.v.s. of finite dimension. It is clear from (3.2) that if K is locally compact (hence complete), then every finite dimensional Hausdorff t.v.s. over K is locally compact. Conversely, if K is complete, then every locally compact Hausdorff t.v.s. over K is of finite dimension (cf. Exercise 3).

3.6

Theorem. *Let K be complete. If $L \neq \{0\}$ is a locally compact Hausdorff t.v.s. over K, then K is locally compact and L is of finite dimension.*

Proof. By (3.1) every one-dimensional subspace of L is complete, hence closed in L and therefore locally compact; it follows that K is locally compact. Now let V be a compact, circled 0-neighborhood in L, and let $\{\lambda_n\}$ be a null sequence in K consisting of non-zero terms. We show first that $\{\lambda_n V: n \in N\}$ is a neighborhood base of 0 in L. Given a 0-neighborhood U, choose a circled 0-neighborhood W such that $W + W \subset U$. Since V is compact, there exist elements $x_i \in V$ $(i = 1, ..., k)$ satisfying $V \subset \bigcup_{i=1}^{k} (x_i + W)$, and there exists $\lambda \in K$, $\lambda \neq 0$, such that $\lambda x_i \in W$ for all i, and $|\lambda| \leq 1$. There exists $n \in N$ for which $|\lambda_n| \leq |\lambda|$, and

$$\lambda_n V \subset \lambda V \subset \bigcup_{i=1}^{k} (\lambda x_i + \lambda W) \subset W + W \subset U$$

shows $\{\lambda_n V: n \in N\}$ to be a neighborhood base of 0.

Let $\rho \in K$ satisfy $0 < |\rho| \leq 1/2$. Since V is compact and ρV is a 0-neighborhood, there exist elements y_l $(l = 1, ..., m)$ in V for which $V \subset \bigcup_{l=1}^{m} (y_l + \rho V)$. We denote by M the smallest subspace of L containing all y_l $(l = 1, ..., m)$ and show that $M = L$, which will complete the proof. Assuming that $M \neq L$, there exists $w \in L \sim M$ and $n_0 \in N$ such that $(w + \lambda_{n_0} V) \cap M = \varnothing$; for M, which is finite dimensional and hence complete by (3.2), is closed in L while $\{w + \lambda_n V: n \in N\}$ is a neighborhood base of w. Let μ be any number in K such that $w + \mu V$ intersects M (such numbers exist since V is radial) and set $\delta = \inf|\mu|$. Clearly, $\delta \geq |\lambda_{n_0}| > 0$. Choose $v_0 \in V$ so that $y = w + \mu_0 v_0 \in M$, where $\delta \leq |\mu_0| \leq 3\delta/2$. By the definition of $\{y_l\}$ there exists l_0, $1 \leq l_0 \leq m$, such that $v_0 = y_{l_0} + \rho v_1$, where $v_1 \in V$, and therefore

$$w = y - \mu_0 v_0 = (y - \mu_0 y_{l_0}) - \mu_0 \rho v_1 \in M + \mu_0 \rho V.$$

This contradicts the definition of δ, since V is circled and since $|\mu_0 \rho| \leq 3\delta/4$; hence the assumption $M \neq L$ is absurd.

4. LINEAR MANIFOLDS AND HYPERPLANES

If L is a vector space, a **linear manifold** (or **affine subspace**) in L is a subset which is a translate of a subspace $M \subset L$, that is, a set F of the form $x_0 + M$ for some $x_0 \in L$. F determines M uniquely while it determines x_0 only mod M: $x_0 + M = x_1 + N$ if and only if $M = N$ and $x_1 - x_0 \in M$. Two linear manifolds $x_0 + M$ and $x_1 + N$ are said to be **parallel** if either $M \subset N$ or $N \subset M$. The **dimension** of a linear manifold is the dimension of the subspace of which it is a translate. A **hyperplane** in L is a maximal proper affine subspace of L; hence the corresponding subspace of a hyperplane is of codimension 1. It is further clear that two hyperplanes in L are parallel if and only if the corresponding subspaces are identical. A hyperplane which is a subspace (i.e., a hyperplane containing 0) is sometimes called a homogeneous hyperplane.

For any vector space L over K, we denote by L^* the **algebraic dual** of L, that is, the (right) vector space (over K) of all linear forms on L.

4.1

A subset $H \subset L$ is a hyperplane if and only if $H = \{x : f(x) = \alpha\}$ for some $\alpha \in K$ and some non-zero $f \in L^*$. f and α are determined by H to within a common factor β, $0 \neq \beta \in K$.

Proof. If $f \in L^*$ is $\neq 0$, then $M = f^{-1}(0)$ is a maximal proper subspace of L; if, moreover, $x_0 \in L$ is such that $f(x_0) = \alpha$, then $H = \{x : f(x) = \alpha\} = x_0 + M$, which shows H to be a hyperplane. Conversely, if H is a hyperplane, then $H = x_0 + M$, where M is a subspace of L such that $\dim L/M = 1$, so that L/M is algebraically isomorphic with K_0. Denote by ϕ the natural map of L onto L/M and by g an isomorphism of L/M onto K_0; then $f = g \circ \phi$ is a linear form $\neq 0$ on L such that $H = \{x : f(x) = \alpha\}$ when $\alpha = f(x_0)$. If $H = \{x : f_1(x) = \alpha_1\}$ is another representation of H, then because of $f_1^{-1}(0) = M$ we must have $f_1 = g_1 \circ \phi$, where g_1 is an isomorphism of L/M onto K_0; if ξ is the element of L/M for which $g(\xi) = 1$ and if $g_1(\xi) = \beta$, then $f_1(x) = f(x)\beta$ for all $x \in L$, thus completing the proof.

Since translations in a t.v.s. L are homeomorphisms, it follows from (2.1) that the closure of an affine subspace F is an affine subspace $\bar{F}$; but $\bar{F}$ need not be a proper subset of L if F is.

4.2

A hyperplane H in a t.v.s. L is either closed or dense in L; $H = \{x : f(x) = \alpha\}$ is closed if and only if f is continuous.

Proof. If a hyperplane $H \subset L$ is not closed, it must be dense in L; otherwise, its closure would be a proper affine subspace of L, contradicting the maximality of H. To prove the second assertion, it is sufficient to show that $f^{-1}(0)$ is closed if and only if f is continuous. If f is continuous, $f^{-1}(0)$ is closed, since $\{0\}$ is closed in K. If $f^{-1}(0)$ is closed in L, then $L/f^{-1}(0)$ is a

Hausdorff t.v.s. by (2.3), of dimension 1; writing $f = g \circ \phi$ as in the preceding proof, (3.1) implies that g, hence f, is continuous.

We point out that, in general, there exist no closed hyperplanes in a t.v.s. L, even if it is Hausdorff (Exercises 6, 7).

5. BOUNDED SETS

A subset A of a t.v.s. L is called **bounded** if for each 0-neighborhood U in L, there exists $\lambda \in K$ such that $A \subset \lambda U$. Since by (1.2) the circled 0-neighborhoods in L form a base at 0, $A \subset L$ is bounded if and only if each 0-neighborhood absorbs A. A **fundamental system** (or **fundamental family**) of bounded sets of L is a family $\mathfrak{B}$ of bounded sets such that every bounded subset of L is contained in a suitable member of $\mathfrak{B}$.

A subset B of a t.v.s. L is called **totally bounded** if for each 0-neighborhood U in L there exists a finite subset $B_0 \subset B$ such that $B \subset B_0 + U$. Recall that a separated uniform space P is called **precompact** if the completion $\tilde{P}$ of P is compact; it follows readily from (1.4) and a well-known characterization of precompact uniform spaces (see Prerequisites) that a subset B of a Hausdorff t.v.s. is precompact if and only if it is totally bounded. (We shall use the term *precompact* exclusively when dealing with Hausdorff spaces.) From the preceding we obtain an alternative characterization of precompact sets: A subset B of a Hausdorff t.v.s. L is precompact if and only if the closure of B in the completion $\tilde{L}$ of L is compact.

5.1

Let L be a t.v.s. over K and let A, B be bounded (respectively, totally bounded) subsets of L. Then the following are bounded (respectively, totally bounded) subsets of L:

(i) *Every subset of A.*

(ii) *The closure $\bar{A}$ of A.*

(iii) *$A \cup B$, $A + B$, and λA for each $\lambda \in K$.*

Moreover, every totally bounded set is bounded. The circled hull of a bounded set is bounded; if K is locally precompact, the circled hull of every totally bounded set in L is totally bounded.

Proof. If A, B are bounded subsets of L, then (i) is trivial and (ii) is clear from (1.3). To prove (iii), let λ_1 and λ_2 be two elements of K such that $A \subset \lambda_1 U$ and $B \subset \lambda_2 U$ for a given circled 0-neighborhood U. Since K is nondiscrete, there exists $\lambda_0 \in K$ such that $|\lambda_0| > \sup(|\lambda_1|, |\lambda_2|)$. We obtain $A \cup B \subset \lambda_0 U$ and $A + B \subset \lambda_0(U + U)$; since by (1.2) $U + U$ runs through a neighborhood base of 0 when U does, it follows that $A \cup B$ and $A + B$ are bounded; the boundedness of λA is trivial. The proof for totally bounded sets A, B is similarly straightforward and will be omitted.

Since $\varnothing$ and every one-point set are clearly bounded, it follows from a repeated application of (iii) that every finite set is bounded. If B is totally bounded and U is a given circled 0-neighborhood, there exists a finite set $B_0 \subset B$ such that $B \subset B_0 + U$. Now $B_0 \subset \lambda_0 U$, where we can assume that $|\lambda_0| \geqq 1$, since U is circled; we obtain $B \subset \lambda_0(U + U)$ and conclude as before that B is bounded. The fact that the circled hull of a bounded set is bounded is clear from (1.3). To prove the final assertion, it is evidently sufficient to show that the circled hull of a finite subset of L is totally bounded, provided that K is locally precompact. In view of (iii), it is hence sufficient to observe that each set Sa is totally bounded where $a \in L$ and $S = \{\lambda: |\lambda| \leqq 1\}$; but this is clear from $(LT)_2$ and the assumed precompactness of S (cf. (5.4) below). This completes the proof.

COROLLARY 1. *The properties of being bounded and of being totally bounded are preserved under the formation of finite sums and unions and under dilatations* $x \to \lambda_0 x + x_0$.

COROLLARY 2. *The range of every Cauchy sequence is bounded.*

COROLLARY 3. *The family of all closed and circled bounded subsets of a t.v.s. L is a fundamental system of bounded sets of L.*

It is clear from the definition of precompactness that a subset of a Hausdorff t.v.s. is compact if and only if it is precompact and complete. We record the following simple facts on compact sets.

5.2

Let L be a Hausdorff t.v.s. over K and let A, B be compact subsets of L. Then $A \cup B$, $A + B$, and λA ($\lambda \in K$) are compact; if K is locally compact, then also the circled hull of A is compact.

Proof. The compactness of $A \cup B$ is immediate from the defining property of compact spaces (each open cover has a finite subcover; cf. Prerequisites); $A + B$ is compact as the image of the compact space $A \times B$ under $(x, y) \to x + y$ which is continuous by $(LT)_1$; the same argument applies to λA by $(LT)_2$. (Another proof consists in observing that $A \cup B$, $A + B$, and λA are precompact and complete.) Finally, the circled hull of A is the continuous image of $S \times A$ (under $(\lambda, x) \to \lambda x$), and hence compact if S is compact.

COROLLARY. *Compactness of subsets of a Hausdorff t.v.s. is preserved under the formation of finite sums and unions and under dilatations.*

The following is a sequential criterion for the boundedness of a subset of a t.v.s. (for a sequential criterion of total boundedness, see Exercise 5). By a **null sequence** in a t.v.s. L, we understand a sequence converging to $0 \in L$.

5.3

A subset A of a t.v.s. L is bounded if and only if for every null sequence $\{\lambda_n\}$ in K and every sequence $\{x_n\}$ in A, $\{\lambda_n x_n\}$ is a null sequence in L.

Proof. Let A be bounded and let V be a given circled 0-neighborhood in L. There exists $\mu \in K$, $\mu \neq 0$ such that $\mu A \subset V$. If $\{\lambda_n\}$ is any null sequence in K, there exists $n_0 \in N$ such that $|\lambda_n| \leq |\mu|$ whenever $n \geq n_0$; hence we obtain $\lambda_n x_n \in V$ for all $n \geq n_0$ and any sequence $\{x_n\}$ in A. Conversely, suppose that A is a subset of L satisfying the condition; if A were not bounded, there would exist a 0-neighborhood U such that A is not contained in $\rho_n U$ for any sequence $\{\rho_n\}$ in K. Since K is non-discrete, we can choose ρ_n so that $|\rho_n| \geq n$ for all $n \in N$, and $x_n \in A \sim \rho_n U$ $(n \in N)$; it would follow that $\rho_n^{-1} x_n \notin U$ for all n, which is contradictory, since $\{\rho_n^{-1}\}$ is a null sequence in K.

5.4

Let L, M be t.v.s. over K and let u be a continuous linear map of L into M. If B is a bounded (respectively, totally bounded) subset of L, $u(B)$ is bounded (respectively, totally bounded) in M.

Proof. If V is any 0-neighborhood in M, then $u^{-1}(V)$ is a 0-neighborhood in L; hence if B is bounded, then $B \subset \lambda u^{-1}(V)$ for a suitable $\lambda \in K$, which implies $u(B) \subset \lambda V$. If B is totally bounded, then $B \subset B_0 + u^{-1}(V)$ for some finite set $B_0 \subset B$, whence $u(B) \subset u(B_0) + V$.

The preceding result will enable us to determine the bounded sets in a product space $\prod_\alpha L_\alpha$. We omit the corresponding result for totally bounded sets.

5.5

If $\{L_\alpha : \alpha \in A\}$ is a family of t.v.s. and if $L = \prod_\alpha L_\alpha$, a subset B of L is bounded if and only if $B \subset \prod_\alpha B_\alpha$, where each B_α $(\alpha \in A)$ is bounded in L_α.

Proof. It is easy to verify from the definition of the product topology that if B_α is bounded in L_α $(\alpha \in A)$, then $\prod_\alpha B_\alpha$ is bounded in L; on the other hand, if B is bounded in L, then $u_\alpha(B)$ is bounded in L_α, since the projection map u_α of L onto L_α is continuous $(\alpha \in A)$, and, clearly, $B \subset \prod_\alpha u_\alpha(B)$.

Thus a fundamental system of bounded sets in $\prod_\alpha L_\alpha$ is obtained by forming all products $\prod_\alpha B_\alpha$, where B_α is any member of a fundamental system of bounded sets in $L_\alpha(\alpha \in A)$. Further, if L is a t.v.s. and M a subspace of L, a set is bounded in M if and only if it is bounded as a subset of L; on the other hand, a bounded subset of L/M is not necessarily the canonical image of a bounded set in L (Chapter IV, Exercises 9, 20).

A t.v.s. L is **quasi-complete** if every bounded, closed subset of L is complete; this notion is of considerable importance for non-metrizable t.v.s. By (5.1), Corollary 2, every quasi-complete t.v.s. is semi-complete; many results on quasi-complete t.v.s. are valid in the presence of semi-completeness, although there are some noteworthy exceptions (Chapter IV, Exercise 21). Note also that in a quasi-complete Hausdorff t.v.s., every precompact subset is relatively compact.

5.6

The product of any number of quasi-complete t.v.s. is quasi-complete.

The proof is immediate from the fact that the product of any number of complete uniform spaces is complete, and from (5.5).

6. METRIZABILITY

A t.v.s. $(L, \mathfrak{T})$ is **metrizable** if its topology $\mathfrak{T}$ is metrizable, that is, if there exists a metric on L whose open balls form a base for $\mathfrak{T}$. We point out that the uniformity generated by such a metric need not be translation-invariant and can hence be distinct from the uniformity associated with $\mathfrak{T}$ by (1.4) (Exercise 13). However, as we have agreed earlier, any uniformity notions to be employed in connection with any t.v.s. (metrizable or not) refer to the uniformity $\mathfrak{N}$ of (1.4).

It is known from the theory of uniform spaces that a separated uniform space is metrizable if and only if its vicinity filter has a countable base. For topological vector spaces, the following more detailed result is available.

6.1

Theorem. *A Hausdorff t.v.s. L is metrizable if and only if it possesses a countable neighborhood base of 0. In this case, there exists a function $x \rightarrow |x|$ on L into $\mathbf{R}$ such that:*

(i) $|\lambda| \leq 1$ *implies* $|\lambda x| \leq |x|$ *for all* $x \in L$.
(ii) $|x + y| \leq |x| + |y|$ *for all* $x \in L$, $y \in L$.
(iii) $|x| = 0$ *is equivalent with* $x = 0$.
(iv) *The metric* $(x, y) \rightarrow |x - y|$ *generates the topology of L.*

We note that (i) implies $|x| = |-x|$ and that (i) and (iii) imply $|x| \geq 0$ for all $x \in L$. Moreover, since the metric $(x, y) \rightarrow |x - y|$ is translation-invariant, it generates also the uniformity of the t.v.s. L.

A real function $x \rightarrow |x|$, defined on a vector space L over K and satisfying (i) through (iii) above, is called a **pseudo-norm** on L. It is clear that a given pseudo-norm on L defines, via the metric $(x, y) \rightarrow |x - y|$, a topology $\mathfrak{T}$ on L satisfying $(LT)_1$; on the other hand, $(LT)_2$ is not necessarily satisfied (Exercise 12). However, if $x \rightarrow |x|$ is a pseudo-norm on L such that $\lambda_n \rightarrow 0$ implies $|\lambda_n x| \rightarrow 0$ for each $x \in L$ and $|x_n| \rightarrow 0$ implies $|\lambda x_n| \rightarrow 0$ for each $\lambda \in K$, then it follows from (i) and the identity

$$\lambda x - \lambda_0 x_0 = \lambda_0 (x - x_0) + (\lambda - \lambda_0) x_0 + (\lambda - \lambda_0)(x - x_0)$$

that the topology $\mathfrak{T}$ defined by $x \rightarrow |x|$ satisfies $(LT)_2$, and hence that $(L, \mathfrak{T})$ is a t.v.s. over K.

Proof of (6.1). Let $\{V_n : n \in N\}$ be a base of circled 0-neighborhoods satisfying

$$V_{n+1} + V_{n+1} \subset V_n \qquad (n \in N). \tag{1}$$

For each non-empty finite subset H of N, define the circled 0-neighborhood V_H by $V_H = \sum_{n \in H} V_n$ and the real number p_H by $p_H = \sum_{n \in H} 2^{-n}$. It follows from (1) by induction on the number of elements of H that these implications hold:

$$p_H < 2^{-n} \Rightarrow n < H \Rightarrow V_H \subset V_n, \tag{2}$$

where $n < H$ means that $n < k$ for all $k \in H$. We define the real-valued function $x \to |x|$ on L by $|x| = 1$ if x is not contained in any V_H, and by

$$|x| = \inf_H \{p_H \colon x \in V_H\}$$

otherwise; the range of this function is contained in the real unit interval. Since each V_H is circled, (i) is satisfied. Let us show next that the triangle inequality (ii) is valid. This is evident for each pair (x, y) such that $|x| + |y| \geq 1$. Hence suppose that $|x| + |y| < 1$. Let $\varepsilon > 0$ be any real number such that $|x| + |y| + 2\varepsilon < 1$; there exist non-empty finite subsets H, K of N such that $x \in V_H$, $y \in V_K$ and $p_H < |x| + \varepsilon$, $p_K < |y| + \varepsilon$. Since $p_H + p_K < 1$, there exists a unique finite subset M of N for which $p_M = p_H + p_K$; by virtue of (1), M has the property that $V_H + V_K \subset V_M$. It follows that $x + y \in V_M$ and hence that

$$|x + y| \leq p_M = p_H + p_K < |x| + |y| + 2\varepsilon,$$

which proves (ii).

For any $\varepsilon > 0$, let $S_\varepsilon = \{x \in L \colon |x| \leq \varepsilon\}$; we assert that

$$S_{2^{-n-1}} \subset V_n \subset S_{2^{-n}} \qquad (n \in N). \tag{3}$$

The inclusion $V_n \subset S_{2^{-n}}$ is obvious since $x \in V_n$ implies $|x| \leq 2^{-n}$. On the other hand, if $|x| \leq 2^{-n-1}$, then there exists H such that $x \in V_H$ and $p_H < 2^{-n}$; hence (2) implies that $x \in V_n$.

It is clear from (3) that (iii) holds, since L is a Hausdorff space and hence $x = 0$ is equivalent with $x \in \bigcap\{V_n \colon n \in N\}$. Moreover, (3) shows that the family $\{S_\varepsilon \colon \varepsilon > 0\}$ is a neighborhood base of 0 in L; since the topology generated by the metric $(x, y) \to |x - y|$ is translation-invariant, (iv) also holds. This completes the proof.

REMARK. It is clear from the preceding proof that on every non-Hausdorff t.v.s. L over K possessing a countable neighborhood base of 0, there exists a real-valued function having properties (i), (ii) and (iv) of (6.1).

If L is a metrizable t.v.s. over K and if $x \to |x|$ is a pseudo-norm generating the topology of L, this pseudo-norm is clearly uniformly continuous; hence it has a unique continuous extension, $\tilde{x} \to |\tilde{x}|$, to the completion $\tilde{L}$ of L. We conclude from (1.5) that this extension, which is obviously a pseudo-norm on $\tilde{L}$, generates the topology of $\tilde{L}$.

Example. Denote by I the real unit interval and by μ Lebesgue measure on I. Further let $\mathscr{L}^p$ $(p > 0)$ be the vector space over R of all real-valued, μ-measurable functions for which $|f|^p$ (where $|f|$ denotes the function $t \to |f(t)|$) is μ-integrable, and let L^p be the quotient space of $\mathscr{L}^p$ over the subspace of μ-null functions. If $p \leq 1$,

$$f \to \int_I |f|^p \, d\mu$$

is a pseudo-norm on L^p, and it is easy to verify that L^p is complete under the corresponding topology. If $p < 1$, L^p is an example of a Hausdorff t.v.s. on which there exists no continuous linear form other than 0 (Exercise 6).

A t.v.s. L is said to be **locally bounded** if L possesses a bounded neighborhood of 0; clearly, such a space has a neighborhood base of 0 consisting of bounded sets. The spaces L^p of the preceding paragraph are locally bounded. We shall encounter further examples in Chapter II, Section 2.

6.2

Every locally bounded Hausdorff t.v.s. is metrizable.

Proof. Let V be a bounded 0-neighborhood in L and let $\{\lambda_n\}$ be a sequence of non-zero elements of K such that $\lim \lambda_n = 0$. If U is any circled neighborhood of 0, there exists $\lambda \in K$ such that $V \subset \lambda U$, since V is bounded; if n is such that $|\lambda_n \lambda| \leq 1$, then $\lambda_n V \subset U$, since U is circled. It follows that $\{\lambda_n V : n \in N\}$ is a 0-neighborhood base, whence L is metrizable by (6.1).

A quasi-complete, locally bounded t.v.s is complete, since it possesses a complete neighborhood of 0. We observe that the converse of (6.2) is false; an example is furnished by the product of a countably infinite number of one-dimensional t.v.s. which is metrizable (see below), but not locally bounded by (5.5).

Clearly, every subspace M of a metrizable t.v.s. is metrizable; if $x \to |x|$ is a pseudo-norm on L generating its topology, the restriction of $x \to |x|$ to M generates the topology of M. Let $L = \prod_n L_n$ be the product of countably many metrizable t.v.s. Since the product topology is metrizable, (6.1) implies that it can be generated by a pseudo-norm. Such a pseudo-norm can be constructed explicitly if, on each factor $L_n (n \in N)$, a generating pseudo-norm $x \to |x|_n$ is given: Writing $x = (x_n)$,

$$x \to |x| = \sum_{n=1}^{\infty} \frac{1}{2^n} \frac{|x_n|_n}{1 + |x_n|_n}$$

is a generating pseudo-norm on L. It is not difficult to verify conditions (i)–(iv) of (6.1); for (i) and (ii), recall that $u \to u/(1 + u)$ is monotone for $u \geq 0$ and that

$$\frac{a + b}{1 + a + b} \leq \frac{a}{1 + a} + \frac{b}{1 + b}$$

for any two real numbers $a, b \geq 0$. We leave it to the reader to verify that $x \to |x|$ generates the product topology on L.

For a quotient space L/M of a metrizable t.v.s. to be metrizable, M must necessarily be closed by (2.3); this condition is also sufficient. In terms of a generating pseudo-norm on L, we prove the following more detailed result.

6.3

The quotient space of a metrizable t.v.s. L over a closed subspace M is metrizable, and if L is complete then L/M is complete. If $x \rightarrow |x|$ is a pseudo-norm generating the topology of L, then (with $\hat{x} = x + M$)

$$\hat{x} \rightarrow |\hat{x}| = \inf\{|x|: x \in \hat{x}\}$$

is a pseudo-norm generating the topology of L/M.

Proof. We note first that $\hat{x} \rightarrow |\hat{x}|$ satisfies (i)–(iii) of (6.1). Clearly, $|\hat{0}| = 0$; if $|\hat{x}| = 0$, then $0 \in \hat{x}$, since M is closed. For (ii), let $\varepsilon > 0$ be given; then $|x| < |\hat{x}| + \varepsilon$, $|y| < |\hat{y}| + \varepsilon$ for suitable $x \in \hat{x}$, $y \in \hat{y}$; now,

$$|\hat{x} + \hat{y}| \leq |x + y| \leq |x| + |y| \leq |\hat{x}| + |\hat{y}| + 2\varepsilon.$$

(i) follows from the corresponding property of $x \rightarrow |x|$ on L, since the quotient map $x \rightarrow \hat{x}$ is linear.

Let $V_n = \{x: |x| < n^{-1}\}(n \in N)$. $\{V_n\}$ is a 0-neighborhood base in L; hence $\{\phi(V_n)\}$ is a 0-neighborhood base in L/M, since the natural map $x \rightarrow \hat{x} = \phi(x)$ is both open and continuous. We set $\hat{V}_n = \{\hat{x}: |\hat{x}| < n^{-1}\}$ and claim that $\hat{V}_n = \phi(V_n)$ for $n \in N$. Clearly, $\phi(V_n) \subset \hat{V}_n$. Conversely, if $\hat{x} \in \hat{V}_n$, there exists $x \in \hat{x}$ such that $x \in V_n$; hence $\phi^{-1}(\hat{V}_n) \subset V_n + M$, which implies $\hat{V}_n \subset \phi(V_n)$. Thus $\hat{x} \rightarrow |\hat{x}|$ generates the topology of L/M.

There remains to show that L/M is complete when L is complete. Given a Cauchy sequence in L/M, there exists a subsequence $\{\hat{x}_n\}$ such that $|\hat{x}_{n+1} - \hat{x}_n| < 2^{-n-1}(n \in N)$. Hence there exist representatives $y_{n+1} \in \hat{x}_{n+1} - \hat{x}_n$ such that $|y_{n+1}| < 2^{-n}$. Let $x_1 \in \hat{x}_1$ be arbitrarily chosen; then $x_n = x_1 + \sum_{v=2}^{n} y_v \in \hat{x}_n$ for all $n \geq 2$. Using condition (ii) of (6.1), it is readily verified that $\{x_n\}$ is a Cauchy sequence in L, hence convergent to some $x \in L$. Since ϕ is continuous, $\{\hat{x}_n\}$ converges in L/M; thus the given Cauchy sequence converges, which shows L/M to be complete.

We point out that if L is a non-metrizable, complete t.v.s. and if M is a closed subspace of L, the quotient space L/M is, in general, not complete (cf. Chapter IV, Exercise 11).

7. COMPLEXIFICATION

In this section we consider vector spaces over a more restricted type of fields K than were admissible so far: We assume that either K is a subfield of R, or else that K is a subfield of C containing the imaginary unit i and invariant under conjugation; in both cases, K is understood to carry the induced absolute value.

Any such field can be written as $K = H + iH$ if it contains the imaginary unit i; $H = K \cap R$ is a subfield of R. If H is, on the other hand, a subfield of R, let us denote by $H(i) = H + iH$ the complex extension of H. If L is a vector space over H, can scalar multiplication in L be extended to $K = H(i)$? If it can, then L possesses an automorphism u such that $u^2 = -e$ (e the identity mapping); namely, $x \to ix$ is such an automorphism. Conversely, if u is an automorphism of L (over H) satisfying $u^2 = -e$, then the definition $(\lambda, \mu \in H)$

$$(\lambda + i\mu)x = \lambda x + \mu u(x) \tag{1}$$

extends scalar multiplication to $K = H + iH$, which can be quickly verified. Similarly, if L is a t.v.s. over H and if u is a (topological) automorphism of L such that $u^2 = -e$, then (1) makes L into a t.v.s. over K.

7.1

If L is a t.v.s. over $H \subset R$, scalar multiplication in L has a continuous extension to $H(i) \times L$ into L if and only if L permits an automorphism u satisfying $u^2 = -e$.

Conversely, if L is a vector space (or t.v.s.) over a field $K = H(i)$ containing i, then the restriction of scalar multiplication to $H \times L$ turns L into a vector space (or t.v.s.) L_0 over H. L_0 will be called the **real underlying space** of (or associated with) L. A **real linear form** on L is a linear form on L_0, and a **real hyperplane** in L is a hyperplane in L_0. Accordingly, a **real subspace** (**real affine subspace**) of L is a subspace (affine subspace) of L_0.

Let L be a vector space over $K = H + iH$ and let $f \in L^*$ be a linear form on L. Then $f = g + ih$, where g, h are uniquely determined real-valued (more precisely, H-valued) functions on L; obviously g and h are real linear forms on L, called the **real** and **imaginary parts** of f, respectively. Since $g(ix) + ih(ix) = f(ix) = if(x) = ig(x) - h(x)$ for all $x \in L$, we have

$$f(x) = g(x) - ig(ix) \qquad (x \in L). \tag{2}$$

Conversely, if g is any real linear form on L, then f, defined by (2), is a member of L^* (verification is left to the reader), and obviously the only one with real part g. Moreover, if L is a t.v.s. over K, then (2) shows that f is continuous if and only if g is continuous. We have proved:

7.2

Let L be a t.v.s. over K and let L_0 be its real underlying space. The mapping $f \to g$ defined by (2) is an isomorphism of $(L^*)_0$ onto $(L_0)^*$, carrying the space of continuous linear forms on L onto the space of continuous linear forms on L_0.

For hyperplanes in L, we have the following result:

7.3

Let L be a t.v.s. over K. Every (closed) hyperplane in L is the intersection of a uniquely determined pencil of (closed) real hyperplanes.

Proof. By (4.1), a hyperplane G in L is of the form $G = \{x : f(x) = \gamma\}$, where $f \in L^*$ and $\gamma = \alpha + i\beta \in K = H + iH$. If g is the real part of f, then, clearly, $G = G_1 \cap G_2$, where $G_1 = \{x : g(x) = \alpha\}$, $G_2 = \{x : g(ix) = -\beta\}$. Since f is determined by G to within a non-zero factor, G_1 and G_2 determine the unique pencil whose intersection is G. Moreover, by (4.2) and (7.2), G_1 and G_2 are closed if and only if G is closed in L.

If L is a vector space over a field $H \subset \mathbf{R}$, there does not always exist an automorphism u of L satisfying $u^2 = -e$; examples are furnished by real vector spaces of finite odd dimension. It is still often desirable, especially for the purposes of spectral theory, to imbed L isomorphically into a vector space over $K = H(i)$; the following procedure will provide such an imbedding. Consider the product $L \times L$ over H. The mapping $u : (x, y) \rightarrow (-y, x)$ is an automorphism (which is topological if L is a t.v.s. over H) of $L \times L$ satisfying $u^2 = -e$; thus scalar multiplication can be extended to $K \times L \times L$ into $L \times L$ by (1). Thus $i(y, 0) = (0, y)$, and if we agree to write $(x, 0) = x$ for all $x \in L$, then each $z \in L \times L$ has a unique representation $z = x + iy$ with $x \in L$, $y \in L$. If L is a t.v.s. over H, then $L \times L$ over K is a t.v.s. such that $(L \times L)_0 = L \oplus iL$. This type of imbedding is called the **complexification** of a vector space (or t.v.s.) defined over a subfield of $\mathbf{R}$.

It can be shown (Exercise 16) that every vector space over a conjugation invariant field $K \subset \mathbf{C}$ such that K contains i, is algebraically isomorphic to the complexification of any one of its maximal properly real subspaces.

EXERCISES

1. Let $\{L_\alpha : \alpha \in A\}$ be a family of Hausdorff t.v.s. over K and denote by $\mathfrak{B}$ the family of subsets of the vector space $L = \prod_\alpha L_\alpha$ obtained by forming all products $V = \prod_\alpha V_\alpha$, where $V_\alpha (\alpha \in A)$ is any member of a 0-neighborhood base in L_α. Let $\mathfrak{T}$ denote the unique translation-invariant topology on L for which $\mathfrak{B}$ is a neighborhood base of 0. Let M be the subspace of L containing exactly those elements $x \in L$ which have only a finite number of non-zero coordinates (M is denoted by $\oplus_\alpha L_\alpha$ and called the **algebraic direct sum** of the family $\{L_\alpha\}$).

   (a) If an infinite number of the spaces L_α are not reduced to $\{0\}$, $(L, \mathfrak{T})$ is not a t.v.s.

   (b) $(M, \mathfrak{T})$ is a Hausdorff t.v.s. which is complete if and only if each L_α is complete.

   (c) A subset of M is bounded in $(M, \mathfrak{T})$ if and only if it is contained in a set of the form $\prod_{\alpha \in H} B_\alpha \times \{0\}$, where $H \subset A$ is finite and B_α is bounded in L_α for $\alpha \in H$.

2. Let L be a t.v.s. which is not a Hausdorff space, and denote by N the closure of $\{0\}$.

(a) The topology of the subspace N is the trivial topology whose only members are N and $\emptyset$. If M is any algebraic complementary subspace of N in L, then $L = M \oplus N$ and M is isomorphic with the Hausdorff t.v.s. associated with L (use (2.4).)

(b) Deduce from this that every t.v.s. L is isomorphic with a dense subspace of a complete t.v.s. over the same field.

(c) Show that a subset A of L is totally bounded if and only if the canonical image of A in L/N is precompact.

3. Give an example of a finite-dimensional t.v.s. L over a (non-complete) field K such that the completion of L is infinite-dimensional over K, and locally compact.

4. Let Q be the rational number field under its usual absolute value and let $L = Q + Q\sqrt{2}$. Show that L, under the topology induced by R, is a t.v.s. over Q not isomorphic with $Q_0 \times Q_0$.

5. Let B be a subset of a t.v.s. such that every sequence in B has a cluster point; then B is totally bounded. (For a given circled 0-neighborhood V, let B_0 be a subset of B such that $x \in B_0$, $y \in B_0$, and $x \neq y$ imply $x - y \notin V$, and which is maximal with respect to this property (Zorn's lemma); then $B \subset B_0 + V$. Show that the assumption "B_0 is infinite" is absurd.)

6. Let $L^p (0 < p < 1)$ be the vector space over R introduced in Section 6, under the topology generated by the pseudo-norm $f \to |f|_1 = \int |f|^p \, d\mu$. Show that L^p is a complete t.v.s. on which there exist no non-zero continuous linear forms. (If $u \neq 0$ is a continuous linear form, then $|u(f)| = 1$ for some $f \in L^p$. Denote by z_s $(0 \leqq s \leqq 1)$ the characteristic function of $[0, s] \subset [0, 1]$; there exists t such that $|fz_t|_1 = |f(1 - z_t)|_1 = \frac{1}{2}|f|_1$. For at least one of the functions fz_t and $f(1 - z_t)$, call it $\frac{1}{2}f_1$, one has $|u(\frac{1}{2}f_1)| \geqq \frac{1}{2}$. Moreover, $|f_1|_1 = 2^{p-1}|f|_1$. By induction, define a sequence $\{f_n\}$ such that $|u(f_n)| \geqq 1$ and $|f_n|_1 = 2^{n(p-1)}|f|_1$.) (M. Day [1], W. Robertson [1]).

7. Let L be a t.v.s. Show these assertions to be equivalent:
(a) Every subspace of finite codimension is dense in L.
(b) There exist no closed hyperplanes in L.
(c) No finite-dimensional subspace has a complementary subspace in L.

8. Construct a decomposition $L = M + N$ of a t.v.s. L such that $M + N$ is an algebraic, but not a topological, direct sum (use Exercise 4 or Exercises 6, 7).

9. The dimension of a complete metrizable t.v.s. over a complete field K is either finite or uncountably infinite (use Baire's theorem).

10. Let $\{L_\alpha : \alpha \in A\}$ be a family of metrizable t.v.s. The product $\prod_\alpha L_\alpha$ is metrizable only if A is countable, and the direct sum $\oplus_\alpha L_\alpha$ (the space $(M, \mathfrak{T})$ of Exercise 1(b)) is metrizable only if A is finite.

11. Deduce from Exercise 10 an example of a complete Hausdorff t.v.s. of countable dimension which is not metrizable.

12. (a) Let L be a vector space over K and let d be a translation-invariant metric on L such that metric space (L, d) is complete. Suppose, in addition, that $\lambda_n \to 0$ implies $d(\lambda_n x, 0) \to 0$ for each $x \in L$ and that $d(x_n, 0) \to 0$ implies $d(\lambda x_n, 0) \to 0$ for each $\lambda \in K$. Show that under the

topology generated by d, L is a complete t.v.s. (Use Baire's theorem to show that $(LT)_2$ holds.)

(b) Let L be the vector space over $\mathbf{R}$ of all real-valued continuous functions on $\mathbf{R}$. Show that $f \to \sup_{t \in \mathbf{R}} |f(t)|/(1 + |f(t)|)$ is a pseudo-norm on L, and that under the topology generated by this pseudo-norm, L is a complete topological group with respect to addition, but not a t.v.s.

13. The metric $d(x, y) = \tan^{-1}|x - y|$ generates the unique topology on $\mathbf{R}$ under which $\mathbf{R}_0$ is a Hausdorff t.v.s., but the uniformity generated by d (under which $\mathbf{R}$ is precompact) is distinct from the uniformity of the t.v.s. $\mathbf{R}_0$.

14. Let d be a metric on a vector space L such that under the topology $\mathfrak{T}$ generated by d, L is a t.v.s., and such that the metric space (L, d) is complete. Then the t.v.s. $(L, \mathfrak{T})$ is complete. (V. L. Klee [1].)

15. Show that on the vector spaces $\mathbf{R}^{2n+1}(n \in N)$ there exists no automorphism u satisfying $u^2 = -e$.

16. Let L be a vector space over a subfield $K = H + iH$ of $\mathbf{C}$, where H is a subfield of $\mathbf{R}$. Call a real subspace N of L properly real if $N \cap iN = \{0\}$. There exists a properly real subspace M of L such that $L = M + iM$. (Use Zorn's lemma.) Cf. Chapter IV, Exercise 3.

17. Every t.v.s. (Hausdorff or not) over $\mathbf{R}$ or $\mathbf{C}$ is connected and locally connected.

18. Find a formula relating the cardinality of a vector space L over K with its dimension. Prove that if $\dim L \geqq \operatorname{card} K$, then $\dim L^* = (\operatorname{card} K)^{\dim L}$.

Chapter II

LOCALLY CONVEX
TOPOLOGICAL VECTOR SPACES

Since convexity will play a central role in all following chapters, *the scalar field K over which vector spaces are defined is from now on assumed to be the real field **R** or the complex field **C***, unless the contrary is expressly stated. In most definitions and results (for example, the Hahn-Banach theorem) we shall not find it necessary to distinguish between the real and complex case. When several vector spaces occur in one statement and no explicit mention of the respective scalar fields is made, the spaces involved are assumed to be defined over the same field K, where either $K = \mathbf{R}$ or $K = \mathbf{C}$. If $K = \mathbf{C}$, $\mathbf{R}$ will be considered a subfield and restriction of scalars to $\mathbf{R}$ will be indicated by the use of the adjective "real" (Chapter I, Section 7). In particular, the symbols $>$ and $\geqq$, when used between scalars, refer to the customary order in $\mathbf{R}$; for example, "$\lambda > 0$" means "$\lambda \in \mathbf{R}$ and $\lambda > 0$".

The theory of general topological vector spaces whose elements have been presented in Chapter I can be extended in several directions (see, e.g., Bourgin [1], Hyers [1], [2], Landsberg [1], [2]); it remains on the whole an unsatisfactory theory, devoid of a great number of valuable results both from the pure and applied viewpoints. The concept of topological vector space, as defined before, is too general to support a rich theory just as, on the other hand, the concept of Banach space is too narrow. The notion on which a satisfying and applicable theory can be built is that of local convexity; it is the purpose of this chapter to acquaint the reader with the elementary properties of topological vector spaces (over $\mathbf{R}$ or $\mathbf{C}$) in which each point has a base of convex neighborhoods.

Section 1 gives some topological properties of convex sets and introduces semi-norms, a useful tool for the analytical description of certain convex sets. Section 2 is devoted to a brief discussion of normable and normed spaces. The literature on such spaces is vast, and the reader ought to be familiar with their elementary theory, which is now part of every first course in abstract analysis; we confine ourselves to some basic results and a review of the most frequent examples of normed spaces, including some facts on Hilbert space that are recorded for later use. Section 3 proves the Hahn-Banach theorem in its two forms called by Bourbaki [7] the geometrical and analytical forms, respectively. This is the central result of the chapter and fundamental for most of what follows later; it lends power to the notion of locally convex space (due to J. von Neumann [1]), defined in Section 4. Continuous semi-norms constitute an analytical alternative for the use of convex circled 0-neighborhoods which is illustrated by the two forms of the Hahn-Banach theorem; but while applications often suggest the use of semi-norms, we feel that their exclusive or even preferred use does not support the geometrical clarity of the subject.

The separation properties of convex sets, all consequences of the geometrical form of the Hahn-Banach theorem, could logically follow Section 4; we have preferred to place them at the end of the chapter so that the reader would first have a survey of the class of spaces in which those separation results are valid. Following a method extremely useful even in general topology (cf. Prerequisites), we hope to give the reader an efficient way to organize the various means of generating new locally convex spaces from those of a given family, by simply distinguishing between projective and inductive topologies. With the exception of spaces of linear mappings and topological tensor products (Chapter III), Sections 5 and 6 of the present chapter give all standard methods for constructing locally convex spaces. It is interesting to observe ((5.4), Corollary 2) that every locally convex space can be obtained as a subspace of a suitable product of Banach spaces. Two classes of spaces particularly frequent in applications are discussed in Sections 7 and 8. Section 9 furnishes the standard separation theorems which are constantly used later. The chapter closes with a rather compressed approach, following Bourbaki [7], to the Krein-Milman theorem. This is a beautiful and important theorem of which everyone interested in topological vector spaces should be aware; however, it has little bearing on the theory to be presented here, and we refer to Klee [3]–[5] for a deep analysis and the many ramifications of this result.

1. CONVEX SETS AND SEMI-NORMS

A subset A of a vector space L is **convex** if $x \in A$, $y \in A$ imply that $\lambda x + (1 - \lambda)y \in A$ for all scalars λ satisfying $0 < \lambda < 1$. The sets $\{\lambda x + (1 - \lambda)y: 0 \leq \lambda \leq 1\}$ and $\{\lambda x + (1 - \lambda)y: 0 < \lambda < 1\}$ are called the closed and open **line segments**, respectively, joining x and y. It is immediate that convexity of a subset $A \subset L$ is preserved under translations: A is convex if (and only if) $x_0 + A$ is convex for every $x_0 \in L$.

1.1

Let A be a convex subset of a t.v.s. L. If x is interior to A and y in the closure of A, the open line segment joining x and y is interior to A.

Proof. Let $\lambda, 0 < \lambda < 1$, be fixed; we have to show that $\lambda x + (1 - \lambda)y \in \mathring{A}$. By a translation if necessary we can arrange that $\lambda x + (1 - \lambda)y = 0$. Now $y = \alpha x$ where $\alpha < 0$. Since $w \to \alpha w$ is a homeomorphism of L by (I, 1.1)* and $x \in \mathring{A}$, $y \in \bar{A}$, there exists a $z \in \mathring{A}$ such that $\alpha z \in A$. Let $\mu = \alpha/(\alpha - 1)$; then $0 < \mu < 1$ and $\mu z + (1 - \mu)\alpha z = 0$. Hence

$$U = \{\mu w + (1 - \mu)\alpha z : w \in \mathring{A}\}$$

is a neighborhood of 0 since $w \to \mu w + (1 - \mu)\alpha z$ is a homeomorphism of L mapping $z \in \mathring{A}$ onto 0. But $w \in \mathring{A}$ and $\alpha z \in A$ imply $U \subset A$, since A is convex; hence $0 \in \mathring{A}$.

1.2

Let L be a t.v.s. and let A and B be convex subsets of L. Then $\mathring{A}$, $\bar{A}$, $A + B$ and $\alpha A (\alpha \in K)$ are convex.

The convexity of $\mathring{A}$ is immediate from (1.1); if λ is fixed, $0 < \lambda < 1$, then $\lambda A + (1 - \lambda)A \subset A$ whence $\lambda \bar{A} + (1 - \lambda)\bar{A} \subset \bar{A}$ by $(LT)_1$ and $(LT)_2$ (Chapter I, Section 1); thus $\bar{A}$ is convex. The proof that $A + B$ and αA are convex is left to the reader.

1.3

If A is convex with non-empty interior, then the closure $\bar{A}$ of A equals the closure of $\mathring{A}$, and the interior $\mathring{A}$ of A equals the interior of $\bar{A}$.

Proof. Since $\mathring{A} \subset A$, $(\bar{\mathring{A}}) \subset \bar{A}$ holds trivially. If A is convex and $\mathring{A}$ non-empty, (1.1) shows that $\bar{A} \subset (\bar{\mathring{A}})$. To prove the second assertion, it suffices to show that $0 \in (\mathring{\bar{A}})$ implies $0 \in \mathring{A}$ if A is convex with non-empty interior. There exists a circled neighborhood V of 0 such that $V \subset \bar{A}$. Since $\bar{A} = (\bar{\mathring{A}})$, 0 is in the closure of $\mathring{A}$; hence $\mathring{A}$ and V intersect. Let $y \in \mathring{A} \cap V$. Since $V \subset \bar{A}$ and V is circled, we have $-y \in \bar{A}$ and it follows now from (1.1) that $0 \in \mathring{A}$, since $0 = \tfrac{1}{2}y + \tfrac{1}{2}(-y)$.

A **cone** C of vertex 0 is a subset of a vector space L invariant under all homothetic maps $x \to \lambda x$ of strictly positive ratio λ; if, in addition, C is convex, then C is called a **convex cone** of vertex 0. Thus a convex cone of vertex 0 is a subset of L such that $C + C \subset C$ and $\lambda C \subset C$ for all $\lambda > 0$. A (convex) cone of vertex x_0 is a set $x_0 + C$, where C is a (convex) cone of vertex 0. It is a simple exercise to show that the interior and the closure of a (convex) cone of vertex 0 in a t.v.s. L are (convex) cones of vertex 0 in L.

---

* Roman numeral refers to chapter number.

For subsets of a vector space L, the properties of being circled or convex are invariant under the formation of arbitrary intersections. Since L has both properties, every subset $A \subset L$ determines a unique smallest subset containing A and having any one or both of these properties, respectively: the circled hull, the convex hull, and the convex, circled hull, of A. The circled hull of A is the set $\{\lambda a: a \in A \text{ and } |\lambda| \leq 1\}$; if $A \neq \varnothing$, the **convex hull** of A is the set $\{\sum \lambda_v a_v\}$, where $\lambda_v > 0$, $\sum \lambda_v = 1$ and $\{a_v\}$ ranges over all nonempty finite subsets of A (Exercise 1). The **convex, circled hull** of A, denoted by ΓA, is the convex hull of the circled hull of A (Exercise 1). By $\Gamma_\alpha A_\alpha$, we denote the convex, circled hull of the union of a family $\{A_\alpha: \alpha \in A\}$.

If L is a t.v.s., the properties of being circled or convex (or both) can be combined with the property of being closed; obviously, the resulting notions are again intersection-invariant. In particular, the **closed, convex hull** (sometimes referred to as the convex closure) of $A \subset L$ is the closure of the convex hull of A, by (1.2); similarly, the **closed, convex, circled hull** of A is the closure of the convex, circled hull of A (Exercise 1).

We turn to the investigation of convex, radial subsets of a vector space L; certainly the convex hull of a radial set is of this type. If M is any radial subset of L, the non-negative real function on L:

$$x \to p_M(x) = \inf\{\lambda > 0: x \in \lambda M\},$$

is called the **gauge**, or **Minkowski functional**, of M. Obviously, if M is a radial set in L and $M \subset N$, then $p_N(x) \leq p_M(x)$ for all $x \in L$, that is $p_N \leq p_M$.

We define a **semi-norm** on L to be the gauge of a radial, circled and convex subset of L; a **norm** is a semi-norm p such that $p(x) = 0$ implies $x = 0$. The following analytical description of semi-norms is often used as a definition.

1.4

A real-valued function p on a vector space L is a semi-norm if and only if

(a) $p(x + y) \leq p(x) + p(y)$ $(x, y \in L)$

(b) $p(\lambda x) = |\lambda| p(x)$ $(x \in L, \lambda \in K)$.

Proof. Let p be a semi-norm on L, that is, let p be p_M, where M is radial, circled, and convex. If $x \in L$, $y \in L$ are given and $\lambda_1 > p(x)$, $\lambda_2 > p(y)$, then $x + y \in \lambda_1 M + \lambda_2 M$. Since M is convex,

$$\lambda_1 M + \lambda_2 M = (\lambda_1 + \lambda_2)\left[\frac{\lambda_1}{\lambda_1 + \lambda_2} M + \frac{\lambda_2}{\lambda_1 + \lambda_2} M\right] \subset (\lambda_1 + \lambda_2)M \ ;$$

this implies $p(x + y) \leq \lambda_1 + \lambda_2$; hence $p(x + y) \leq p(x) + p(y)$. For (b), observe that $\lambda x \in \mu M$ is equivalent with $|\lambda| x \in \mu M$, since M is circled; hence if $\lambda \neq 0$,

$$p(\lambda x) = \inf\{\mu > 0: x \in |\lambda|^{-1} \mu M\} = \inf_{\mu > 0}\{|\lambda|\mu: x \in \mu M\} = |\lambda| p(x) \ ;$$

this proves (b), since $p(0) = 0$.

Conversely, assume that p is a function satisfying (a) and (b), and let $M = \{x: p(x) < 1\}$. Clearly, M is radial and circled, and it follows from (a) and (b) that M is convex. We show that $p = p_M$. It follows from (b) that $\{x: p(x) < \lambda\} = \lambda M$ for every $\lambda > 0$; hence if $p(x) = \alpha$, then $x \in \lambda M$ for all $\lambda > \alpha$ but for no $\lambda < \alpha$, which proves that $p(x) = \inf\{\lambda > 0: x \in \lambda M\} = p_M(x)$.

Simple examples show that the gauge function p of a radial set $M \subset L$ does not determine M; however, if M is convex and circled, we have the following result, whose proof is similar to that of (1.4) and will be omitted.

1.5

Let M be a radial, convex, circled subset of L; for the semi-norm p on L to be the gauge of M, it is necessary and sufficient that $M_0 \subset M \subset M_1$, where $M_0 = \{x: p(x) < 1\}$ and $M_1 = \{x: p(x) \leqq 1\}$.

If L is a topological vector space, the continuity of a semi-norm p on L is governed by the following relationship.

1.6

Let p be a semi-norm on the t.v.s. L. These properties of p are equivalent:

(a) *p is continuous at $0 \in L$.*
(b) *$M_0 = \{x: p(x) < 1\}$ is open in L.*
(c) *p is uniformly continuous on L.*

Proof. (a) $\Rightarrow$ (c), since by (1.4), $|p(x) - p(y)| \leqq p(x - y)$ for all $x, y \in L$. (c) $\Rightarrow$ (b), since $M_0 = p^{-1}[(-\infty, 1)]$. (b) $\Rightarrow$ (a), since $\varepsilon M_0 = \{x: p(x) < \varepsilon\}$ for all $\varepsilon > 0$.

A subset of a t.v.s. L that is closed, convex, and has non-empty interior is called a **convex body** in L. Thus if p is a continuous semi-norm on L, $M_1 = \{x: p(x) \leqq 1\}$ is a convex body in L.

2. NORMED AND NORMABLE SPACES

By the definition given in Section 1, a norm p on a vector space L (over R or C) is the gauge of a convex, circled, radial set which contains no subspace of L other than $\{0\}$; frequently a norm is denoted by $\| \ \|$. We recall from (1.4) that a norm $\| \ \|$ on L is characterized by these analytical properties:

(i) $\|\lambda x\| = |\lambda| \ \|x\|$ *for all $\lambda \in K$, $x \in L$.*
(ii) $\|x + y\| \leqq \|x\| + \|y\|$ *for all $x \in L$, $y \in L$.*
(iii) $\|x\| = 0$ *implies $x = 0$.*

We define a **normed space** to be a pair $(L, \| \ \|)$, with the understanding that L carries the topology generated by the metric $(x, y) \to \|x - y\|$. Under this topology, L is a t.v.s. (This is clear from the discussion following (I, 6.1), since a norm is also a special case of a pseudonorm (not only of a semi-norm); in view of this, it is possible to define normed spaces over arbitrary, non-discrete valued fields (Exercise 5), but we do not follow this usage.) By contrast, a **normable space** is a t.v.s. L whose topology can be obtained from a norm $\| \ \|$ on L via the metric $(x, y) \to \|x - y\|$; such a norm is, of course, not unique (Exercise 5). A complete normed space is called a **Banach space**, or briefly (B)-space. A norm preserving isomorphism of one normed space onto another is called a **norm isomorphism**, and two normed spaces are called **norm isomorphic** if there exists a norm isomorphism between them. The set $\{x: \|x\| \leq 1\}$ is the (closed) **unit ball** of $(L, \| \ \|)$.

There is a simple necessary and sufficient condition for a (necessarily Hausdorff) t.v.s. to be normable; the result is due to Kolmogoroff [1].

2.1

A Hausdorff t.v.s. L is normable if and only if L possesses a bounded, convex neighborhood of 0.

Proof. The condition is necessary, for if $x \to \|x\|$ generates the topology of L, $V_1 = \{x: \|x\| \leq 1\}$ is a convex neighborhood of 0 which is bounded, since, by (i) above, the multiples $\{n^{-1}V_1\}(n \in N)$ form a 0-neighborhood base in L. Conversely, if V is a convex, bounded 0-neighborhood in L, there exists a circled neighborhood contained in V whose convex hull U is bounded (since it is contained in V). Clearly, the gauge p of U is a norm on L. Now the boundedness of U implies that $\{n^{-1}U\}(n \in N)$ is a 0-neighborhood base, whence it follows that p generates the topology of L.

The completion $\tilde{L}$ of a normable space L is normable, for if V is a bounded, convex 0-neighborhood in L, its closure $\overline{V}$ in $\tilde{L}$ is bounded by (I, 1.5) and convex by (1.2). If (L, p) is a normed space, then p, which is uniformly continuous on L by (1.6), has a unique continuous extension $\bar{p}$ to $\tilde{L}$ that generates the topology of $\tilde{L}$; $(\tilde{L}, \bar{p})$ is a Banach space. It is obvious that a subspace of a normable space is normable and that a closed subspace of a Banach space is a Banach space. However,

2.2

The product of a family of normable spaces is normable if and only if the number of factors $\neq \{0\}$ is finite.

Proof. This follows quickly from (2.1), since, by (I, 5.5), a 0-neighborhood in the product $\prod_\alpha L_\alpha$ can be bounded if and only if the number of factors $L_\alpha \neq \{0\}$ is finite.

REMARK. A norm generating the topology of the product of a finite family of normed spaces can be constructed from the given norms in a variety of ways (Exercise 4).

2.3

The quotient space of a normable (and complete) t.v.s. L over a closed subspace M is normable (and complete). If $(L, \| \ \|)$ is a normed space, the norm $\hat{x} \to \|\hat{x}\| = \inf\{\|x\|: x \in \hat{x}\}$ generates the topology of L/M.

Proof. Since M is closed, L/M is Hausdorff by (I, 2.3); since the natural map ϕ of L onto L/M is linear, open, and continuous, $\phi(V)$ is a convex 0-neighborhood in L/M which is bounded by (I, 5.4) if V is a bounded, convex 0-neighborhood in L; thus L/M is normable by (2.1). By (I, 6.3) L/M is complete when L is, and the pseudonorm $\hat{x} \to \|\hat{x}\|$, which is easily seen to be a norm, generates the topology of L/M.

It is immediate that the bounded sets in a normed space L are exactly those subsets on which $x \to \|x\|$ is bounded. Thus if L, N are normed spaces over K, and u is a continuous linear map on L into N, it follows from (I, 5.4) that the number

$$\|u\| = \sup\{\|u(x)\|: x \in L, \|x\| \leq 1\}$$

is finite. It is easy to show that $u \to \|u\|$ is a norm on the vector space $\mathscr{L}(L, N)$ over K of all continuous linear maps on L into N. $\mathscr{L}(L, N)$ is a Banach space under this norm if N is a Banach space; in particular, if N is taken to be the one-dimensional Banach space $(K_0, | \ |)$ (cf. Chapter I, Section 3) then $L' = \mathscr{L}(L, K_0)$, endowed with the above norm, is a Banach space called the **strong dual** of L.

Examples

The following examples are intended to present some principal types of Banach spaces occurring in analysis. As normed spaces in general, these spaces have been widely covered in the literature (e.g., Day [2], Dunford-Schwartz [1], Köthe [5]), to which we refer for details.

1. Let X be a non-empty set. Denote by $B(X)$ the vector space over $K(K = R$ or $C)$ of all K-valued bounded functions; under the norm $f \to \|f\| = \sup\{|f(t)|: t \in X\}$, $B(X)$ is a Banach space. If X_0 is any subset of X, the subset of all $f \in B(X)$ vanishing on X_0 is a closed subspace.

If Σ is a σ-algebra of subsets of X (cf. Halmos [1]), let $M(X, \Sigma)$ be the set of all Σ-measurable functions in $B(X)$; $M(X, \Sigma)$ is a closed subspace of $B(X)$.

If X is a topological space and $\mathscr{C}(X)$ the set of all continuous functions in $B(X)$, $\mathscr{C}(X)$ is a closed subspace of $B(X)$.

An example of particular importance is the space $\mathscr{C}(X)$ when X is a compact (Hausdorff) space. Using the fact that all (except the second) of the preceding spaces are vector lattices of a particular type, it can be proved (Chapter V, Theorem 8.5) that each of them is norm-isomorphic to a space $\mathscr{C}(X)$, where X is a suitable compact space.

2. Let (X, Σ, μ) be a measure space in the sense of Halmos [1], so that μ is a non-negative (possibly infinity-valued), countably additive set function defined on the σ-algebra Σ of subsets of X. Denote by $\mathscr{L}^p$ the set of all Σ-measurable, scalar-valued functions f for which $|f|^p$ ($1 \leq p < \infty$) is μ-integrable; the well-known inequalities of Hölder and Minkowski (cf. Halmos, l.c.) show that $\mathscr{L}^p$ is a vector space and that $f \to (\int |f|^p d\mu)^{1/p}$ is a semi-norm on $\mathscr{L}^p$. If $\mathscr{N}_\mu$ denotes the subspace of $\mathscr{L}^p$ consisting of all μ-null functions and $[f]$ the equivalence class of $f \in \mathscr{L}^p$ mod $\mathscr{N}_\mu$, then

$$[f] \to (\int |f|^p d\mu)^{1/p}$$

is a norm on the quotient space $\mathscr{L}^p/\mathscr{N}_\mu$, which thus becomes a Banach space usually denoted by $L^p(\mu)$.

$\mathscr{L}^\infty$ commonly denotes the vector space of μ-essentially bounded Σ-measurable functions on X; a Σ-measurable function is μ-essentially bounded if its equivalence class mod $\mathscr{N}_\mu$ contains a bounded function. Thus $\mathscr{L}^\infty/\mathscr{N}_\mu$ is algebraically isomorphic with $M(X, \Sigma)/(\mathscr{N}_\mu \cap M(X, \Sigma))$; the latter quotient of $M(X, \Sigma)$ is a Banach space usually denoted by $L^\infty(\mu)$. $L^\infty(\mu)$ is again norm-isomorphic to $\mathscr{C}(X)$ for a suitable compact space X.

3. Let X be a compact space. Each continuous linear form $f \to \mu_0(f)$ on $\mathscr{C}(X)$ is called a **Radon measure** on X (Bourbaki [9], Chapter III). For each μ_0, there exists a unique regular signed (respectively, complex) Borel measure μ on X in the sense of Halmos [1] such that $\mu_0(f) = \int f d\mu$ for all $f \in \mathscr{C}(X)$. The correspondence $\mu_0 \to \mu$ is a norm isomorphism of the strong dual $\mathscr{M}(X)$ of $\mathscr{C}(X)$ onto the Banach space of all finite signed (respectively, complex) regular Borel measures on X, the norm $\|\mu\|$ being the total variation of μ. Because of this correspondence, one frequently writes $\mu_0(f) = \int f d\mu_0$.

Returning to the general case where (X, Σ, μ) is an arbitrary measure space (Example 2, above), let us note that for $1 < p < +\infty$, the strong dual of $L^p(\mu)$ can be identified with $L^q(\mu)$, where $p^{-1} + q^{-1} = 1$, in the sense that the correspondence $[g] \to ([f] \to \int fg \, d\mu)$ is a norm isomorphism of $L^q(\mu)$ onto the strong dual of $L^p(\mu)$. In the same fashion, the strong dual of $L^1(\mu)$ can be identified with $L^\infty(\mu)$ whenever μ is totally σ-finite. (For a complete discussion of the duality between $L^p(\mu)$ and $L^q(\mu)$, see Kelley-Namioka [1], 14. M.)

4. Let Z_0 denote the open unit disk in the complex plane. Denote by $H^p(1 \leq p < +\infty)$ the subspace of C^{Z_0} consisting of all functions which are holomorphic on Z_0 and for which

$$\|f\|_p = \sup_{0 < r < 1} \left(\int_0^{2\pi} |f(re^{it})|^p dt \right)^{1/p}$$

is finite; $f \to \|f\|_p$ is a norm on H^p under which H^p is a complex Banach space. Similarly, the space of all bounded holomorphic functions on Z_0 is a complex Banach space H^∞ under the norm $f \to \|f\|_\infty = \sup\{|f(\zeta)|: \zeta \in Z_0\}$; for details we refer to Hoffman [1].

The preceding examples can be substantially extended by considering Banach spaces of functions taking their values in an arbitrary (B)-space F.

5. An especially important class of Banach spaces is the class of Hilbert spaces. The presence of an inner product distinguishes Hilbert spaces quite drastically from general Banach spaces; the theory of Hilbert spaces, and in particular of their linear transformations, is elaborate and the literature is very extensive. For later use, we record here only the definition and the most elementary properties of Hilbert space; see also Bourbaki ([8], chap. V), Halmos [2], and Sz.-Nagy [1].

Let H be a vector space over C and let $(x, y) \rightarrow [x, y]$ be a complex-valued function on $H \times H$ such that the following conditions are satisfied (α^* denoting the complex conjugate of $\alpha \in C$):

(i) *For each $y \in H$, $x \rightarrow [x, y]$ is a linear form on H.*
(ii) *$[x, y] = [y, x]^*$ for all $x \in H$, $y \in H$.*
(iii) *$[x, x] \geqq 0$ for all $x \in H$.*
(iv) *$[x, x] = 0$ implies $x = 0$.*

The mapping $(x, y) \rightarrow [x, y]$ is called a **positive definite Hermitian form** (or **inner product**); $x \rightarrow \sqrt{[x, x]}$ is a norm $\| \ \|$ on H, and $(H, \| \ \|)$ is called a **pre-Hilbert** (or **inner product**) **space**. The inner product satisfies **Schwarz' inequality**: $|[x, y]| \leqq \|x\| \ \|y\|$. If the normed space $(H, \| \ \|)$ is complete (hence a Banach space), it is called a **Hilbert space**. The corresponding notion of real inner product space or real Hilbert space, respectively, is obtained by assuming $(x, y) \rightarrow [x, y]$ to be real valued and H to be a real vector space. A function on $H \times H$ satisfying (i) through (iii) but not necessarily (iv) is called a **positive semi-definite Hermitian form**; in these circumstances, $x \rightarrow \sqrt{[x, x]} = p(x)$ is a semi-norm on H, and the quotient space $H/p^{-1}(0)$ is an inner product space under $(\hat{x}, \hat{y}) \rightarrow [x, y]$, where $x \rightarrow \hat{x}$ denotes the canonical map of H onto $H/p^{-1}(0)$.

It is clear that the property of being an inner product (respectively, Hilbert) space is inherited by subspaces (respectively, by closed subspaces). More important, every closed subspace M of an inner product space H possesses a (topologically) complementary subspace: the subspace $M^{\perp} = \{x \in H: [x, y] = 0$ for all $y \in M\}$, called the subspace of H **orthogonal** to M, satisfies the relation $H = M \oplus M^{\perp}$. The projection of H on to M that vanishes on $M^{\perp}$ is called the **orthogonal projection** of H onto M. Hence for every closed subspace M of H, the quotient space H/M, being norm isomorphic with $M^{\perp}$, is an inner product space. A subset $\{x_{\alpha}: \alpha \in A\}$ of H is **orthonormal** if $[x_{\alpha}, x_{\beta}] = \delta_{\alpha\beta}$ for all $\alpha, \beta \in A$; any total orthonormal subset is called an **orthonormal basis** of H (cf. Chapter III, Section 9 and Exercise 23). The existence of orthonormal bases in every complete inner product space H is implied by Zorn's lemma, and it can be shown that every orthonormal basis of H has the same cardinality d; d is called the **Hilbert dimension** of H.

Every Hilbert space H is self-dual in the following sense: If f is a continuous linear form on H, there exists a unique element $z \in H$ such that $f(x) = [x, z](x \in H)$; $f \rightarrow z$ is a norm-preserving, additive map of the strong dual H' onto H under which αf is mapped onto $\alpha^* z (\alpha \in C)$, and which is therefore called **conjugate-linear**. ($f \to z$ is a norm isomorphism if H is a real Hilbert space.) This has an immediate consequence: If H_1, H_2 are Hilbert spaces and u is a continuous linear map of H_1 into H_2, then the identity $[u(x), y]_2 = [x, u^*(y)]_1$ on $H_1 \times H_2$ defines a continuous linear map u^* of H_2 into H_1, called the **conjugate** of u. It is easy to see that $\|u\| = \|u^*\|$ and that $u \to u^*$ is a conjugate-linear map of $\mathscr{L}(H_1, H_2)$ onto $\mathscr{L}(H_2, H_1)$.

The most important concrete examples of Hilbert spaces are the spaces $L^2(\mu)$ (Example 2 above) with inner product $\int fg^* \, d\mu$; special instances of the latter are the spaces l_d^2 (or $l^2(A)$), defined to be the subspace of C^A (or of R^A in the real case) of all families $\{\xi_\alpha : \alpha \in A\}$ for which $\sum_\alpha |\xi_\alpha|^2 < +\infty$, A being a set of cardinality d. The inner product on these spaces is defined to be $[\xi, \eta] = \sum_\alpha \xi_\alpha \eta_\alpha^*$; for each pair (ξ, η) the family $\{\xi_\alpha \eta_\alpha^* : \alpha \in A\}$, which has at most countably many non-zero members, is summable by Schwarz' inequality. (Cf. Chapter III, Exercise 23.) Every Hilbert space of Hilbert dimension d is isomorphic with l_d^2; if $\{x_\alpha : \alpha \in A\}$ is any orthonormal basis of H, the mapping $x \to \{[x, x_\alpha] : \alpha \in A\}$ is an isomorphism of H onto l_d^2.

Finally, the method used in constructing the spaces $l^2(A)$ can be applied to any family $\{H_\alpha : \alpha \in A\}$ of Hilbert spaces. Consider the subspace H of $\prod_\alpha H_\alpha$ consisting of all elements (x_α) such that the family $\{\|x_\alpha\|^2 : \alpha \in A\}$ is summable; then $[x, y] = \sum_\alpha [x_\alpha, y_\alpha]$ defines an inner product under which H becomes a Hilbert space; H is called the **Hilbert direct sum** of the family $\{H_\alpha : \alpha \in A\}$. In particular, if A is finite, then $H = \prod_\alpha H_\alpha$ and the topology of the Hilbert direct sum is the product topology (cf. Exercise 4); hence each finite product of Hilbert spaces is a Hilbert space in a natural way.

3. THE HAHN-BANACH THEOREM

In the preceding section we have seen that the strong dual of a Banach space is a Banach space; however, we could not assert that this space always contains elements other than 0. We have also seen (Chapter I, Exercise 6) that there exist metrizable t.v.s. on which 0 is the only continuous linear form. It is the purpose of this section to establish a theorem guaranteeing that on a large class of t.v.s. (Section 4), including the normable spaces, there exist sufficiently many continuous linear forms to distinguish points. This result, the theorem of Hahn-Banach, is undoubtedly one of the most important and far-reaching theorems in functional analysis.

We begin by establishing a lemma that contains the core of the Hahn-Banach theorem.

LEMMA. *Let L be a Hausdorff t.v.s. over R of dimension at least 2. If B is an open, convex set in L not containing 0, there exists a one-dimensional subspace of L not intersecting B.*

Proof. Let E be any fixed two-dimensional subspace of L. If $E \cap B = \varnothing$, the result is immediate; so let us assume that $B_1 = E \cap B$ is non-empty. B_1 is a convex, open subset of E not containing 0. By (I, 3.2) we can identify E with R_0^2 (the Euclidean plane). Project B_1 onto a subset of the unit circle C of E by the mapping

$$f: (x, y) \rightarrow \left(\frac{x}{r}, \frac{y}{r}\right), \quad r = (x^2 + y^2)^{\frac{1}{2}}.$$

Since B_1, being convex, is connected, $f(B_1)$ is connected, for f is continuous on B_1; moreover, $f(B_1)$ is an open subset of C. Hence $f(B_1)$ is an open arc on C which subtends an angle $\leq \pi$ at 0; otherwise, there would exist points in B_1 whose images under f are diametrical, contradicting the hypothesis $0 \notin B_1$, since B_1 is convex. Consequently, there exists a straight line in E passing through 0 and not intersecting B_1.

The following theorem is sometimes called Mazur's theorem (cf. Day [2]), and is virtually the geometrical form of the Hahn-Banach theorem (Bourbaki [7]).

3.1

Theorem. *Let L be a t.v.s., let M be a linear manifold in L, and let A be a non-empty convex, open subset of L, not intersecting M. There exists a closed hyperplane in L, containing M and not intersecting A.*

Proof. After a translation, if necessary, we can have $0 \in M$, so that M is a subspace of L. Consider the set $\mathfrak{M}$ of all closed real subspaces of L that contain M and do not intersect A; $\mathfrak{M}$ is non-empty, since $\overline{M} \in \mathfrak{M}$.

Order $\mathfrak{M}$ by inclusion $\subset$. If $\{M_\alpha\}$ is a totally ordered subset of $\mathfrak{M}$, the closure of $\bigcup_\alpha M_\alpha$ is clearly its least upper bound; hence by Zorn's lemma there exists a maximal element H_0 of $\mathfrak{M}$. If L_0 denotes the real underlying space of L (Chapter I, Section 7), the quotient space L_0/H_0 is Hausdorff by (I, 2.3), for H_0 is closed. Because of $A \neq \varnothing$, L_0/H_0 has dimension ≥ 1; suppose that L_0/H_0 is of dimension ≥ 2. Since the natural map ϕ of L_0 onto L_0/H_0 is linear and open, $B = \phi(A)$ is a convex, open subset of L_0/H_0, not containing 0, since H_0 does not intersect A. Hence by the preceding lemma, there exists a one-dimensional subspace N of L_0/H_0 not intersecting B; this implies that $H = \phi^{-1}(N)$ is a closed subspace of L_0 containing H_0 properly and not intersecting A. This contradicts the maximality of H_0 in $\mathfrak{M}$; hence L_0/H_0 has dimension 1, and H_0 is a closed, real hyperplane containing M and not intersecting A. This completes the proof when L is a t.v.s. over R.

If L is a t.v.s. over C, then $M = iM$ (assuming $0 \in M$), since M is a subspace of L. Consequently $H_1 = H_0 \cap iH_0$, which is a closed hyperplane in L not intersecting A, contains M, and the proof is complete.

COROLLARY. *If L is a t.v.s., there exists a continuous linear form $f \neq 0$ on L if and only if L contains a non-empty convex, open subset $A \neq L$.*

Proof. If $f \neq 0$ is a continuous linear form on L, the subset $A = \{x: |f(x)| < 1\}$ is $\neq L$, convex, and open. Conversely, if the convex set $A \subset L$ is open and $x_0 \notin A$, x_0 is contained in a closed hyperplane (not intersecting A) by (3.1) which by (I, 4.2) implies the existence of a non-zero continuous linear form on L.

We deduce now from (3.1) its analytic equivalent, the theorem of Hahn-Banach; for a more general form valid in real vector spaces, see Exercise 6.

3.2

Theorem. *Let L be a vector space, let p be a semi-norm on L, and let M be a subspace of L. If f is a linear form on M such that $|f(x)| \leq p(x)$ for all $x \in M$, there exists a linear form f_1 extending f to L and such that $|f_1(x)| \leq p(x)$ for all $x \in L$.*

Proof. Since the case $f = 0$ is trivial, we assume that $f(x) \neq 0$ for some $x \in M$. By (I, 1.2), the convex, circled sets $V_n = \{x \in L: p(x) < n^{-1}\}$, $n \in N$, form a 0-neighborhood base for a topology $\mathfrak{T}$ under which L is a t.v.s. Define $H = \{x \in M: f(x) = 1\}$; then H is a hyperplane in M and a linear manifold in L. Let $A = V_1$; A is open in $L(\mathfrak{T})$ by (1.6) and $A \cap H = \varnothing$, since $p(x) \geq 1$ for $x \in H$. By (3.1) there exists a hyperplane H_1 in L, containing H and not intersecting A. Since $H_1 \cap M \neq M$ (for $0 \notin H_1$) and $H_1 \supset H$, it follows that $H_1 \cap M = H$, since H and $H_1 \cap M$ are both hyperplanes in M. By (I, 4.1), we have $H_1 = \{x: f_1(x) = 1\}$ for some linear form f_1 on L, since $0 \notin H_1$. Now $H = H_1 \cap M$ implies that $f(x) = f_1(x)$ for all $x \in M$; that is, f_1 is an extension of f to L. From $H_1 \cap A = \varnothing$, it follows that $|f_1(x)| \leq p(x)$ for all $x \in L$, thus completing the proof.

COROLLARY. *If $(L, \| \ \|)$ is a normed space, M is a subspace of L, and f is a linear form on M such that $|f(x)| \leq \|x\|$ $(x \in M)$, then f has a linear extension f_1 to L satisfying $|f_1(x)| \leq \|x\|$ $(x \in L)$.*

This is the classical form of the theorem for normed spaces.

4. LOCALLY CONVEX SPACES

A topological vector space E over R or C will be called **locally convex** if it is a Hausdorff space such that every neighborhood of any $x \in E$ contains a convex neighborhood of x. Equivalently, E is a locally convex t.v.s. or briefly locally convex space (**l.c.s.**) if the convex neighborhoods of 0 form a base at 0 with intersection $\{0\}$. A topology on a vector space over R or C, not necessarily Hausdorff but satisfying $(LT)_1$ and $(LT)_2$ (Chapter I, Section 1) and possessing a base of convex 0-neighborhoods, is called a **locally convex topology**. It will be convenient to have this distinction, since the majority of the valuable results produced by convexity (such as Corollary 1 of (4.2), below) holds only in Hausdorff spaces, while it is sometimes necessary (such as in the proof of (3.2), above) to consider locally convex topologies that are not separated.

By the **topological dual** (or briefly **dual**) of a t.v.s. L, we understand the vector space L' of continuous linear forms on L; L' is a subspace of the algebraic dual L^* of L. If E is a l.c.s., its dual E' separates points in E; that is, for any two elements $x, y \in E$, $x \neq y$, there exists an $f \in E'$ such that $f(x) \neq f(y)$. (Equivalently, for every non-zero $x \in E$ there exists $f \in E'$ with $f(x) \neq 0$.) This important result is an immediate consequence of (3.1), and formally contained in (4.2), Corollary 1.

If E is a vector space, a locally convex topology on E can geometrically be defined by selecting a filter base $\mathfrak{B}$ in E, consisting of radial, convex, circled sets and such that $V \in \mathfrak{B}$ implies $\frac{1}{2}V \in \mathfrak{B}$; since $\frac{1}{2}V + \frac{1}{2}V = V$ by the convexity of each V, the corollary of (I, 1.2) implies that $\mathfrak{B}$ is a 0-neighborhood base for a unique locally convex topology. Conversely, every l.c. topology on E can be so defined; for example, the family of all closed, convex, circled 0-neighborhoods is a base at 0.

Analytically a locally convex topology on E is determined by an arbitrary family $\{p_\alpha: \alpha \in A\}$ of semi-norms as follows: For each $\alpha \in A$, let $U_\alpha = \{x \in E: p_\alpha(x) \leqq 1\}$ and consider the family $\{n^{-1}U\}$, where n runs through all positive integers and U ranges over all finite intersections of sets U_α ($\alpha \in A$). This family $\mathfrak{B}$ satisfies the conditions indicated above and hence is a base at 0 for a locally convex topology $\mathfrak{T}$ on E, called the topology **generated** by the family $\{p_\alpha\}$; equivalently, $\{p_\alpha\}$ is said to be a **generating family** of semi-norms for $\mathfrak{T}$. Conversely, every locally convex topology on E is generated by a suitable family of semi-norms; it suffices to take the gauge functions of a family of convex, circled 0-neighborhoods whose positive multiples form a subbase at 0. It is clear from (1.6) that every member of a generating family of semi-norms is continuous for $\mathfrak{T}$, and it is easy to see that $\mathfrak{T}$ is Hausdorff if and only if for each non-zero $x \in E$ and each family $\mathscr{P}$ of semi-norms generating $\mathfrak{T}$ there exists $p \in \mathscr{P}$ such that $p(x) > 0$. We can leave it to the reader to prove that, for a given l.c. topology $\mathfrak{T}$, the smallest cardinality of a base at 0 is identical with the smallest cardinality of a generating family of semi-norms, except when the latter is 1.

It is a direct consequence of the definitions (Chapter I, Section 2) that induced, quotient, and product topologies of locally convex topologies are locally convex; accordingly, subspaces, separated quotients, and products of l.c.s. are again l.c.s. These will be discussed in the subsequent sections. Here we confine ourselves to a few simple facts concerning metrizable l.c.s.

In view of (I, 6.1) a l.c.s. is metrizable if and only if it possesses a countable base at 0 consisting of convex, circled sets, and hence a base which consists of the members of a decreasing sequence $\{U_n\}$ of convex, circled sets. Equivalently, a l.c.s. is metrizable if and only if its topology is generated by a countable family of semi-norms, and hence by a sequence of semi-norms

$\{p_n\}$ which is increasing. A complete metrizable l.c.s. is called a **Fréchet-space**, or briefly (F)-space. Clearly, every complete normable space (and hence every Banach space) is an (F)-space; the simplest example of an (F)-space which is not normable is furnished by the space K_0^N of all numerical sequences under the product topology (K_0^N is not normable by (I, 5.5) and (2.2)). It follows from the results of Chapter I, Section 6, that every closed subspace and every separated quotient of an (F)-space is an (F)-space, and so is every countable product of (F)-spaces.

As a simple example for the definition of a locally convex topology by families of semi-norms, let E be any vector space with algebraic dual E^*, and suppose that M is a subset of E^* ($M \neq \varnothing$). The semi-norms $x \to |f(x)|$ ($f \in M$) generate a locally convex topology on E under which E is a l.c.s. if and only if M separates points in E.

4.1

The completion $\tilde{E}$ of a l.c.s. E is a l.c.s., whose topology is generated by the continuous extensions to $\tilde{E}$ of the members of any generating family of semi-norms on E.

Proof. If p is any member of a family $\mathscr{P}$ of generating semi-norms on E, p has a unique continuous extension $\bar{p}$ to $\tilde{E}$ by (1.6). If $U = \{x \in E: p(x) \leq 1\}$, then $\bar{U} = \{x \in \tilde{E}: \bar{p}(x) \leq 1\}$ is the closure of U in $\tilde{E}$. It follows from (I, 1.5) that $\tilde{E}$ is a l.c.s. (since $\bar{U}$ is convex) and that $\{\bar{p}: p \in \mathscr{P}\}$ is a generating family of semi-norms on $\tilde{E}$.

The following consequence of the Hahn-Banach theorem reflects a basic property of locally convex topologies:

4.2

Theorem. *Let E be a t.v.s. whose topology is locally convex. If f is a linear form, defined and continuous on a subspace M of E, then f has a continuous linear extension to E.*

Proof. Since f is continuous on M, $V = \{x: |f(x)| \leq 1\}$ is a 0-neighborhood in M. There exists a convex, circled 0-neighborhood U in E such that $U \cap M \subset V$; the gauge p of U is a continuous semi-norm on E such that $|f(x)| \leq p(x)$ on M, since $U \cap M \subset V$. By (3.2) there exists an extension f_1 of f to E such that $|f_1(x)| \leq p(x)$ on E; f_1 is continuous, since $|f_1(x) - f_1(y)| \leq \varepsilon$ whenever $x - y \in \varepsilon U$ ($\varepsilon > 0$).

COROLLARY 1. *Given n ($n \in N$) linearly independent elements x_ν of a l.c.s. E, there exist n continuous linear forms f_μ on E such that $f_\mu(x_\nu) = \delta_{\mu\nu}$ ($\mu, \nu = 1, ..., n$).*

Proof. Denote by M the n-dimensional subspace with basis $\{x_\nu\}$ in E. By (I, 3.4), the linear forms g_μ ($\mu = 1, ..., n$) on M which are determined by $g_\mu(x_\nu) = \delta_{\mu\nu}$, where $\delta_{\mu\mu} = 1$ and $\delta_{\mu\nu} = 0$ for $\mu \neq \nu$ ($\mu, \nu = 1, ..., n$), are continuous. Any set $\{f_\mu\}$ of continuous extensions of the respective forms g_μ to E has the required properties.

COROLLARY 2. *Any finite-dimensional subspace M of a l.c.s. E has a complementary subspace.*

Proof. Let M have dimension n and let $\{f_\mu\}$ be n continuous linear forms on E whose restrictions to M are linearly independent (cf. Corollary 1). Then $N = \bigcap\limits_{\mu=1}^{n} f_\mu^{-1}(0)$ is a closed subspace of E and an algebraic complementary subspace of M, and the assertion follows from (I, 3.5).

If E is a t.v.s. whose topology is locally convex, the definition of boundedness (Chapter I, Section 5) implies that the convex hull of a bounded subset of E is bounded. In particular, the family of all closed, convex, and circled bounded subsets is a fundamental system of bounded sets in E.

4.3

In every locally convex space, the convex hull and the convex, circled hull of a precompact subset is precompact.

Proof. Since the circled hull of a precompact set is clearly precompact (cf. Chapter I, 5.1), it is enough to prove the assertion for convex hulls; the reader will notice, however, that (with the obvious modifications) the following proof also applies to convex, circled hulls. Observe first that the convex hull P of a finite set $\{a_i: i = 1, ..., n\}$ is compact, for P is the image of the compact simplex $\{(\lambda_1, ..., \lambda_n): \lambda_i \geq 0, \sum\limits_{1}^{n}\lambda_i = 1\} \subset R^n$ under the continuous map $(\lambda_1, ..., \lambda_n) \to \sum\lambda_i a_i$. (This is a special case of (10.2) below.) Now let $B \subset E$ be precompact, C the convex hull of B, V an arbitrary convex neighborhood of 0 in E. Supposing B to be non-empty, there exist elements $a_i \in B$ $(i = 1, ..., n)$ such that $B \subset \bigcup\limits_{1}^{n}(a_i + V)$. The convex hull P of $\{a_i\}$ is compact, and $C \subset P + V$, since $P + V$ is convex and contains B; hence we have $P \subset \bigcup\limits_{1}^{m}(b_j + V)$ for a suitable finite subset $\{b_j: j = 1, ..., m\}$ of $P \subset C$, and it follows that $C \subset \bigcup\limits_{1}^{m}(b_j + 2V)$, which shows C to be precompact.

COROLLARY. *If E is a quasi-complete l.c.s., then the closed, convex hull and the closed, convex, circled hull of every precompact subset of E are compact.*

However, if E is not quasi-complete, then the closed, convex hull of a compact subset of E can fail to be compact (Exercise 27).

For the construction of locally convex spaces from those of a given class a substantially more general approach proves fruitful than we have discussed in Chapter I, Section 2, for arbitrary topological vector spaces; the two following sections are concerned with two basic methods of generating locally convex spaces.

5. PROJECTIVE TOPOLOGIES

Let E and E_α ($\alpha \in A$) be vector spaces over K, let f_α be a linear map on E into E_α, and let $\mathfrak{T}_\alpha$ be a locally convex topology on E_α ($\alpha \in A$). The **projective topology** $\mathfrak{T}$ on E with respect to the family $\{(E_\alpha, \mathfrak{T}_\alpha, f_\alpha): \alpha \in A\}$ is the coarsest topology on E for which each of the mappings f_α ($\alpha \in A$) on E into $(E_\alpha, \mathfrak{T}_\alpha)$ is continuous.

Clearly, $\mathfrak{T}$ is the upper bound (in the lattice of topologies on E) of the topologies $f_\alpha^{-1}(\mathfrak{T}_\alpha)$ ($\alpha \in A$); if $x \in E$ and $x_\alpha = f_\alpha(x) \in E_\alpha$, a $\mathfrak{T}$-neighborhood base of x is given by all intersections $\bigcap\limits_{\alpha \in H} f_\alpha^{-1}(U_\alpha)$, where U_α is any neighborhood of x_α with respect to $\mathfrak{T}_\alpha$, and H is any finite subset of A. Since the f_α are linear maps and the $\mathfrak{T}_\alpha$ are locally convex topologies on the respective spaces, E_α, $\mathfrak{T}$ is a translation-invariant topology on E with a base of convex 0-neighborhoods satisfying conditions (a) and (b) of (I, 1.2); hence $\mathfrak{T}$ is a locally convex topology on E.

5.1

The projective topology on E with respect to the family $\{(E_\alpha, \mathfrak{T}_\alpha, f_\alpha): \alpha \in A\}$ is a Hausdorff topology if and only if for each non-zero $x \in E$, there exists an $\alpha \in A$ and a 0-neighborhood $U_\alpha \subset E_\alpha$ such that $f_\alpha(x) \notin U_\alpha$.

Proof. If $\mathfrak{T}$ is Hausdorff and $0 \neq x \in E$, there exists a 0-neighborhood U in $(E, \mathfrak{T})$ not containing x; since there exist 0-neighborhhods $U_\alpha \subset E_\alpha$ and a finite subset $H \subset A$ with $\bigcap\limits_{\alpha \in H} f_\alpha^{-1}(U_\alpha) \subset U$, we must have $f_\alpha(x) \notin U_\alpha$ for some $\alpha \in H$. Conversely, $f_\alpha(x) \notin U_\alpha$ implies $x \notin f_\alpha^{-1}(U_\alpha)$ which shows $\mathfrak{T}$ to be a Hausdorff topology.

5.2

A mapping u of a topological space F into E, where E is endowed with the projective topology defined by the family $\{(E_\alpha, \mathfrak{T}_\alpha, f_\alpha): \alpha \in A\}$, is continuous if and only if for each $\alpha \in A$, $f_\alpha \circ u$ is continuous on F into $(E_\alpha, \mathfrak{T}_\alpha)$.

Proof. If u is continuous, then, clearly, so is each $f_\alpha \circ u$ ($\alpha \in A$). Conversely, let G_α be any open subset of E_α; then $u^{-1}[f_\alpha^{-1}(G_\alpha)]$ is an open subset of F. Now each open subset G of E is the union of a suitable family of finite intersections of sets $f_\alpha^{-1}(G_\alpha)$, whence it follows that $u^{-1}(G)$ is open in F, and hence u is continuous.

The reader will have noticed that no vector space concepts are needed in the preceding result; in fact, it reflects a general property of projective topologies (cf. Prerequisites).

We proceed to enumerate the most important examples of projective locally convex topologies.

Subspaces. Let M be a subspace of the l.c.s. $(E, \mathfrak{T})$; the topology of M (i.e., the topology induced on M by $\mathfrak{T}$) is the projective topology on M with respect to the canonical imbedding $M \to E$.

Products. Let $\{(E_\alpha, \mathfrak{T}_\alpha): \alpha \in A\}$ be a family of t.v.s., each $\mathfrak{T}_\alpha$ being a locally convex topology. The product topology $\mathfrak{T}$ on $E = \prod_\alpha E_\alpha$ (Chapter I, Section 2) is evidently locally convex; $\mathfrak{T}$ is the projective topology on E with respect to the projections $E \to E_\alpha$ ($\alpha \in A$). In particular, the product of any family of l.c.s. is a l.c.s.

Upper Bounds. Let $\{\mathfrak{T}_\alpha: \alpha \in A\}$ be a family of locally convex topologies on a vector space E; their least upper bound $\mathfrak{T}$ (in the lattice of topologies on E) is a locally convex topology which is a projective topology; in fact, $\mathfrak{T}$ is the projective topology with respect to the family $\{(E, \mathfrak{T}_\alpha, e): \alpha \in A\}$, where e is the identity map of E.

Weak Topologies. Let E be a vector space over K and let F be a subset of E^* that is non-empty. Set $E_f = K_0$ for every $f \in F$; the projective topology on E with respect to the family $\{(E_f, f): f \in F\}$ is called the **weak topology** generated by F, and is denoted by $\sigma(E, F)$. Since F can be replaced by its linear hull in E^* without changing the corresponding projective topology, F can be assumed to be a subspace of E^*. By (5.1), E is a l.c.s. under $\sigma(E, F)$ if and only if F separates points in E.

In particular, when $(E, \mathfrak{T})$ is a locally convex space, then its dual E' separates points in E by (4.2), Corollary 1; $\sigma(E, E')$ is called the **weak topology** of E (associated with $\mathfrak{T}$ if this distinction is necessary). On the other hand, E' is a l.c.s. under $\sigma(E', E)$, called the **weak dual** of $(E, \mathfrak{T})$; here E is to be viewed as a subspace of $(E')^*$.

Projective Limits. Let A be an index set directed under a (reflexive, transitive, anti-symmetric) relation " $\leqq$ ", let $\{E_\alpha: \alpha \in A\}$ be a family of l.c.s. over K, and denote, for $\alpha \leqq \beta$, by $g_{\alpha\beta}$ a continuous linear map of E_β into E_α. Let E be the subspace of $\prod_\alpha E_\alpha$ whose elements $x = (x_\alpha)$ satisfy the relation $x_\alpha = g_{\alpha\beta}(x_\beta)$ whenever $\alpha \leqq \beta$; E is called the **projective limit** of the family $\{E_\alpha: \alpha \in A\}$ with respect to the mappings $g_{\alpha\beta}(\alpha, \beta \in A; \alpha \leqq \beta)$, and denoted by $\varprojlim g_{\alpha\beta}E_\beta$. It is evident that the topology of E is the projective topology on E with respect to the family $\{(E_\alpha, \mathfrak{T}_\alpha, f_\alpha): \alpha \in A\}$, where $\mathfrak{T}_\alpha$ denotes the topology of E_α, and f_α denotes the restriction to E of the projection map p_α of $\prod_\beta E_\beta$ onto E_α.

5.3

The projective limit of a family of quasi-complete (respectively, complete) locally convex spaces is quasi-complete (respectively, complete).

Proof. Let $E = \varprojlim g_{\alpha\beta}E_\beta$, $F = \prod_\alpha E_\alpha$. If every E_α is complete, then F is complete; if every E_α is quasi-complete, so is F by (I, 5.6). Hence the proposition will be proved when we show that E is a closed subspace of F. Denote, for each pair $(\alpha, \beta) \in A \times A$ such that $\alpha \leq \beta$, by $V_{\alpha\beta}$ the subspace $\{x: x_\alpha - g_{\alpha\beta}(x_\beta) = 0\}$ of F. Since E_α is Hausdorff and $V_{\alpha\beta}$ is the null space of the continuous linear map $p_\alpha - g_{\alpha\beta} \circ p_\beta$ of F into E_α, $V_{\alpha\beta}$ is closed; thus $E = \bigcap_{\alpha \leq \beta} V_{\alpha\beta}$ is closed in F.

The product $\prod_\alpha E_\alpha$ of a family $\{E_\alpha: \alpha \in A\}$ of l.c.s. is itself an example of a projective limit. If $\{H\}$ denotes the family of all non-empty finite subsets of A, ordered by inclusion, $E_H = \prod_{\alpha \in H} E_\alpha$ and $g_{H,\Lambda}$ denotes the projection of E_Λ onto E_H when $H \subset \Lambda$, then $\prod_\alpha E_\alpha = \varprojlim g_{H,\Lambda} E_\Lambda$. Other examples of projective limits are provided by the duals of inductive limits (Chapter IV, Section 4), for which concrete examples will be given in Section 6. In the proof of (5.4) below we construct, for every complete locally convex space E, a projective limit of Banach spaces to which E is isomorphic. Finally, we point out that, in general, there is nothing to prevent a projective limit of l.c.s. from being $\{0\}$; but it can be shown (Exercise 10) that if A is countable and certain additional conditions are satisfied, then $p_\alpha(E) = E_\alpha$ $(\alpha \in A)$.

5.4

Every complete l.c.s. E is isomorphic to a projective limit of a family of Banach spaces; this family can be so chosen that its cardinality equals the cardinality of a given 0-neighborhood base in E.

Proof. Let $\{U_\alpha: \alpha \in A\}$ denote a given base of convex, circled neighborhoods of 0 in E. A is directed under the relation $\alpha \leq \beta$, defined by "$\alpha \leq \beta$ if $U_\beta \subset U_\alpha$". Denote by p_α the gauge of U_α and set $F_\alpha = E/V_\alpha$, where $V_\alpha = p_\alpha^{-1}(0)$ $(\alpha \in A)$. If x_α is the equivalence class of $x \in E$ mod V_α, then $x_\alpha \to \|x_\alpha\| = p_\alpha(x)$ is a norm on F_α generating a topology $\mathfrak{T}_\alpha$ which is coarser than the topology of the quotient space E/V_α. If $\alpha \leq \beta$, every equivalence class mod V_β, x_β, is contained in a unique equivalence class mod V_α, x_α, since $V_\beta \subset V_\alpha$; the mapping $g_{\alpha\beta}: x_\beta \to x_\alpha$ is linear, and continuous from $(F_\beta, \mathfrak{T}_\beta)$ onto $(F_\alpha, \mathfrak{T}_\alpha)$, since $\|x_\alpha\| \leq \|x_\beta\|$.

Let us form the projective limit $F = \varprojlim g_{\alpha\beta} F_\beta(\mathfrak{T}_\beta)$. The mapping $x \to (x_\alpha)$ of E into F is clearly linear, and one–to–one, since E is Hausdorff. We show that this mapping is onto F. Let H be any non-empty finite subset of A and let $z = (z_\alpha)$ be any fixed element of F. There exists $x_H \in E$ such that the equivalence class of x_H mod V_α is z_α for every $\alpha \in H$ (if $\beta \in A$ is such that $\beta \geq \alpha$ for all $\alpha \in H$, then any $x \in E$ with $x_\beta = z_\beta$ will do). Now, if $\alpha \in H_1$ and $\alpha \in H_2$, then $x_{H_1} - x_{H_2} \in U_\alpha$ which shows the filter of sections of $\{x_H\}$, where for each finite $H \neq \varnothing$ an x_H has been selected as above, to be a Cauchy filter in E. Since E is complete, $\{x_H\}$ has a limit y in E for which $y_\alpha = z_\alpha (\alpha \in A)$, because $x \to (x_\alpha)$ is clearly continuous. Moreover, $x \to (x_\alpha)$ is readily seen to be a homeomorphism, and hence an isomorphism of E onto F. Now F is a dense subspace (proof!) of the projective limit $\varprojlim \bar{g}_{\alpha\beta} \tilde{F}_\beta$, where $\tilde{F}_\alpha$ $(\alpha \in A)$ is the completion of $(F_\alpha, \mathfrak{T}_\alpha)$ and $\bar{g}_{\alpha\beta}$ the continuous extension of $g_{\alpha\beta}$ $(\alpha \leq \beta)$ to $\tilde{F}_\beta$

with values in $\tilde{F}_\alpha$. But F, being isomorphic with E, is complete, and hence $F = \varprojlim \bar{g}_{\alpha\beta}\tilde{F}_\beta$, which completes the proof of the theorem.

COROLLARY 1. *Every Fréchet space is isomorphic with a projective limit of a sequence of Banach spaces.*

COROLLARY 2. *Every locally convex space is isomorphic with a subspace of a product of Banach spaces.*

6. INDUCTIVE TOPOLOGIES

Let E and E_α ($\alpha \in A$) be vector spaces over K, let g_α be a linear mapping of E_α into E, and let $\mathfrak{T}_\alpha$ be a locally convex topology on E_α ($\alpha \in A$). The **inductive topology** on E with respect to the family $\{(E_\alpha, \mathfrak{T}_\alpha, g_\alpha): \alpha \in A\}$ is the finest locally convex topology for which each of the mappings g_α ($\alpha \in A$) is continuous on $(E_\alpha, \mathfrak{T}_\alpha)$ into E.

To see that this topology is well defined, we note that the class $\mathscr{T}$ of l.c. topologies on E for which all g_α are continuous is not empty; the trivial topology (whose only open sets are $\varnothing$ and E) is a member of $\mathscr{T}$. Now the upper bound $\mathfrak{T}$ of $\mathscr{T}$ (in the lattice of topologies on E) is clearly a locally convex topology for which all g_α are continuous, and hence $\mathfrak{T}$ is the topology whose existence was to be verified. (As an upper bound, $\mathfrak{T}$ is also a projective topology, namely the projective topology with respect to the family $\mathscr{T}$ and the identity map on E.) $\mathfrak{T}$ need not be separated, even if all $\mathfrak{T}_\alpha$ are. (We leave it to the reader to construct an example.) A 0-neighborhood base for $\mathfrak{T}$ is given by the family $\{U\}$ of all radial, convex, circled subsets of E such that for each $\alpha \in A$, $g_\alpha^{-1}(U)$ is a 0-neighborhood in $(E_\alpha, \mathfrak{T}_\alpha)$. If E is the linear hull of $\bigcup g_\alpha(E_\alpha)$, such a base can be obtained by forming all sets of the form $\Gamma_\alpha g_\alpha(U_\alpha)$, where U_α is any member of a 0-neighborhood base in $(E_\alpha, \mathfrak{T}_\alpha)$.

6.1

A linear map v on a vector space E into a l.c.s. F is continuous for an inductive topology on E if and only if each map $v \circ g_\alpha$ ($\alpha \in A$) is continuous on $(E_\alpha, \mathfrak{T}_\alpha)$ into F.

Proof. The condition is clearly necessary. Conversely, let each $v \circ g_\alpha$ be continuous ($\alpha \in A$) and let W be a convex, circled 0-neighborhood in F. Then $(v \circ g_\alpha)^{-1}(W) = g_\alpha^{-1}[v^{-1}(W)]$ is a neighborhood of 0 in E_α ($\alpha \in A$), which implies that $v^{-1}(W)$ is a neighborhood of 0 for the inductive topology on E.

We consider now the most important instances of inductive topologies:

Quotient Spaces. If E is a l.c.s. and M is a subspace of E, it is immediate from the discussion in Chapter I, Section 2, that the topology of E/M is locally convex. If $\mathfrak{T}$ denotes the topology of E, the quotient topology is the inductive topology with respect to the family $\{(E, \mathfrak{T}, \phi)\}$, where ϕ is the natural (or canonical) map of E onto E/M. By (I, 2.3) this topology is Hausdorff if and only if M is closed in E. The quotient topology can be generated by a family of semi-norms derived from certain generating families of semi-norms on E through a process analogous to that used in (2.3) (Exercise 8).

Locally Convex Direct Sums. If $\{E_\alpha : \alpha \in A\}$ is a family of vector spaces over K, the algebraic direct sum $\oplus_\alpha E_\alpha$ (Chapter I, Exercise 1) is defined to be the subspace of $\prod_\alpha E_\alpha$ for whose elements x all but a finite number of the projections $x_\alpha = p_\alpha(x)$ are 0. Denote by g_α ($\alpha \in A$) the injection map (or canonical imbedding) $E_\alpha \to \oplus_\beta E_\beta$. The **locally convex direct sum** of the family $\{E_\alpha(\mathfrak{T}_\alpha):$ $\alpha \in A\}$ of l.c.s. is defined to be $\oplus_\alpha E_\alpha$ under the inductive topology with respect to the family $\{(E_\alpha, \mathfrak{T}_\alpha, g_\alpha): \alpha \in A\}$ and, when reference to the topologies is desired, denoted by $E(\mathfrak{T}) = \oplus_\alpha E_\alpha(\mathfrak{T}_\alpha)$. Since $\mathfrak{T}$ is finer than the topology induced on E by $\prod_\alpha E_\alpha(\mathfrak{T}_\alpha)$, $\mathfrak{T}$ is a Hausdorff topology and hence $E(\mathfrak{T})$ is a l.c.s. From the remarks made above it follows that a 0-neighborhood base of $\oplus_\alpha E_\alpha(\mathfrak{T}_\alpha)$ is provided by all sets of the form $U = \Gamma_\alpha g_\alpha(U_\alpha)$; that is

$$U = \{\sum \lambda_\alpha g_\alpha(x_\alpha): \sum |\lambda_\alpha| \leq 1, x_\alpha \in U_\alpha\}, \qquad (*)$$

where $\{U_\alpha : \alpha \in A\}$ is any family of respective 0-neighborhoods in the spaces E_α. For simplicity of notation, we shall often write x_α in place of $g_\alpha(x_\alpha)$, thus identifying E_α with its canonical image $g_\alpha(E_\alpha)$ in $\oplus_\beta E_\beta$. (Note that each g_α is an isomorphism of $E_\alpha(\mathfrak{T}_\alpha)$ into $\oplus_\beta E_\beta(\mathfrak{T}_\beta)$.)

6.2

The locally convex direct sum $\oplus_\alpha E_\alpha$ of a family of l.c.s. is complete if and only if each summand E_α is complete.

Proof. Let $E = \oplus_\alpha E_\alpha$ be complete. Since each of the projections p_α of E onto E_α is continuous, every summand E_α is closed in E and hence is complete.

Conversely, suppose that each E_α is complete ($\alpha \in A$). Denote by $\mathfrak{T}_1$ the unique translation-invariant topology on E for which a 0-neighborhood base is given by the sets $E \cap V$, where $V = \prod_\alpha V_\alpha$ and V_α is any 0-neighborhood in E_α. Then $\mathfrak{T}_1$ is evidently coarser than the locally convex direct sum topology $\mathfrak{T}$ on E, and $(E, \mathfrak{T}_1)$ is a t.v.s. (Chapter I, Exercise 1) which is complete. (In fact, the sets V form a 0-neighborhood base in $F = \prod_\alpha E_\alpha$ for a unique translation-invariant topology, under which F is easily seen to be a complete topological group with respect to addition. If $z \in F$ is in the closure of E, then for each V there exists $x \in E$ satisfying $x - z \in V$; this implies $z \in E$, and hence E is a closed subgroup of F.) To prove that $(E, \mathfrak{T})$ is complete it suffices, in view of (I, 1.6), to show that the $\mathfrak{T}$-closures $\overline{U}$ are $\mathfrak{T}_1$-closed, where $U = \Gamma_\alpha U_\alpha$ and U_α is any convex, circled 0-neighborhood in E_α. Let U be given, and let $\mathfrak{G}$ be a $\mathfrak{T}_1$-Cauchy filter on $\overline{U}$ with $\mathfrak{T}_1$-limit $x = (x_\alpha)$. Denote by H the finite set of indices $\{\alpha: x_\alpha \neq 0\}$, and write $E_H = \oplus\{E_\alpha: \alpha \in H\}$ (setting $E_H = \{0\}$ if $H = \varnothing$), and $E_B = \oplus\{E_\beta: \beta \in A \sim H\}$; finally, let p be the projection $E \to E_H$ that vanishes on E_B. Clearly, p is $\mathfrak{T}_1$-continuous and $\mathfrak{T}$-continuous, and satisfies $p(\overline{U}) \subset \overline{U}$. Thus $p(\mathfrak{G})$ is a filter base on $\overline{U} \cap E_H$ that $\mathfrak{T}_1$-converges to $x = p(x)$. Since, H being finite, the topologies $\mathfrak{T}$ and $\mathfrak{T}_1$ agree on E_H, $p(\mathfrak{G})$ also $\mathfrak{T}$-converges to x; it follows that $x \in \overline{U}$, completing the proof.

We remark that the locally convex sum of a finite family of l.c.s. is identical with their product.

Example. Let E be a vector space over K and let $\{x_\alpha : \alpha \in A\}$ be a basis of E. Obviously E is isomorphic with the algebraic direct sum $\oplus_\alpha K_\alpha$, where $K_\alpha = K_0 (\alpha \in A)$ and K_0 is the one-dimensional vector space associated with K. By (I, 3.4), the imbedding map on K_α into E (K_α being endowed with the topology of K) is continuous for any topology on E under which E is a t.v.s.; hence the locally convex direct sum topology $\mathfrak{T}$ on $E = \oplus_\alpha K_\alpha$ is the *finest locally convex topology* on E. $\mathfrak{T}$ is consequently a Hausdorff topology under which E is complete; it is generated by the family of all semi-norms on E. Equivalently, a 0-neighborhood base in $E(\mathfrak{T})$ is formed by the family of all radial, convex, circled subsets of E (hence $\mathfrak{T}$ is sometimes called the convex core topology). For further properties of this topology see Exercise 7.

6.3

A subset B of a locally convex direct sum $\oplus\{E_\alpha : \alpha \in A\}$ is bounded if and only if there exists a finite subset $H \subset A$ such that $p_\alpha(B) = \{0\}$ for $\alpha \notin H$, and $p_\alpha(B)$ is bounded in E_α if $\alpha \in H$.

Proof. Let B be bounded in $\oplus_\alpha E_\alpha$. Since, as we have noted above, the projection p_α of the direct sum onto the subspace E_α is continuous, $p_\alpha(B)$ is bounded for all $\alpha \in A$ by (I, 5.4). Suppose there is an infinite subset $B \subset A$ such that $p_\alpha(B) \neq \{0\}$ whenever $\alpha \in B$; then B contains a sequence $\{\alpha_n\}$ of distinct indices. There exists a sequence $\{y^{(n)}\} \subset B$ such that $y_{\alpha_n}^{(n)} \neq 0$ for all $n \in N$ and hence, since E_{α_n} is Hausdorff, such that $y_{\alpha_n}^{(n)} \notin n V_n$, where V_n is a suitable circled 0-neighborhood in E_{α_n}. Now if U is a 0-neighborhood of type (*) in $\oplus_\alpha E_\alpha$ such that $U_{\alpha_n} = V_n$ for all n, then $n^{-1} y^{(n)} \notin U$ for any $n \in N$, which contradicts the boundedness of B by (I, 5.3). Conversely, it is clear that the condition is sufficient for a set B to be bounded.

COROLLARY. *The l.c. direct sum of a family of quasi-complete l.c.s. is quasi-complete.*

Inductive Limits. Let $\{E_\alpha : \alpha \in A\}$ be a family of l.c.s. over K, where A is an index set directed under a (reflexive, transitive, anti-symmetric) relation " $\leqq$ " and denote, whenever $\alpha \leqq \beta$, by $h_{\beta\alpha}$ a continuous linear map of E_α into E_β. Set $F = \oplus_\alpha E_\alpha$ and denote (g_α being the canonical imbedding of E_α in F) by H the subspace of F generated by the ranges of the linear maps $g_\alpha - g_\beta \circ h_{\beta\alpha}$ of E_α into F, where (α, β) runs through all pairs such that $\alpha \leqq \beta$. If the quotient space F/H is Hausdorff (equivalently, if H is closed in F), the l.c.s. F/H is called the **inductive limit** of the family $\{E\alpha: \alpha \in A\}$ with respect to the mappings $h_{\beta\alpha}$, and is denoted by $\lim_{\longrightarrow} h_{\beta\alpha}E_\alpha$. It appears to be unknown whether H is necessarily closed in F.

The requirements for the construction of an inductive limit are often realized in the following special form: $\{E_\alpha: \alpha \in A\}$ is a family of subspaces of a vector space E such that $E_\alpha \neq E_\beta$ for $\alpha \neq \beta$, directed under inclusion and satisfying $E = \bigcup_\alpha E_\alpha$; then A is directed under "$\alpha \leq \beta$ if $E_\alpha \subset E_\beta$". Moreover, on each E_α ($\alpha \in A$) a Hausdorff locally convex topology $\mathfrak{T}_\alpha$ is given such that, whenever $\alpha \leq \beta$, the topology induced by $\mathfrak{T}_\beta$ on E_α is coarser than $\mathfrak{T}_\alpha$. Denoting by $g_\alpha(\alpha \in A)$ the canonical imbedding of E_α into E and by $h_{\beta\alpha}$ the canonical imbedding of E_α into E_β ($\alpha \leq \beta$), and supposing that the inductive topology $\mathfrak{T}$ on E with respect to the family $\{(E_\alpha, \mathfrak{T}_\alpha, g_\alpha): \alpha \in A\}$ is Hausdorff, it is easy to see that the inductive limit $\lim_{\longrightarrow} h_{\beta\alpha}E_\alpha$ exists and is isomorphic with $E(\mathfrak{T})$. In these circumstances, $E(\mathfrak{T})$ is called the **inductive limit** of the family $\{E_\alpha(\mathfrak{T}_\alpha): \alpha \in A\}$ of subspaces. An inductive limit of a family of subspaces is **strict** if $\mathfrak{T}_\beta$ induces $\mathfrak{T}_\alpha$ on E_α whenever $\alpha \leq \beta$.

Examples

1. The locally convex direct sum of a family $\{E_\alpha: \alpha \in A\}$ of l.c.s. is itself an example of an inductive limit. If $\{H\}$ denotes the family of all non-empty finite subsets of A, ordered by inclusion, $E_H = \bigoplus_{\alpha \in H} E_\alpha$ and $h_{\Lambda,H}$ the canonical imbedding of E_H into E_Λ when $H \subset \Lambda$, then $\bigoplus_\alpha E_\alpha = \lim_{\longrightarrow} h_{\Lambda,H}E_H$.

2. Let R^n ($n \in N$ fixed) be represented as the union of countably many compact subsets $G_m(m \in N)$ such that G_m is contained in the interior of G_{m+1} for all m. The vector space $\mathscr{D}(G_m)$ over C of all complex-valued functions, infinitely differentiable on R^n and supported by G_m, is a Fréchet-space under the topology of uniform convergence in all derivatives; a generating family of semi-norms $p_k(k = 0, 1, 2, \ldots)$ is given by

$$f \to p_k(f) = \sup|D^k f|,$$

the sup being taken over all $t \in R^n$ and all derivatives of order k. If $\mathscr{D}$ denotes the vector space of all complex-valued infinitely differentiable functions on R^n whose support is compact (but arbitrarily variable with f), then the inductive topology $\mathfrak{T}$ on $\mathscr{D}$ with respect to the sequence $\{\mathscr{D}(G_m)\}$ of subspaces is separated, for $\mathfrak{T}$ is finer than the topology of uniform convergence on compact subsets of R^n, which is a Hausdorff topology. Thus $\mathscr{D}(\mathfrak{T})$ is the strict inductive limit of a sequence of subspaces; its dual $\mathscr{D}'$ is the space of complex distributions on R^n (L. Schwartz [1]).

3. Let X be a locally compact space and let E be the vector space of all continuous, complex-valued functions on X with compact support. For any fixed compact subset $C \subset X$, denote by $(E_C, \mathfrak{T}_C)$ the Banach

space of functions in E that are supported by C, with the uniform norm generating $\mathfrak{T}_C$. Ordering the family of compact subsets of X by inclusion and denoting by g_C the canonical imbedding of E_C into E, the inductive topology $\mathfrak{T}$ on E is readily seen to be Hausdorff: $\mathfrak{T}$ is finer than the topology of compact convergence. Hence $E(\mathfrak{T})$ is the inductive limit of the subspaces $(E_C, \mathfrak{T}_C)$; the dual $E(\mathfrak{T})'$ is the space of complex Radon measures on X (Bourbaki [9], chap. III).

The preceding definitions of inductive limit, and even of strict inductive limit of a family of subspaces, are too general to ensure a great number of interesting results (cf. Komura [1]). The situation is different for the strict inductive limit of a sequence $\{E_n : n \in N\}$ of subspaces of E, with N under its natural order. We prove some of the most important results which are due to Dieudonné and Schwartz [1], and Köthe [2].

The strict inductive limit of an increasing sequence of (B)-spaces will be called an (**LB**)-**space**, and that of (F)-spaces an (**LF**)-**space**. Example 2 above is an (LF)-space. Example 3 is an (LB)-space, provided the locally compact space X is countable at infinity (i.e., a countable union of compact sub-spaces).

6.4

If $\{E_n(\mathfrak{T}_n) : n \in N\}$ is an increasing sequence of l.c.s. such that the topology $\mathfrak{T}_{n+1}$ induces $\mathfrak{T}_n$ for all n and if the vector space E is the union of the subspaces E_n ($n \in N$), then the inductive topology on E with respect to the canonical imbeddings $E_n \to E$ is separated and induces $\mathfrak{T}_n$ on E_n ($n \in N$).

The proof is based on this lemma:

LEMMA. *If E is a l.c.s., M a subspace of E, and U a convex, circled 0-neighborhood in M, there exists a convex, circled 0-neighborhood V in E with $U = V \cap M$. If $x_0 \in E$ is not in $\overline{M}$, then V can be chosen so that, in addition, $x_0 \notin V$.*

Proof. Let W be a convex, circled 0-neighborhood in E such that $W \cap M \subset U$. Then $V = \Gamma(W \cup U)$ satisfies $V \cap M = U$. Obviously $U \subset V \cap M$; if $z \in V \cap M$, then $z = \lambda w + \mu u$, where $w \in W$, $u \in U$, $|\lambda| + |\mu| \leq 1$, and $\lambda w = z - \mu u \in M$ implies either $\lambda = 0$ or $w \in M$; in both cases we have $z \in U$, whence $V \cap M \subset U$. If x_0 is not in the closure $\overline{M}$ of M, then in the preceding construction W can be so chosen that $(x_0 + W) \cap M$ is empty, whence it follows that $x_0 \notin V$, for $x_0 = \lambda w + \mu u \in V$ would imply $x_0 - \lambda w \in M$ and $x_0 - \lambda w \in x_0 + W$, which contradicts the choice of W.

Proof of (6.4). Let n be fixed and let V_n be a convex, circled neighborhood of 0 in $E_n(\mathfrak{T}_n)$. Using the lemma above, we can by induction construct a sequence $\{V_{n+k}\}(k = 1, 2, \ldots)$ of subsets of E such that V_{n+k} is a convex, circled 0-neighborhood in E_{n+k} and $V_{n+k+1} \cap E_{n+k} = V_{n+k}$ for all $k \geq 0$. Clearly, $V = \bigcup_{k \geq 0} V_{n+k}$ is a 0-neighborhood for the inductive topology $\mathfrak{T}$ on E such that $V \cap E_n = V_n$. It follows that the topology on E_n induced by $\mathfrak{T}$ is both finer and coarser than $\mathfrak{T}_n$, and hence is identical with $\mathfrak{T}_n$. Since $E = \bigcup_n E_n$, it is also clear that $\mathfrak{T}$ is separated.

A similar construction enables us to determine all bounded subsets in $E(\mathfrak{T})$ when $E(\mathfrak{T}) = \lim_{\longrightarrow} E_n(\mathfrak{T}_n)$ is the strict inductive limit of a sequence of subspaces.

6.5

Let $E(\mathfrak{T}) = \lim_{\longrightarrow} E_n(\mathfrak{T}_n)$ and let E_n be closed in $E_{n+1}(\mathfrak{T}_{n+1})$ $(n \in N)$. A subset $B \subset E$ is bounded in $E(\mathfrak{T})$ if and only if for some $n \in N$, B is a bounded subset of $E_n(\mathfrak{T}_n)$.

Proof. By (6.4) the condition is clearly sufficient for B to be bounded. Conversely, assume that B is bounded but not contained in E_n for any $n \in N$. There exists an increasing sequence $\{k_1, k_2, \ldots\} \subset N$ and a sequence $\{x_n\} \subset B$ such that $x_n \in E_{k_{n+1}}$, but $x_n \notin E_{k_n}$ $(n \in N)$. Using the lemma in (6.4), we construct inductively a sequence $\{V_{k_n}\}$ of convex, circled 0-neighborhoods in E_{k_n}, respectively, such that $n^{-1}x_n \notin V_{k_{n+1}}$ and $V_{k_{n+1}} \cap E_{k_n} = V_{k_n}$ for all $n \in N$. Again $V = \bigcup_{n=1}^{\infty} V_{k_n}$ is a 0-neighborhood in $E(\mathfrak{T})$, but $n^{-1}x_n \notin V$ $(n \in N)$, which is impossible by (I, 5.3); hence the assumption that B be not contained in any E_n is absurd.

6.6

The strict inductive limit of a sequence of complete locally convex spaces is complete.

Proof. Let $E(\mathfrak{T}) = \lim_{\longrightarrow} E_n(\mathfrak{T}_n)$ be a strict inductive limit of complete l.c.s. If $\mathfrak{F}$ is a Cauchy filter in $E(\mathfrak{T})$ and $\mathfrak{U}$ is the neighborhood filter of 0, then $\mathfrak{F} + \mathfrak{U} = \{F + U: F \in \mathfrak{F}, U \in \mathfrak{U}\}$ is a Cauchy filter base in $E(\mathfrak{T})$ which converges if and only if $\mathfrak{F}$ converges. We show that there must exist an $n_0 \in N$ for which the trace of $\mathfrak{F} + \mathfrak{U}$ on E_{n_0} is a filter base; if so, this trace converges in $E_{n_0}(\mathfrak{T}_{n_0})$, since E_{n_0} is complete, and hence $\mathfrak{F}$ converges in E. Otherwise, there exists a sequence $F_n \in \mathfrak{F}$ $(n \in N)$ and a decreasing sequence of convex, circled 0-neighborhoods W_n in $E(\mathfrak{T})$ such that $(F_n + W_n) \cap E_n = \varnothing$ for every n. Now $U = \Gamma \bigcup_{n=1}^{\infty} (W_n \cap E_n)$ is a 0-neighborhood in $E(\mathfrak{T})$; we show that $(F_n + U) \cap E_n = \varnothing$ for all n. Let $y \in E_n \cap (F_n + U)$; then $y = z_n + \sum_{i=1}^{p} \lambda_i x_i$, where $\sum |\lambda_i| \leq 1$, $x_i \in W_i \cap E_i$ $(i = 1, \ldots, p)$ and $z_n \in F_n$, hence

$$y - \sum_{i \leq n} \lambda_i x_i = z_n + \sum_{i > n} \lambda_i x_i.$$

Since $W_i \subset W_n$ for $i > n$ and W_n is circled and convex, the right-hand member of the last equality is in $F_n + W_n$, while the left-hand member is in E_n, which is impossible; thus $(F_n + U) \cap E_n = \emptyset$ for all n. Since $\mathfrak{F}$ is a Cauchy filter, there exists $F \in \mathfrak{F}$ such that $F - F \subset U$. Let $w \in F$, then $w \in E_k$ for some $k \in N$. Let $v \in F_k \cap F$; then $w = v + (w - v) \in v + (F - F) \subset F_k + U$, which is contradictory.

COROLLARY. *Every space of type* (LB) *or* (LF) *is complete.*

7. BARRELED SPACES

In this and the following section, we discuss the elementary properties of two types of locally convex spaces that occur frequently in applications, and whose importance is largely due to the fact that they include all Fréchet (and hence Banach) spaces and that their defining properties are invariant under the formation of inductive topologies.

A **barrel** (tonneau) in a t.v.s. E is a subset which is radial, convex, circled, and closed. A l.c.s. E is **barreled** (tonnelé) if each barrel in E is a neighborhood of 0. Equivalently, a barreled space is a l.c.s. in which the family of all barrels forms a neighborhood base at 0 (or on which each semi-norm that is semi-continuous from below, is continuous).

7.1

Every locally convex space which is a Baire space is barreled.

Proof. Let D be a barrel in E; since D is radial and circled, $E = \bigcup_{n \in N} nD$. Since E is a Baire space, there exists $n_0 \in N$ such that $n_0 D$ (which is closed) has an interior point; hence D has an interior point y. Since D is circled, $-y \in D$; hence $0 = \frac{1}{2}y + \frac{1}{2}(-y)$ is interior to D by (1.1) because D is convex.

COROLLARY. *Every Banach space and every Fréchet space is barreled.*

The property of being barreled is, in general, not inherited by projective topologies; for instance, there exist (non-complete) normed spaces which are not barreled (Exercise 14), and even a closed subspace of a barreled space is, in general, not barreled (Chapter IV, Exercise 10). The same is true for projective limits (cf. (5.4)). However, it can be shown that the completion of a barreled space is barreled (Exercise 15), and that the product of any family of barreled spaces is barreled (Chapter IV, Section 4). Moreover, any inductive topology inherits this property.

7.2

If $\mathfrak{T}$ is the inductive topology on E with respect to a family of barreled spaces (and corresponding linear maps), then each barrel in E is a 0-neighborhood for $\mathfrak{T}$.

Proof. Let $\mathfrak{T}$ be the inductive topology on E with respect to the family $\{(E_\alpha, \mathfrak{T}_\alpha, g_\alpha): \alpha \in A\}$, where $\mathfrak{T}_\alpha$ ($\alpha \in A$) is a barreled l.c. topology on E_α. If D

is a barrel in $E(\mathfrak{T})$, then $g_\alpha^{-1}(D)$ is a barrel D_α in $E_\alpha(\mathfrak{T}_\alpha)$ for each $\alpha \in A$, and hence a neighborhood of 0; it follows that D is a 0-neighborhood in $E(\mathfrak{T})$.

COROLLARY 1. *Every separated quotient of a barreled space is barreled; the locally convex direct sum and the inductive limit of a family of barreled spaces are barreled.*

COROLLARY 2. *Every space of type* (LB) *or* (LF) *is barreled.*

Since a space of type (LF) (a strict inductive limit of a sequence of Fréchet spaces) is not a Baire space, there exist barreled spaces which are not Baire spaces. Examples of such spaces are given in Section 6.

8. BORNOLOGICAL SPACES

A locally convex space E is **bornological** if every circled, convex subset $A \subset E$ that absorbs every bounded set in E is a neighborhood of 0. Equivalently, a bornological space is a l.c.s. on which each semi-norm that is bounded on bounded sets, is continuous.

8.1

Every metrizable l.c.s. is bornological.

Proof. If E is metrizable, there exists a countable 0-neighborhood base $\{V_n : n \in N\}$ by (I, 6.1), which can be chosen to be decreasing. Let A be a convex, circled subset of E that absorbs every bounded set; we must have $V_n \subset nA$ for some $n \in N$. For if this were false, there would exist elements $x_n \in V_n$ such that $x_n \notin nA$ $(n \in N)$; since $\{x_n\}$ is a null sequence, it is bounded by (I, 5.1), Corollary 2, and hence absorbed by A, which is contradictory.

It can be shown (Chapter IV, Exercise 20, and Köthe [5], §28.4) that a closed subspace of a bornological space is not necessarily bornological. It is not known whether every product of bornological spaces is bornological, but the answer to this question depends only on the cardinality of the set of factor spaces (Exercise 19). Thus since $K_0^{\aleph_0}$ is bornological, every countable product of bornological spaces is bornological; more generally, the theorem of Mackey-Ulam (Köthe [5], §28.8) asserts that every product of d bornological spaces is bornological if d is smaller than the smallest strongly inaccessible cardinal. (It is not known if strongly inaccessible cardinals exist; a cardinal d_0 is called strongly inaccessible if (a) $d_0 > \aleph_0$ (b) $\sum \{d_\alpha : \alpha \in A\} < d_0$ whenever card $A < d_0$ and $d_\alpha < d_0$ for all $\alpha \in A$ (c) $d < d_0$ implies $2^d < d_0$. For details on strongly inaccessible cardinals see, e.g., Gillman-Jerison [1].) In particular, it follows from the Mackey-Ulam theorem that K_0^d is bornological when $d = \aleph$ or $d = 2^\aleph$, where $\aleph$ is the cardinality of the continuum.

We note from (8.1) that every Fréchet space (and hence every Banach space) is bornological. Moreover, the property of being bornological is preserved under the formation of inductive topologies.

8.2

Let $\mathfrak{T}$ be the inductive topology on E with respect to a family of bornological spaces (and corresponding linear maps); each convex, circled subset of E absorbing all bounded sets in $E(\mathfrak{T})$ is a 0-neighborhood for $\mathfrak{T}$.

Proof. Let A be such a subset of E and let $\mathfrak{T}$ be the inductive topology with respect to the family $\{(E_\alpha, \mathfrak{T}_\alpha, g_\alpha): \alpha \in A\}$. If B_α is bounded in $(E_\alpha, \mathfrak{T}_\alpha)$, then $g_\alpha(B_\alpha)$ is bounded in $E(\mathfrak{T})$ by (I, 5.4); hence A absorbs $g_\alpha(B_\alpha)$, whence $g_\alpha^{-1}(A)$ absorbs B_α, and so $g_\alpha^{-1}(A)$ is a 0-neighborhood in $(E_\alpha, \mathfrak{T}_\alpha)$. Since this holds for all $\alpha \in A$, A is a 0-neighborhood in $E(\mathfrak{T})$.

COROLLARY 1. *Every separated quotient of a bornological space is bornological; the locally convex direct sum and the inductive limit of any family of bornological spaces is bornological.*

In conjunction with (8.1) we obtain:

COROLLARY 2. *Every space of type* (LB) *or* (LF) *is bornological.*

Essentially by virtue of (I, 5.3), bornological spaces E have the interesting property that continuity of a linear map u into a l.c.s. F is equivalent to the sequential continuity of u, which is in turn equivalent to u being bounded on bounded sets. This latter property, stating that the continuous linear maps of E into any l.c.s. are exactly those linear maps that preserve boundedness, actually characterizes bornological spaces (see Exercise 18).

8.3

Let $E(\mathfrak{T})$ be bornological, let F be any l.c.s., and let u be a linear map on E into F. These assertions are equivalent:

(a) *u is continuous.*
(b) *$\{u(x_n)\}$ is a null sequence for every null sequence $\{x_n\} \subset E$.*
(c) *$u(B)$ is bounded for every bounded subset $B \subset E$.*

Proof. (a) $\Rightarrow$ (b) is obvious. (b) $\Rightarrow$ (c): If $\{u(x_n)\}$ $(x_n \in B, n \in N)$ is any sequence of elements of $u(B)$, then $\{\lambda_n x_n\}$ is a null sequence in E for every null sequence of scalars $\lambda_n \in K$, by (I, 5.3); hence $\{\lambda_n u(x_n)\}$ is a null sequence in F by (b), and repeated application of (I, 5.3) shows $u(B)$ to be bounded in F. (c) $\Rightarrow$ (a): If B is any bounded set in E, and V is a given convex, circled 0-neighborhood in F, then V absorbs $u(B)$; hence $u^{-1}(V)$ absorbs B. Since B was arbitrary and E is bornological, $u^{-1}(V)$ is a 0-neighborhood for $\mathfrak{T}$. This holds for any given V which implies (since the topologies of E and F are translation-invariant) the continuity of u.

Let $E(\mathfrak{T})$ be any l.c.s., $\mathfrak{B}$ the family of all bounded subsets of $E(\mathfrak{T})$. The class $\mathscr{T}$ of separated l.c. topologies on E for each of which every $B \in \mathfrak{B}$ is bounded is non-empty, since $\mathfrak{T} \in \mathscr{T}$. The upper bound $\mathfrak{T}_0$ of $\mathscr{T}$ (in the lattice of topologies on E) is a projective topology (Section 5), and the finest l.c.

topology on E whose family of bounded sets is identical with $\mathfrak{B}$. Clearly, $E(\mathfrak{T}_0)$ is bornological; $E(\mathfrak{T}_0)$ is called the **bornological space associated with** $E(\mathfrak{T})$. Thus $E(\mathfrak{T})$ is bornological if and only if $\mathfrak{T} = \mathfrak{T}_0$.

The bornological space $E(\mathfrak{T}_0)$ associated with $E(\mathfrak{T})$ can also be defined as follows: Let $\mathfrak{B}$ denote the family of all closed, convex, and circled bounded subsets of $E(\mathfrak{T})$, ordered by inclusion. For a fixed $B \in \mathfrak{B}$, consider the subspace $E_B = \bigcup_{n \in N} nB$ of E; if p_B denotes the gauge function* of B in E_B, p_B is a norm (since B is bounded and $E(\mathfrak{T})$ is Hausdorff) and (E_B, p_B) is a normed space whose topology is finer than the topology induced by $\mathfrak{T}$. (If B is complete, it follows from (I, 1.6) that (E_B, p_B) is a Banach space.) Let g_B denote the canonical imbedding of E_B into E; it is evident that the inductive topology on E with respect to the family $\{(E_B, p_B, g_B): B \in \mathfrak{B}\}$ is the bornological topology $\mathfrak{T}_0$ associated with $\mathfrak{T}$. Hence:

8.4

Every bornological space E is the inductive limit of a family of normed spaces (and of Banach spaces if E is quasi-complete); the cardinality of this family can be chosen as the cardinality of any fundamental system of bounded sets in E.

COROLLARY. *Every quasi-complete bornological space is barreled.*

This is immediate from (7.2), Corollary 1. Since there exist normed spaces which are not barreled (Exercise 14), a bornological space is not necessarily barreled; conversely, Nachbin [1] and Shirota [1] have given examples of barreled spaces that are not bornological.

We conclude this section with a remark that refines the last corollary and follows easily from the preceding discussion, and which will be needed later on (Chapter III, Section 3).

8.5

Let E be any l.c.s. and let D be a barrel in E. Then D absorbs each bounded subset $B \subset E$ that is convex, circled, and complete.

Proof. (E_B, p_B) is a Banach (hence barreled) space whose topology is finer than the topology induced on E_B by E. Thus $D \cap E_B$ is a barrel in (E_B, p_B), which implies that D absorbs B.

9. SEPARATION OF CONVEX SETS

Let E be a vector space over K and let $H = \{x: f(x) = \alpha\}$ be a real hyperplane in E; the four convex sets, $F_\alpha = \{x: f(x) \leqq \alpha\}$, $F^\alpha = \{x: f(x) \geqq \alpha\}$, $G_\alpha = \{x: f(x) < \alpha\}$, $G^\alpha = \{x: f(x) > \alpha\}$, are called the **semi-spaces** determined

---

* For $B = \varnothing$, set $E_B = \{0\}$ and $p_B = 0$.

by H. (Note that F^α and F_α are closed, G^α and G_α are open for the finest locally convex topology on E.) If E is a t.v.s. and H is a closed real hyperplane in E (equivalently, f is a continuous real linear form $\neq 0$), then F^α and F_α are called **closed semi-spaces**, and G^α, G_α are called **open semi-spaces**. Two non-empty subsets A, B of E, are said to be **separated** (respectively, **strictly separated**) by the real hyperplane H if either $A \subset F_\alpha$ and $B \subset F^\alpha$, or $B \subset F_\alpha$ and $A \subset F^\alpha$ (respectively, if either $A \subset G_\alpha$ and $B \subset G^\alpha$, or $B \subset G_\alpha$ and $A \subset G^\alpha$). If A is a subset of the t.v.s. E, a closed real hyperplane H is called a **supporting hyperplane** of A if $A \cap H \neq \varnothing$ and if A is contained in one of the closed semi-spaces determined by H.

Theorem (3.1) is a separation theorem; it asserts that every convex open set $A \neq \varnothing$ and affine subspace M, not intersecting A, in a t.v.s. can be separated by a closed real hyperplane. We derive from (3.1) two more separation theorems (for the second of which it will be important that E is a l.c.s.) that have become standard tools of the theory.

9.1

(FIRST SEPARATION THEOREM). *Let A be a convex subset of a t.v.s. E, such that $\mathring{A} \neq \varnothing$ and let B be a non-empty convex subset of E not intersecting the interior $\mathring{A}$ of A. There exists a closed real hyperplane H separating A and B; if A and B are both open, H separates A and B strictly.*

Proof. $\mathring{A}$ is convex by (1.2) and so is $\mathring{A} - B$, which is open and does not contain 0, since $\mathring{A} \cap B = \varnothing$. Hence by (3.1), there exists a closed real hyperplane H_0 containing the subspace $\{0\}$, $H_0 = \{x : f(x) = 0\}$, and disjoint from $\mathring{A} - B$. Now $f(\mathring{A} - B)$ is convex and hence is an interval in R and does not contain 0; we have, after a change of sign in f if necessary, $f(\mathring{A} - B) > 0$. Thus if $\alpha = \inf f(\mathring{A})$, $H = \{x : f(x) = \alpha\}$ separates $\mathring{A}$ and B: $\mathring{A} \subset F^\alpha$, $B \subset F_\alpha$. Since $A \subset \bar{A}$ and $\bar{A} = (\mathring{\bar{A}})$ by (1.3), we have $A \subset F^\alpha$, since F^α is closed in E; thus H separates A and B. If A and B are open sets, $f(A)$ and $f(B)$ are open intervals in R. For, $f = g \circ \phi$ where ϕ is the natural map of E_0 onto E_0/H_0, which is open (E_0 denoting the underlying real space of E), and g is an isomorphism of E_0/H_0 onto R_0 (Chapter I, Section 4). Hence $A \subset G^\alpha$, $B \subset G_\alpha$ in this case, so that H separates A and B strictly.

COROLLARY. *Let C be a convex body in E. Every boundary point of C is contained in at least one supporting hyperplane of C, and C is the intersection of the closed semi-spaces which contain C and are determined by the supporting hyperplanes of C.*

Proof. To see that each boundary point x_0 of C is contained in at least one supporting hyperplane, it will do to apply (9.1) with $A = C$, $B = \{x_0\}$. To prove the second assertion, we need the lemma that no supporting hyperplane of C contains an interior point of C. Assuming this to be true, suppose that $y \notin C$; we have to show that there exists a closed semi-space containing C, but not containing y. Let $x \in \mathring{C}$; the open segment joining x and y contains exactly one boundary point x_0. There exists a supporting hyperplane H passing through x_0; H does not contain y, or else H would contain x. It is clear that one of the closed semi-spaces determined by H contains C but not y. We prove the lemma:

LEMMA. *If C is a convex body in a t.v.s. E, no supporting hyperplane of C contains an interior point of C.*

Assume that $x \in H \cap \mathring{C}$, where $H = \{x: f(x) = \alpha\}$ is a supporting hyperplane of C such that $C \subset F_\alpha$. There exists $y \in \mathring{C}$ with $f(y) < \alpha$, since H cannot contain $\mathring{C}$. Now $f[x + \varepsilon(x - y)] > f(x) = \alpha$ for every $\varepsilon > 0$; since $x \in \mathring{C}$, $x + \varepsilon(x - y) \in C$ for some $\varepsilon > 0$. This contradicts $C \subset F_\alpha$; hence the assumption $H \cap \mathring{C} \neq \varnothing$ is absurd.

9.2

(SECOND SEPARATION THEOREM). *Let A, B be non-empty, disjoint convex subsets of a l.c.s. such that A is closed and B is compact. There exists a closed real hyperplane in E strictly separating A and B.*

Proof. We shall show that there exists a convex, open 0-neighborhood V in E such that the sets $A + V$ and $B + V$ are disjoint; since these are open, convex subsets of E, the assertion will follow at once from (9.1).

It suffices to prove the existence of a convex, circled, open 0-neighborhood W for which $(A + W) \cap B = \varnothing$; then $V = \frac{1}{2}W$ will satisfy the requirement above. Denote by $\mathfrak{U}$ the filter base of all open, convex, circled neighborhoods of 0 in E and assume that $A + U$ intersects B for each $U \in \mathfrak{U}$; then $\{(A + U) \cap B: U \in \mathfrak{U}\}$ is a filter base in B which has a contact point $x_0 \in B$, since B is compact. Hence $x_0 \in \overline{A + U} \subset A + 2U$ for each $U \in \mathfrak{U}$, whence $x_0 \in \bigcap\{A + U: U \in \mathfrak{U}\} = \overline{A} = A$, since A is closed; this is contradictory.

Since sets containing exactly one point are compact, we obtain

COROLLARY 1. *Every non-empty closed, convex subset of a locally convex space is the intersection of all closed semi-spaces containing it.*

This implies a very important property of convex sets in locally convex spaces:

COROLLARY 2. *In every l.c.s. $E(\mathfrak{T})$, the $\mathfrak{T}$-closure and the weak (that is, $\sigma(E, E')$-) closure of any convex set are identical.*

Proof. Since $\mathfrak{T}$ is finer than $\sigma(E, E')$, every $\sigma(E, E')$-closed subset of E is $\mathfrak{T}$-closed. Conversely, every convex, $\mathfrak{T}$-closed subset of E is weakly closed, since it is the intersection of a family of semi-spaces $F_\alpha = \{x: f(x) \leq \alpha\}$, and each F_α is weakly closed.

For non-convex sets in an infinite-dimensional l.c.s., the weak closure is, in general, larger than the $\mathfrak{T}$-closure (Exercise 22).

10. COMPACT CONVEX SETS

For compact, convex subsets of a locally convex space, a number of strong separation results can be established that will lead to the theorem of Krein-Milman. The theorem asserts that each compact, non-empty convex set contains extreme points (for the definition, see below) and is, in fact, the convex closure of its set of extreme points. We begin with a sharpening of Corollary 1 of (9.2); our proofs follow Bourbaki [7].

10.1

If C is a non-empty, compact, convex subset of a l.c.s. E, then for each closed real hyperplane H in E there exist at least one and, at most, two supporting hyperplanes of C parallel to H. Moreover, C is the intersection of the closed semi-spaces that contain C and are determined by its hyperplanes of support.

Proof. Let $H = \{x: f(x) = \gamma\}$ be any closed real hyperplane in E. Since the restriction of f to C is a continuous real-valued function, there exist points $x_0 \in C$ and $x_1 \in C$ such that $f(x_0) = \alpha = \inf f(C)$ and $f(x_1) = \beta = \sup f(C)$. It is clear that $H_0 = \{x: f(x) = \alpha\}$ and $H_1 = \{x: f(x) = \beta\}$ are (not necessarily distinct) supporting hyperplanes of C, and that there exist no further supporting hyperplanes parallel to H. To prove the second assertion, let $y \notin C$; by (9.2), there exists a closed real hyperplane H strictly separating $\{y\}$ and C. Evidently there exists a hyperplane H_1 parallel to H and supporting C, and such that y is contained in that open semi-space determined by H_1 which does not intersect C.

COROLLARY. *If E is a l.c.s., E_0' is the space of all continuous real linear forms on E, and C is a compact, convex subset of E, then*

$$C = \bigcap \{f^{-1}[f(C)]: f \in E_0'\}.$$

10.2

The convex hull of a finite family of compact, convex subsets of a Hausdorff t.v.s. is compact.

Proof. In fact, if A_i ($i = 1, \ldots, n$) are non-empty convex subsets of E, it is readily verified that their convex hull is $A = \{\sum_{i=1}^{n} \lambda_i a_i: a_i \in A_i, \lambda_i \geqq 0, \sum_{i=1}^{n} \lambda_i = 1$ ($i = 1, \ldots, n$)\}. Thus if $P \subset R^n$ is the compact set $\{(\lambda_1, \ldots, \lambda_n): \lambda_i \geqq 0, \sum \lambda_i = 1\}$, A is the continuous image of the set $P \times \prod_i A_i \subset R^n \times E^n$; hence A is compact if each A_i is ($i = 1, \ldots, n$).

We need now a generalization of the concept of a supporting hyperplane for convex sets. If A is a convex subset of a l.c.s. E, a closed, real linear manifold is said to **support** A if $M \cap A \neq \emptyset$, and if every closed segment $S \subset A$ belongs to M whenever the corresponding open segment S_0 intersects M. In other words, M supports A if $S \subset A$ and $S_0 \cap M \neq \emptyset$ together imply

$S \subset M$, supposing in addition that $M \cap A \neq \emptyset$. An **extreme point** of A is a point $x_0 \in A$ such that $\{x_0\}$ supports A.

Examples. Every vertex of a convex polyhedron A in $\mathbf{R}^3$ is an extreme point of A; every straight line containing an edge of A is a supporting manifold; every plane containing a face of A is a supporting hyperplane. For an infinite-dimensional example of extreme points, see Exercise 29.

The following theorem, asserting the existence of an extreme point in every hyperplane supporting a compact, convex set, is the final step toward the Krein-Milman theorem:

10.3

If C is a compact, convex subset of a locally convex space, every closed real hyperplane supporting C contains at least one extreme point of C.

Proof. Let H be a closed real hyperplane supporting C and denote by $\mathfrak{M}$ the family of all closed real linear manifolds contained in H and supporting C. $\mathfrak{M}$ is inductively ordered under downward inclusion $\supset$; for, if $\{M_\alpha : \alpha \in A\}$ is a totally ordered subfamily, then $M = \bigcap_\alpha M_\alpha$ will be its lower bound in $\mathfrak{M}$, provided that $M \cap C \neq \emptyset$. But the family $\{M_\alpha \cap C : \alpha \in A\}$, again totally ordered under $\supset$, is a filter base consisting of closed subsets of C; since C is compact, it follows that $M \cap C = (\bigcap_\alpha M_\alpha) \cap C = \bigcap_\alpha (M_\alpha \cap C)$ is not empty. Hence by Zorn's lemma, there exists a minimal element $M_0 \in \mathfrak{M}$. If $M_0 = \{x_0\}$, then x_0 is an extreme point of C contained in H; we shall show that the assumption dim $M_0 \geqq 1$ contradicts the minimality of M_0. Now $C_0 = C \cap M_0$ is a compact, convex subset of the affine subspace M_0, and if the dimension of M_0 is $\geqq 1$, (10.1) implies that there exists a closed hyperplane M_1 in M_0 such that M_1 supports C_0. We claim that $M_1 \in \mathfrak{M}$, for if S is a closed segment, $S \subset C$, and the corresponding open segment S_0 intersects M_1, then S_0 intersects M_0 and hence we have $S \subset M_0$ which, in turn, implies $S \subset C_0$ and therefore $S \subset M_1$. Thus $M_1 \in \mathfrak{M}$, which contradicts the minimality of M_0 in $\mathfrak{M}$.

10.4

Theorem. (Krein-Milman). *Every compact, convex subset of a locally convex space is the closed, convex hull of its set of extreme points.*

Proof. If $C \neq \emptyset$ is convex and compact and B is the closed, convex hull of the set of extreme points of C, then clearly $B \subset C$. On the other hand, if $f \neq 0$ is a continuous real linear form on E and $f(C) = [\alpha, \beta]$ there exist, by (10.3), extreme points of C in the supporting hyperplanes $f^{-1}(\alpha)$ and $f^{-1}(\beta)$, whence it follows that $f(C) \subset f(B)$. Thus $f^{-1}[f(C)] \subset f^{-1}[f(B)]$ for each $f \in E'_0$, which implies $C \subset B$ by the corollary of (10.1).

The following supplement of (10.4) is due to Milman [1].

10.5

If A is a compact subset of a locally convex space such that the closed, convex hull C of A is compact, then each extreme point of C is an element of A.

Proof. Let x_0 be an extreme point of C, and V any closed convex 0-neighborhood. There exist points $y_i \in A$ $(i = 1, ..., n)$, with $A \subset \bigcup_i (y_i + V)$. Denote by W_i the closed, convex hull of $A \cap (y_i + V)$. Since W_i (which are subsets of C) are compact, by (10.2) so is the convex hull of their union which is, therefore, identical with C. Hence $x_0 = \sum_{i=1}^{n} \lambda_i w_i$, where $w_i \in W_i$, $\lambda_i \geq 0$ $(i = 1, ..., n)$ and $\sum_{i=1}^{n} \lambda_i = 1$. x_0 being an extreme point of C, it follows that $x_0 = w_i$ for some i; hence $x_0 \in W_i \subset y_i + V$, and since $y_i \in A$, it follows that $x_0 \in A + V$. Since V is an arbitrary member of a 0-neighborhood base and A is closed, it follows that $x_0 \in A$.

COROLLARY. *If C is a compact, convex subset of a l.c.s. and $\mathscr{E}$ is the set of its extreme points, then $\bar{\mathscr{E}}$ is the minimal closed subset of C whose convex closure equals C.*

However, in general, $\mathscr{E}$ is dense in C (Exercise 29).

EXERCISES

1. Let E be a vector space and let $A \neq \varnothing$ be a subset of E. The convex hull (the convex, circled hull) of A consists of all finite sums $\sum_{1}^{n} \lambda_i x_i$ such that $x_i \in A$, $\lambda_i \geq 0$ and $\sum_{1}^{n} \lambda_i = 1$ (such that $\sum_{1}^{n} |\lambda_i| \leq 1$); the convex, circled hull ΓA is the convex hull of the circled hull of A. If E is a t.v.s., the convex hull of an open subset is open, and the closed, convex, circled hull of A is the closure of ΓA.

2. A real-valued function ϕ on a convex subset of a vector space E is **convex** if $\lambda, \mu > 0$ and $\lambda + \mu = 1$ imply that $\phi(\lambda x + \mu y) \leq \lambda \phi(x) + \mu \phi(y)$. If E is a t.v.s. and ϕ is a convex function on E, show that these assertions are equivalent:

   (a) ϕ is continuous on E.
   (b) ϕ is upper semi-continuous on E.
   (c) There exists a non-empty, convex, open subset of E on which ϕ is bounded above.

Deduce from this that there exists a continuous linear form $f \neq 0$ on E if and only if there exists a non-constant convex function on E which is upper semi-continuous. (Use the corollary of (3.1).)

3. Show that each convex, radial subset of a finite-dimensional vector space E is a 0-neighborhood for the unique separated topology on E under which E is a t.v.s. (Chapter I, Section 3).

4. A real-valued function ψ on a vector space E is **sublinear** if it is

convex and positive homogeneous (i.e., if $\psi(\lambda x) = \lambda\psi(x)$ for all $x \in E$ and all $\lambda \geqq 0$).

(a) Every sublinear function on $\mathbf{R}^n$ is continuous; examples are $x \to \sup x_i$, $x \to [\sum |x_i|^q]^{1/q}$ $(q \geqq 1)$ $(i = 1, ..., n)$. (Use Exercises 2, 3.)

(b) If ψ is a sublinear function on $\mathbf{R}^n$ such that $\psi(x_1, ..., x_n) \geqq 0$ whenever $x_i \geqq 0$ $(i = 1, ..., n)$ and $\psi(x_1, ..., x_n) \leqq 0$ if all $x_i \leqq 0$, and if p_i are continuous semi-norms on a t.v.s. E, then $\psi(p_1, ..., p_n)$ is a continuous semi-norm on E.

(c) Assume that ψ is a sublinear function on $\mathbf{R}^n$, as in (b), having the additional property that $x_i \geqq 0$ $(i = 1, ..., n)$ and $\psi(x_1, ..., x_n) = 0$ imply $x_i = 0$ $(i = 1, ..., n)$. Show that if (E_i, p_i) are n normed spaces, then $(x_1, ..., x_n) \to \psi[p_1(x_1), ..., p_n(x_n)]$ is a norm on $\prod_i E_i$ generating the product topology.

5. Let L be a vector space over a complete, non-discrete valuated field K (not necessarily $\mathbf{R}$ or $\mathbf{C}$), and call (L, p) a normed space over K if p satisfies conditions (i) through (iii) of Section 2. Find out to what extent the results of Section 2 can be carried over to this more general case. Show that the topologies generated on L by two such norms p_1, p_2 are identical if and only if there exist constants $c, C > 0$ such that $cp_1(x) \leqq p_2(x) \leqq Cp_1(x)$ for all $x \in L$.

6. Let E be a vector space over $\mathbf{R}$, let M be a subspace of E, and let g be a linear form on M such that $g(x) \leqq p(x)$ $(x \in M)$, where p is a sublinear function on E. There exists a linear form f on E extending g and such that $f(x) \leqq p(x)$ for all $x \in E$. (Observe that the linear forms on $E \times \mathbf{R}$ are the maps $(x, t) \to h(x) + \alpha t$ with $h \in E^*$ and $\alpha \in \mathbf{R}$. Consider the linear manifold $H_0 = \{(x, t): g(x) - t = 1\}$ and the convex cone $C = \{(x, t): p(x) \leqq t\}$ in $E \times \mathbf{R}$, and prove the existence of a hyperplane $H \subset E \times \mathbf{R}$ such that $H \supset H_0$ and $H \cap C = \varnothing$. Cf. proof of (3.2).) Show that this form of the Hahn-Banach theorem, (3.1), and (3.2) imply each other.

7. Denote by E an infinite-dimensional vector space and by $\mathfrak{T}$ the finest locally convex topology on E (Example following (6.2)). Show that $E(\mathfrak{T})$ is a l.c.s. having these properties:

(a) Every linear map u on $E(\mathfrak{T})$ into any l.c.s. F is continuous; hence $E(\mathfrak{T})' = E^*$, every subspace is closed, and every algebraic direct sum decomposition of E is topological.

(b) A subset $B \subset E$ is bounded if and only if it is contained in a finite-dimensional subspace and bounded there; a subset of E is sequentially closed if and only if its intersection with each finite-dimensional subspace is closed.

(c) If E has a countable basis, a convex subset of E is closed if (and only if) its intersection with each finite-dimensional subspace is closed.

(d) $E(\mathfrak{T})$ is complete and not metrizable.

Show also that the property of carrying the finest locally convex topology is inherited by quotients, by inductive limits and by subspaces, but not by infinite products.

8. A family P of semi-norms on a vector space E is **directed** if it is directed for the usual order $\leqq$, defined by "$p(x) \leqq q(x)$ for all $x \in E$".

(a) Let P be a family of semi-norms on E, and let $U_{p,n} = \{x: p(x) < n^{-1}\}$ for $p \in P$, $n \in N$. For $\{U_{p,n}: p \in P, n \in N\}$ to be a 0-neighborhood base of a locally convex topology on E, it is necessary and sufficient that the family $\{cp: c > 0, p \in P\}$ of semi-norms be directed.

(b) If P_0 is a family of semi-norms generating the topology of the l.c.s. E, the family P of the suprema of non-empty finite subsets of P_0 is directed and generates the topology of E.

(c) Let M be a subspace of E. For a given semi-norm p on E, define $\hat{p}(\hat{x}) = \inf\{p(x): x \in \hat{x}\}(\hat{x} \in E/M)$; $\hat{p}$ is a semi-norm on E/M. If P is a directed family of semi-norms on E, the family $\{\hat{p}: p \in P\}$ generates on E/M the quotient of the topology generated by the family P.

9. Let $E(\mathfrak{T}) = \varprojlim g_{\alpha\beta}E_\beta(\mathfrak{T}_\beta)$ be the projective limit of a family of l.c.s. and suppose that $g_{\alpha\gamma} = g_{\alpha\beta} \circ g_{\beta\gamma}$ whenever $\alpha \leqq \beta \leqq \gamma$. Prove that $E(\mathfrak{T})$ is isomorphic with $\varprojlim g_{\delta,\varepsilon}E_\varepsilon(\mathfrak{T}_\varepsilon)$ $(\delta, \varepsilon \in B)$ if B is a cofinal subset of A. A corresponding result holds for inductive limits if $h_{\gamma\alpha} = h_{\gamma\beta} \circ h_{\beta\alpha}$ whenever $\alpha \leqq \beta \leqq \gamma$.

10. If $E(\mathfrak{T}) = \varprojlim g_{mn}E_n(\mathfrak{T}_n)$ is the projective limit of a sequence of l.c.s. such that $g_{mp} = g_{mn} \circ g_{np}$ and $E_m = g_{mn}(E_n)$ whenever $m \leqq n \leqq p$, then $f_n(E) = E_n$ (f_n denoting the projection of $\prod_k E_k$ onto E_n). The result carries over to projective limits of countable families (use Exercise 9).

11. The l.c. direct sum of an infinite family of locally convex spaces, each not reduced to $\{0\}$, is not metrizable. (Consider the completion of the l.c. direct sum of a countable subfamily and use Baire's theorem.)

12. Show that if $E(\mathfrak{T})$ is the locally convex direct sum of a denumerable family of l.c.s., $\mathfrak{T}$ is identical with the topology defined in Chapter I, Exercise 1.

13. If E is a metrizable l.c.s. which possesses a countable fundamental system of bounded sets, then E is normable. (Observe that the completion of E is the union of countably many bounded subsets.) Give an example of a non-metrizable l.c.s. that possesses a countable fundamental system of bounded sets. (Use (6.3).)

14. Let E be the vector space over R of all continuous real-valued functions f on $[0, 1]$ that vanish in a neighborhood (depending on f) of $t = 0$, under the uniform topology. Show that $D = \{f: n|f(n^{-1})| \leqq 1, n \in N\}$ is a barrel in E but not a neighborhood of 0, thus exhibiting a normable (hence bornological) space which is not barreled.

15. Let E be a l.c.s., $\tilde{E}$ its completion. (a) If E is barreled, $\tilde{E}$ is barreled; (b) if E is bornological, $\tilde{E}$ is barreled. (For (b), use (8.5).)

16. Prove the following generalization of (8.1): Let L be a metrizable t.v.s. over a non-discrete, valuated field K (not necessarily R or C); each circled subset of L that absorbs every null sequence in L is a neighborhood of 0.

17. Let E, F be l.c.s., where E is bornological, and let u be a linear map of E into F. If for each null sequence $\{x_n\} \subset E$, the sequence $\{u(x_n)\}$ is bounded in F, then $\lim x_n = 0$ implies $\lim u(x_n) = 0$. Use this result to derive a more general form of (8.3).

18. A l.c.s. E is bornological if and only if every linear map u on E into any Banach space F such that u carries bounded sets onto bounded

sets is continuous. (To establish the sufficiency of the condition, consider the bornological space E_0 associated with E, and show that the identity map of E onto E_0 is continuous by using (5.2) and (5.4), Corollary 2.)

19. Let $\{E_\alpha\colon \alpha \in A\}$ be a family of l.c.s. over K.

(a) Assume that K_0^A is bornological. If u is a linear map on $E = \prod_\alpha E_\alpha$ into a l.c.s. F such that u carries bounded sets onto bounded sets and the restriction of u to each of the subspaces $\dot{E}_\alpha = \{x \in E\colon x_\beta = 0$ for $\beta \neq \alpha\}$ of $E (\alpha \in A)$ vanishes, then $u = 0$. (Consider the restriction of u to the bornological space $\prod_\alpha K x_\alpha$ for each $x = (x_\alpha) \in E$, and use (8.3).)

(b) Let F be a Banach space and let u be a linear map of $E = \prod_\alpha E_\alpha$ into F transforming bounded sets into bounded sets. Show that u must vanish on all but a finite number of the subspaces $\dot{E}_\alpha$.

(c) Using (a) and (b), show that if K_0^A is bornological and u is a linear map on $\prod_\alpha E_\alpha$ into a Banach space F such that u maps bounded sets onto bounded sets, there exists a finite subset $H \subset A$ such that $\prod_\alpha E_\alpha = \prod_{\beta \in H} E_\beta \oplus G$ and $u(G) = \{0\}$.

(d) Using (c) and Exercise 18, show that $\prod_\alpha E_\alpha$ is bornological if E_α $(\alpha \in A)$ and K_0^A are bornological. Deduce from this that the product of a countable number of bornological spaces is bornological.

(e) If $\prod_\alpha E_\alpha$ is bornological, then $\prod_{\beta \in B} E_\beta = G$ is bornological for any subset $B \subset A$. (Observe that G is isomorphic with a quotient space of $\prod_\alpha E_\alpha$.)

20. Let E be a vector space and let A and B be non-empty convex subsets of E such that $A \cap B = \varnothing$ and A has a core point x_0 (i.e., a point x_0 such that $A - x_0$ is radial). There exists a real hyperplane in E separating A and B. (Note that a core point of A is an interior point of A for the finest locally convex topology on E.)

21. Let A, B be non-empty, non-intersecting, convex subsets of a vector space E. Show that there exist convex subsets C, D of E such that $A \subset C, B \subset D, C \cap D = \varnothing$ and $C \cup D = E$. (Use Zorn's lemma.)

22. Let $E = l^2$, the Hilbert space of square summable sequences $x = (x_1, x_2, \ldots)$ with $\|x\| = (\sum_n |x_n|^2)^{\frac{1}{2}}$. Show that the weak closure of the sphere $S = \{x\colon \|x\| = 1\}$, which is closed in E, is the ball $B = \{x\colon \|x\| \leq 1\}$.

23. Show that in a finite-dimensional l.c.s., each pair of non-empty, non-intersecting, closed, convex subsets is separated by a real hyperplane. (Represent one of the sets as the union of an increasing sequence of compact convex subsets.) Show by an example in R_0^2 that in this result, "separated" cannot be replaced by "strictly separated".

24. Let E be a l.c.s and let C be a closed, convex cone of vertex 0 in E such that $C \neq E$.. Show that C is the intersection of the closed semi-spaces containing it and determined by the supporting hyperplanes of C.

25. Let E be a l.c.s. over R, let C be a convex cone of vertex 0 in E, and let C' be the subset of the dual E' whose elements are non-negative on C. C' is a convex cone (of vertex 0) in E' which separates points in E if and only if $\bar{C} \cap (-\bar{C}) = \{0\}$.

26. Show that in a finite-dimensional l.c.s. the convex hull of a compact set is compact.

27. By establishing the following propositions show that in an infinite-dimensional l.c.s., the convex hull of a compact subset is not necessarily closed, and the closed, convex hull of a compact subset is not necessarily compact.

(a) Denote by X the family of all real-valued continuous functions on the unit interval $[0, 1]$ and by E the product space R_0^X; E is a l.c.s. For each fixed $t \in [0, 1]$, let $\phi_t \in E$ be the "evaluation map" $f \to f(t)$. The set $K_1 = \{\phi_t : t \in [0, 1]\}$ is compact in E.

(b) The element $\phi \in E$ given by the Riemann integral $f \to \int_0^1 f(t)\, dt$ is in the closure of the convex hull C of K_1, but $\phi \notin C$; hence C is not closed in E.

(c) Denote by F the smallest subspace of E that contains K_1; C is closed in F and hence is the closed, convex hull of K_1 in F, but is not compact.

28. Let E be the Banach space of real null sequences $x = (x_1, x_2, \cdots)$ with $\|x\| = \sup_n |x_n|$. The unit ball $B = \{x : \|x\| \leq 1\}$ in E is closed and convex, but has no extreme points.

29. Let E be the Hilbert space l^2 over R. Denote by C the subset of E determined by $\sum_{n=1}^{\infty} (2^n x_n)^2 \leq 1$. Then C is a convex and compact set, and the closure of its subset $\mathscr{E}$ of extreme points. (Let E_n be the subspace of E determined by $x_k = 0$ for $k > n$; the boundary points of the ellipsoid $C \cap E_n$ form a set $\mathscr{E}_n$ consisting entirely of extreme points of C. Show that $\bigcup_{n=1}^{\infty} \mathscr{E}_n$ is dense in C. The example is due to Poulsen [1].)

30. A convex cone C of vertex 0 in a l.c.s. E has **compact base** if there exists a real affine subspace N of E, $0 \notin N$, such that $N \cap C$ is compact, non-empty, and C is the smallest cone of vertex 0 containing $N \cap C$. A ray $R = \{\lambda x_0 : \lambda \geq 0\}$, $0 \neq x_0 \in C$, is **extreme** if $x \in R$, $y \in C$, and $x - y \in C$ imply $y \in R$. Show that a convex cone with compact base is closed, satisfies $C \cap -C = \{0\}$, and is the closed, convex hull of the set of its extreme rays. (For more general results in this direction, see Klee [5].)

Chapter III

LINEAR MAPPINGS

The notion of linear mapping has been used frequently before and is obviously indispensable for any discussion of topological vector spaces. But the accent in this chapter is on vector spaces whose elements are vector-valued functions, especially linear mappings. The study of such spaces and their topologies forms the natural background for much of what follows in this book, in particular, duality (Chapter IV) and spectral theory (Appendix); it also leads, via spaces of bilinear maps and topological tensor products, to the important class of nuclear spaces.

The first two sections, concerned with topological homomorphisms, Banach's classical theorem and its close relative, the closed graph theorem, appear to be somewhat isolated from the general theory; however, it will become evident in Chapter IV (Section 8) that for locally convex spaces, these results find their proper place in the general framework of duality. Revealing an intimate relationship with the concept of completeness, this deeper analysis will eliminate the dominating role of category in the proofs of these theorems and lead to what is probably the natural bound of their validity. Thus from a purely esthetical point of view, one is tempted to defer the discussion of the homomorphism and closed graph theorems until the tool of duality is fully available. For the benefit of the reader who is interested in a quick approach, we give the classical versions with their direct proofs here. The other fact in favor of an independent treatment (namely, the fact that the classical proofs can dispense with local convexity) is of little practical importance. Section 3 discusses topologizing spaces of vector-valued functions, boundedness, and the most frequent types of $\mathfrak{S}$-topologies. The section on equicontinuity that follows is fundamental; here also the celebrated principle of uniform boundedness and the Banach-Steinhaus theorem have their natural places. Spaces of bilinear mappings and forms (Section 5) are not only an interesting class of vector-valued function spaces but furnish the background of the theory of topological tensor products of which the elements are presented in Section 6 (barring the use of duality). This, in turn, leads naturally to nuclear mappings and spaces, an important class of locally convex spaces that are beyond the reach of Banach space theory. The comparatively recent results in this area are practically all due to Grothendieck [13]; it is perhaps of interest to the expert how many of the basic results on these spaces can be obtained without the use of duality or abstract measure theory. The final section discusses the approximation problem, with some emphasis on Banach spaces, and briefly the basis problem. It becomes apparent here that duality is hardly dispensable, but the results on strong duals and adjoints in (B)-spaces used here are elementary, so we have decided to place this discussion before Chapter IV despite some technical inconveniences.

1. CONTINUOUS LINEAR MAPS AND TOPOLOGICAL HOMOMORPHISMS

If L and M are t.v.s. over K and u is a linear map (an algebraic homomorphism) of L into M, then u is continuous if and only if u is continuous at $0 \in L$; for if V is a given 0-neighborhood in M, and U is a 0-neighborhood in L such that $u(U) \subset V$, then $x - y \in U$ implies $u(x - y) = u(x) - u(y) \in V$. Hence if u is continuous at 0, it is even uniformly continuous on L into M for the respective uniformities (Chapter I, Section 1). Thus if u is continuous on L into M with M separated and complete, then u has a unique continuous extension $\bar{u}$, with values in M, to any t.v.s. $\bar{L}$ of which L is a dense subspace (in particular, to the completion $\tilde{L}$ of L if L is separated); it is easy to see that $\bar{u}$ is linear. We supplement these simple facts by a statement in terms of semi-norms.

1.1

Let the topologies of L and M be locally convex and let $\mathscr{P}$ be a family of semi-norms generating the topology of L. A linear map u of L into M is continuous if and only if for each continuous semi-norm q on M, there exists a finite subset $\{p_i: i = 1, ..., n\}$ of $\mathscr{P}$ and a number $c > 0$ such that $q[u(x)] \leq c \sup_i p_i(x)$ for all $x \in L$.

Proof. The condition is necessary. Let V be the 0-neighborhood $\{y: q(y) \leq 1\}$, where q is a given continuous semi-norm on M. Since u is continuous and $\mathscr{P}$ generates the topology of L (Chapter II, Section 4), there exist 0-neighborhoods $U_i = \{x: p_i(x) \leq \varepsilon_i\}$ ($\varepsilon_i > 0$, $p_i \in \mathscr{P}$; $i = 1, ..., n$) such that $u(\bigcap_i U_i) \subset V$. Hence, letting $\varepsilon = \min_i \varepsilon_i$, the relation $\sup_i p_i(x) \leq \varepsilon$ implies $u(x) \in V$; thus $q[u(x)] \leq 1$. Clearly, then, $q[u(x)] \leq \varepsilon^{-1} \sup_i p_i(x)$ for all $x \in L$.

The condition is sufficient. If V is a given convex circled 0-neighborhood in M, its gauge function q is a continuous semi-norm on M. Thus if $q[u(x)] \leq$

$c \sup_i p_i(x)$, where $c > 0$ and $p_i \in \mathscr{P}$ $(i = 1, ..., n)$, it follows that $u(U) \subset \bar{V}$ for the 0-neighborhood $U = \{x: cp_i(x) \leq 1, i = 1, ..., n\}$ in L.

COROLLARY. *If u is a linear map on a normed space $(L, \| \ \|)$ into a normed space $(M, \| \ \|)$, u is continuous if and only if $\|u(x)\| \leq c\|x\|$ for some $c > 0$ and all $x \in L$.*

A continuous linear map on L into M, where L and M are t.v.s. over K, is called a **topological homomorphism** (or, briefly, homomorphism when no confusion is likely to occur) if for each open subset $G \subset L$, the image $u(G)$ is an open subset of $u(L)$ (for the topology induced by M). Examples of topological homomorphisms are, for any subspaces H and N of L, the canonical (quotient) map $\phi: L \to L/N$ and the canonical imbedding $\psi: H \to L$. With the aid of these two mappings, every linear map u of L into M can be "canonically" decomposed:

$$L \underset{\phi}{\to} L/N \underset{u_0}{\to} u(L) \underset{\psi}{\to} M.$$

Here $N = u^{-1}(0)$ is the null space of u, and u_0 is the algebraic isomorphism which maps each equivalence class $\hat{x}$ of L mod N onto the common image $u(x)$ $(x \in \hat{x})$ of this class under u. Hence $u = \psi \circ u_0 \circ \phi$, and we call the bijective map u_0 **associated** with u. We leave it to the reader to verify that u is an open map if and only if u_0 is open and that u is a continuous map if and only if u_0 is continuous.

1.2

Let L and M be t.v.s. and let u be a linear map of L into M. These assertions are equivalent:

(a) *u is a topological homomorphism.*
(b) *For every neighborhood base $\mathfrak{U}$ of 0 in L, $u(\mathfrak{U})$ is a neighborhood base of 0 in $u(L)$.*
(c) *The map u_0 associated with u is an isomorphism.*

Proof. (a) $\Rightarrow$ (b): Since u is open, every element of $u(\mathfrak{U})$ is a 0-neighborhood, and $u(\mathfrak{U})$ is a base at $0 \in u(L)$, since u is continuous. (b) $\Rightarrow$ (c): Since $\phi(\mathfrak{U})$ is a 0-neighborhood base in L/N, $N = u^{-1}(0)$, for any 0-neighborhood base in L, u_0 has property (b) and is consequently an isomorphism. (c) $\Rightarrow$ (a): Since ϕ, u_0, ψ are all continuous and open, so is $u = \psi \circ u_0 \circ \phi$, and hence is a topological homomorphism.

1.3

Let u be a linear map on L whose range is a finite-dimensional Hausdorff t.v.s. These assertions are equivalent:

(a) *u is continuous.*
(b) *$u^{-1}(0)$ is closed in L.*
(c) *u is a topological homomorphism.*

Proof. (a) $\Rightarrow$ (b): Since $u(L)$ is Hausdorff, $\{0\}$ is closed and thus $u^{-1}(0)$ is closed if u is continuous. (b) $\Rightarrow$ (c): If $u^{-1}(0)$ is closed, $L/u^{-1}(0)$ is a Hausdorff t.v.s. of finite dimension, whence by (I, 3.4), u_0 is an isomorphism; it follows from (1.2) that u is a topological homomorphism. (c) $\Rightarrow$ (a) is clear.

COROLLARY. *Every continuous linear form on a t.v.s. L is a topological homomorphism.*

This fact has been used implicitly in the proof of (II, 9.1).

2. BANACH'S HOMOMORPHISM THEOREM

It follows from (1.3) that every continuous linear map with finite-dimensional separated range is a topological homomorphism; the question arises for what, if any, larger class of t.v.s. it is true that a continuous linear mapping is automatically open (hence a homomorphism). We shall see that this holds for all mappings of one Fréchet space onto another, and in certain more general cases. For a deeper study of this subject, the reader is referred to Chapter IV, Section 8. We first prove a classical result of Banach ([1], chap. III, theor. 3) for which we need the following lemma:

Let L, M be metric t.v.s. whose respective metrics d, δ are given by pseudo-norms (Chapter I, Section 6): $d(x_1, x_2) = |x_1 - x_2|$ and $\delta(y_1, y_2) = |y_1 - y_2|$. We denote by $S_r = \{x \in L: |x| \leq r\}$ and $S_\rho = \{y \in M: |y| \leq \rho\}$, respectively, the closed balls of center 0 and radius r, ρ.

LEMMA. *Let L be complete and let u be a continuous linear map of L into M satisfying*

(P): *For every $r > 0$, there exists $\rho = \rho(r) > 0$ such that $\overline{u(S_r)} \supset S_\rho$.*

Then $u(S_t) \supset S_\rho$ for each $t > r$.

Proof. Let r and t, $t > r > 0$ be fixed and denote by $\{r_n\}$ a sequence of positive real numbers such that $r_1 = r$ and $\sum_1^\infty r_n = t$. Let $\{\rho_n\}$ be a null sequence of numbers > 0 such that $\rho_1 = \rho$ and for each $n \in N$, ρ_n satisfies $\overline{u(S_{r_n})} \supset S_{\rho_n}$. For each $y \in S_\rho$, we must establish the existence of $z \in S_t$ with $u(z) = y$.

We define inductively a sequence $\{x_n: n = 0, 1, \ldots\}$ such that for all $n \geq 1$:

(i) $|x_n - x_{n-1}| \leq r_n$.

(ii) $|u(x_n) - y| \leq \rho_{n+1}$.

Set $x_0 = 0$ and assume that $x_1, x_2, \ldots, x_{k-1}$ have been selected to satisfy (i) and (ii) ($k \geq 1$). By property (P), the set $u(x_{k-1} + S_{r_k})$ is dense with respect to $u(x_{k-1}) + S_{\rho_k}$. From (ii) we conclude that $y \in u(x_{k-1}) + S_{\rho_k}$; thus there exists x_k satisfying $|x_k - x_{k-1}| \leq r_k$ and $|u(x_k) - y| \leq \rho_{k+1}$.

Since $\sum_1^\infty r_n$ converges, $\{x_n\}$ is a Cauchy sequence in the complete space L and thus converges to some $z \in L$. Clearly, $|z| \leq t$, and $u(z) = y$ follows from the continuity of u and (ii), since $\{\rho_n\}$ was chosen to be a null sequence.

Let us point out that the results and, up to minor modifications, the proofs of this section, with the exception of (2.2), are valid for topological vector spaces L over an arbitrary, non-discrete valued field K (Chapter I). Also the following remark may not be amiss. A Baire space is, by definition, a topological space in which every non-empty open subset is not meager. This implies that every t.v.s. L over K which is non-meager (of second category) in itself, is a Baire space. Otherwise, there would exist a meager, non-empty, open subset of L, and hence a meager 0-neighborhood U. Since L is a countable union of homothetic images of U (hence of meager subsets), we arrive at a contradiction.

2.1

Theorem. *Let L, M be complete, metrizable t.v.s. and let u be a continuous linear map of L with range dense in M. Then either $u(L)$ is meager (of first category) in M, or else $u(L) = M$ and u is a topological homomorphism.*

Proof. Suppose that $u(L)$ is not meager in M. As in the preceding lemma, we can assume the topologies of L and M to be generated by pseudonorms by (I, 6.1), and we continue to use the notation of the lemma. The family $\{S_r : r > 0\}$ is a 0-neighborhood base in L. For fixed r, let $U = S_r$, $V = S_{r/2}$; then $V + V \subset U$ and $u(L) = \bigcup_1^\infty nu(V)$, since V is radial. Let us denote the closure of a set A in $u(L)$ by $[A]^-$. Since, by assumption, $u(L)$ is a Baire space, there exists $n \in N$ such that $[nu(V)]^-$ has an interior point; hence $[u(V)]^-$ has an interior point by (I, 1.1). Now

$$[u(V)]^- + [u(V)]^- \subset [u(V) + u(V)]^- = [u(V + V)]^- \subset [u(U)]^- ;$$

thus $[u(U)]^-$ is a 0-neighborhood in $u(L)$, since 0 is interior to $[u(V)]^- + [u(V)]^-$. Hence there exists $\rho > 0$ such that $u(L) \cap S_\rho \subset [u(U)]^-$, and the lemma above implies that $u(L) \cap S_\rho \subset u(S_{r+\varepsilon})$ for every $\varepsilon > 0$. Thus $\{u(S_t) : t > 0\}$ is a neighborhood base of 0 in $u(L)$, whence by (1.2), u is a topological homomorphism. Therefore u_0 is an isomorphism of the space $L/u^{-1}(0)$ which is complete by (I, 6.3), onto $u(L)$, whence it follows that $u(L) = M$.

COROLLARY 1. *A continuous linear map u of a complete, metrizable t.v.s. L into another such space, M, is a topological homomorphism if and only if $u(L)$ is closed in M.*

Proof. The necessity of the condition is immediate, since $u(L)$, being isomorphic with $L/u^{-1}(0)$, is complete and hence closed in M. Conversely, if $u(L)$ is closed in M, then it is complete and metrizable and hence can replace M in (2.1).

COROLLARY 2. *Let L be a complete, metrizable t.v.s. for both of the top-ologies* $\mathfrak{T}_1$ *and* $\mathfrak{T}_2$, *and suppose that* $\mathfrak{T}_1$ *is finer than* $\mathfrak{T}_2$. *Then* $\mathfrak{T}_1$ *and* $\mathfrak{T}_2$ *are identical.*

This is immediate from Corollary 1, since the identity map is continuous from $(L, \mathfrak{T}_1)$ onto $(L, \mathfrak{T}_2)$.

COROLLARY 3. *If a complete, metrizable t.v.s. L is the direct algebraic sum of two closed subspaces M and N, the sum is topological:* $L = M \oplus N$.

Proof. Since M and N are complete and metrizable, so is $M \times N$; whence it follows that the continuous mapping $(x_1, x_2) \to x_1 + x_2$ of $M \times N$ onto L is an isomorphism (Chapter I, Section 2).

With our present tools, we can extend Corollary 1 somewhat beyond the metrizable case. The following extension is due to Dieudonné-Schwartz [1].

2.2

Let E be a locally convex space of type (LF) *and let F be a locally convex space of type* (F) *or* (LF). *Every continuous linear map u of E onto F is a topo-logical homomorphism.*

Proof. Let $E = \lim_{\longrightarrow} E_n$ be an (LF)-space and let $F = \lim_{\longrightarrow} F_n$ be an (LF)-space; the case where F is a Fréchet space can be formally subsumed under the following proof by letting $F_n = F (n \in N)$. For all $m, n \in N$, set $G_{m,n} = E_m \cap u^{-1}(F_n)$; as a closed subspace of E_m, $G_{m,n}$ is complete and metrizable. Since $u(E) = F$ and $u(G_{m,n}) = u(E_m) \cap F_n$, it follows that $\bigcup\limits_{m=1}^{\infty} u(G_{m,n}) = F_n$ for each fixed n. Since F_n is a Baire space, there exists m_0 (depending on n) such that $u(G_{m_0,n})$ is non-meager in F_n; it follows from (2.1) that $u(G_{m_0,n}) = F_n$. If U is any 0-neighborhood in E, $U \cap G_{m_0,n}$ is a 0-neighborhood in $G_{m_0,n}$ by (II, 6.4); hence $u(U \cap G_{m_0,n})$ is a 0-neighborhood in F_n and a fortiori $u(U) \cap F_n$ is a neighborhood of 0 in F_n. Since this holds for all $n \in N$, it follows that $u(U)$ is a 0-neighborhood in F, and hence u is a homomorphism.

Another direct consequence of Banach's theorem (2.1) is the following frequently used result, called the **closed graph theorem**.

2.3

Theorem. *If L and M are complete, metrizable t.v.s., a linear mapping of L into M is continuous if and only if its graph is closed in* $L \times M$.

Proof. Recall that the graph of u is the subset $G = \{(x, u(x)): x \in L\}$ of $L \times M$. Clearly, if u is continuous, then G is closed in the product space $L \times M$. Conversely, if G is closed, it is (since u is linear) a complete, metrizable subspace of $L \times M$. The mapping $(x, u(x)) \to x$ of G onto L is biunivocal, linear, and continuous, and hence an isomorphism by (2.1). It follows that

$x \to (x, u(x))$ is continuous, whence u is continuous by definition of the product topology on $L \times M$.

3. SPACES OF LINEAR MAPPINGS

Let F be a vector space over K, let T be a set, and let $\mathfrak{S}$ be a family of subsets of T directed under set-theoretic inclusion $\subset$. (Whenever the letter $\mathfrak{S}$ is used in the following, it will denote a family of sets with this property.) A subfamily $\mathfrak{S}_1$ of $\mathfrak{S}$ is **fundamental** (with respect to $\mathfrak{S}$) if it is cofinal with $\mathfrak{S}$ under inclusion (that is, if each member of $\mathfrak{S}$ is contained in some member of $\mathfrak{S}_1$). Consider the vector space F^T, product of T (more precisely, of card T) copies of F; as a set, F^T is the collection of all mappings of T into F. Suppose in addition that F is a t.v.s., and let $\mathfrak{B}$ be a neighborhood base of 0 in F. When S runs through $\mathfrak{S}$, V through $\mathfrak{B}$, the family

$$M(S, V) = \{f : f(S) \subset V\} \qquad (*)$$

is a 0-neighborhood base in F^T for a unique translation-invariant topology, called the **topology of uniform convergence on the sets** $S \in \mathfrak{S}$, or, briefly, the $\mathfrak{S}$**-topology.** For if $V_3 \subset V_1 \cap V_2$ and $S_1 \cup S_2 \subset S_3$, then $M(S_3, V_3)$ is contained in $M(S_1, V_1) \cap M(S_2, V_2)$; hence the sets (*) form a filter base in F^T which has the additional property that $M(S, V) + M(S, V) \subset M(S, U)$ whenever $V + V \subset U$. In (*), the family $\mathfrak{S}$ can evidently be replaced by any fundamental subfamily, and likewise we note that the $\mathfrak{S}$-topology does not depend on the particular choice of the neighborhood base $\mathfrak{B}$ of 0 in F. Now we have to settle the question under what conditions F^T, or a subspace G of F^T, is a t.v.s. for a given $\mathfrak{S}$-topology.

3.1

A (vector) subspace $G \subset F^T$ *is a t.v.s. under an* $\mathfrak{S}$-*topology if and only if for each* $f \in G$ *and* $S \in \mathfrak{S}$, $f(S)$ *is bounded in* F.

Proof. The sets $M(S, V) \cap G$ $(S \in \mathfrak{S}, V \in \mathfrak{B})$ form a base of the 0-neighborhood filter of the topology $\mathfrak{T}$ induced on G by the $\mathfrak{S}$-topology; in what follows we shall denote these sets again by $M(S, V)$ with the understanding that now $M(S, V) = \{f \in G : f(S) \subset V\}$. For $(G, \mathfrak{T})$ to be a t.v.s., by (I, 1.2) it is necessary and sufficient that the 0-neighborhood filter have a base consisting of radial and circled sets, since from the remark made above, it follows that condition (a) of (I, 1.2) is satisfied. Since $M(S, \lambda V) = \lambda M(S, V)$ for each $\lambda \neq 0$, $M(S, V)$ is circled if V is circled; thus let $\mathfrak{B}$ consist of circled sets.

Now suppose that for each $S \in \mathfrak{S}, f \in G$, the set $f(S)$ is bounded in F. Then for given f, S and V, there exists $\lambda > 0$ such that $f(S) \subset \lambda V$ and hence $f \in M(S, \lambda V) = \lambda M(S, V)$; it follows that $M(S, V)$ is radial. Conversely, if $\mathfrak{T}$ is a vector space topology on G, each $M(S, V)$ is necessarily radial; thus for given f, S, and V, there exists $\lambda > 0$ with $f \in \lambda M(S, V) = M(S, \lambda V)$. Hence $f(S) \subset \lambda V$, which shows $f(S)$ to be bounded in F.

3.2

Let F be a locally convex space, let T be a topological space, and let $\mathfrak{S}$ be a family of subsets of T whose union is dense. If G is a subspace of F^T whose elements are continuous on T and bounded on each $S \in \mathfrak{S}$, then G is a locally convex space under the $\mathfrak{S}$-topology.

Proof. If $\mathfrak{B}$ is a 0-neighborhood base in F consisting of convex sets, then each $M(S, V)$ is convex; hence by (3.1) the $\mathfrak{S}$-topology is locally convex. There remains to show that the $\mathfrak{S}$-topology is Hausdorff. Let $f \in G$ and $f \neq 0$; since f is continuous and $\bigcup\{S: S \in \mathfrak{S}\}$ is dense, there exists $t_0 \in S_0 \in \mathfrak{S}$ such that $f(t_0) \neq 0$. Since F is a Hausdorff space (Chapter II, Section 4), we have $f(t_0) \notin V_0$ for a suitable $V_0 \in \mathfrak{B}$. It follows that $f \notin M(S_0, V_0)$, and hence the $\mathfrak{S}$-topology is a Hausdorff topology on G.

If T is itself a t.v.s., and each $S \in \mathfrak{S}$ is bounded, and G is a vector space of continuous linear maps into F, then the assumption that each $f(S)$ be bounded is automatically satisfied by (I, 5.4), and for the conclusion of (3.2) to hold, it suffices that the linear hull of $\bigcup\{S: S \in \mathfrak{S}\}$ be dense. It is convenient to have a term for this: A subset of a t.v.s. L is **total** in L if its linear hull is dense in L. With this notation, we obtain the following corollary of (3.2).

COROLLARY. *Let E be a t.v.s., let F be a l.c.s., and let $\mathfrak{S}$ be a family of bounded subsets of E whose union is total in E. Then the vector space $\mathscr{L}(E, F)$ of all continuous linear mappings of E into F is a locally convex space under the $\mathfrak{S}$-topology.*

Endowed with an $\mathfrak{S}$-topology, the space $\mathscr{L}(E, F)$ is sometimes denoted by $\mathscr{L}_{\mathfrak{S}}(E, F)$. It is no restriction of generality to suppose E separated, for if E_0 is the Hausdorff t.v.s. associated with E (Chapter I, Section 2), then $\mathscr{L}(E, F)$ is algebraically isomorphic with $\mathscr{L}(E_0, F)$ (Exercise 5). Moreover, if E is Hausdorff and F is complete, $\mathscr{L}(E, F)$ is algebraically isomorphic with $\mathscr{L}(\tilde{E}, F)$, where $\tilde{E}$ denotes the completion of E (Exercise 5). We shall see later that every locally convex topology on a vector space E is an $\mathfrak{S}$-topology, where $\mathfrak{S}$ is a suitable family of subsets of the algebraic dual E^* (Chapter IV, Section 1).

Examples

1. Let T be a given set, let F be any t.v.s., and let $\mathfrak{S}$ be the family of all finite subsets of T. Under the $\mathfrak{S}$-topology, F^T is isomorphic with the topological product of T copies of F.

2. Let T be a Hausdorff topological space, let F be a l.c.s., and let $\mathfrak{S}$ be the family of all compact subsets of T. Under the $\mathfrak{S}$-topology (called the topology of *compact* convergence), the space of all continuous functions on T into F is a l.c.s.

3. Let E be a l.c.s. over K with dual E'; the weak dual $(E', \sigma(E', E))$ is the space $\mathscr{L}(E, K_0)$ under the $\mathfrak{S}$-topology, with $\mathfrak{S}$ the family of all finite subsets of E.

4. Let E, F be l.c.s. The following $\mathfrak{S}$-topologies are of special importance on $\mathscr{L}(E, F)$:

a.   The topology of *simple* (or pointwise) convergence: $\mathfrak{S}$ the family
      of all finite subsets of E.
b.   The topology of *convex, circled, compact* convergence: $\mathfrak{S}$ the
      family of all convex, circled, compact subsets of E.
c.   The topology of *precompact* convergence: $\mathfrak{S}$ the family of all
      precompact subsets of E.
d.   The topology of *bounded* convergence: $\mathfrak{S}$ the family of all bounded
      subsets of E.

The families $\mathfrak{S}$ in the preceding examples have the property that the union
of their members is E; such a family $\mathfrak{S}$ is said to **cover** E.

A family $\mathfrak{S} \neq \{\varnothing\}$ of bounded subsets of a l.c.s. E is called **saturated**
if (1) it contains arbitrary subsets of each of its members, (2) it contains
all scalar multiples of each of its members, and (3) it contains the closed,
convex, circled hull of the union of each finite subfamily. Thus the families
$\mathfrak{S}$ of Example 4c and 4d are saturated, and 4a and 4b are not saturated
unless $E = \{0\}$. Since the family of all bounded subsets of E is saturated
and since the intersection of any non-empty collection of saturated families is
saturated, a given family $\mathfrak{S}$ of bounded sets in E determines a smallest
saturated family $\overline{\mathfrak{S}}$ containing it; $\overline{\mathfrak{S}}$ is called the **saturated hull** of $\mathfrak{S}$. E and
F being locally convex, it is clear that for each family $\mathfrak{S}$ of bounded subsets
of E, the $\mathfrak{S}$-topology and the $\overline{\mathfrak{S}}$-topology are identical on $\mathscr{L}(E, F)$. (Cf.
Exercise 7.)

To supplement the corollary of (3.2), we note that if $\{p_\alpha : \alpha \in A\}$ is a family
of semi-norms generating the topology of F, the family of semi-norms

$$u \to p_{S,\alpha}(u) = \sup_{x \in S} p_\alpha[u(x)]$$

($S \in \mathfrak{S}, \alpha \in A$) generates the $\mathfrak{S}$-topology on $\mathscr{L}(E, F)$. In particular, if E and
F are normed spaces, the norm

$$u \to \|u\| = \sup\{\|u(x)\| : \|x\| \leq 1\}$$

generates the topology of bounded convergence on $\mathscr{L}(E, F)$ (cf. Chapter II,
Section 2).

Returning to a more general setting, let E, F be Hausdorff t.v.s. over K,
let $\mathfrak{S}$ be a (directed) family of bounded subsets of E, and let $\mathscr{L}(E, F)$ be the
vector space over K of all continuous linear maps on E into F. We turn our
attention to the subsets of $\mathscr{L}(E, F)$ that are bounded for the $\mathfrak{S}$-topology.

3.3

Let H be a subset of $\mathscr{L}(E, F)$. The following assertions are equivalent:

(a) *H is bounded for the $\mathfrak{S}$-topology.*
(b) *For each 0-neighborhood V in F, $\bigcap_{u \in H} u^{-1}(V)$ absorbs every $S \in \mathfrak{S}$.*
(c) *For each $S \in \mathfrak{S}$, $\bigcup_{u \in H} u(S)$ is bounded in F.*

Proof. (a) $\Rightarrow$ (b): We can assume V to be circled. If H is bounded, it is
absorbed by each $M(S, V)$; hence $u(S) \subset \lambda V$ for all $u \in H$ and some $\lambda > 0$,

which implies $S \subset \lambda \bigcap_{u \in H} u^{-1}(V)$. (b) $\Rightarrow$ (c): If $S \in \mathfrak{S}$ and a circled 0-neighbor-hood V in F are given, then $S \subset \lambda \bigcap_{u \in H} u^{-1}(V)$ implies $u(S) \subset \lambda V$ for all $u \in H$; hence $\bigcup_{u \in H} u(S)$ is bounded in F. (c) $\Rightarrow$ (a): For given S and V, the existence of λ such that $u(S) \subset \lambda V$ for all $u \in H$ implies $H \subset \lambda M(S, V)$; hence H is bounded for the $\mathfrak{S}$-topology.

A subset of $\mathscr{L}(E, F)$ is **simply bounded** if it is bounded for the topology of simple convergence (Examples 1 and 4a above). It is important to know conditions under which simply bounded subsets are bounded for finer $\mathfrak{S}$-topologies on $\mathscr{L}(E, F)$.

3.4

Let E, F be l.c.s. and let $\mathfrak{S}$ be the family of all convex, circled subsets of E that are bounded and complete. Each simply bounded subset of $\mathscr{L}(E, F)$ is bounded for the $\mathfrak{S}$-topology.

Proof. If H is simply bounded in $\mathscr{L}(E, F)$ and V is a closed, convex, circled 0-neighborhood in F, then $D = \bigcap_{u \in H} u^{-1}(V)$ is a closed, convex, circled subset of E which is radial by (3.3)(b), and hence a barrel; thus by (II, 8.5) D absorbs every $S \in \mathfrak{S}$, which implies, again by (3.3), that H is bounded for the $\mathfrak{S}$-topology.

COROLLARY. *If E, F are l.c.s. and E is quasi-complete, then the respective families of bounded subsets of $\mathscr{L}(E, F)$ are identical for all $\mathfrak{S}$-topologies such that $\mathfrak{S}$ is a family of bounded sets covering E.*

Proof. When E is quasi-complete, the family $\mathfrak{S}$ of (3.4) is a fundamental system of bounded sets in E; in other words, the $\mathfrak{S}$-topology of (3.4) is the topology of bounded convergence. The assertion is now immediate.

4. EQUICONTINUITY. THE PRINCIPLE OF UNIFORM BOUNDEDNESS AND THE BANACH-STEINHAUS THEOREM

If T is a topological space and F is a uniform space, a set $H \subset F^T$ is **equicontinuous at** $t_0 \in T$ if for each vicinity (entourage) $N \subset F \times F$, there exists a neighborhood $U(t_0)$ of t_0 such that $[f(t), f(t_0)] \in N$ whenever $t \in U(t_0)$ and $f \in H$; H is **equicontinuous** if it is equicontinuous at each $t \in T$. If T is a uniform space as well and if for each vicinity N in F there exists a vicinity M in T such that $(t_1, t_2) \in M$ implies $[f(t_1), f(t_2)] \in N$ for all $f \in H$, then H is called **uniformly equicontinuous.** It is at once clear that if $T = E$ is a t.v.s., and if F is a t.v.s., a set H of linear mappings of E into F is uniformly equicontinuous (for the unique translation-invariant uniformities associated with the topologies of E and F, respectively (Chapter I, Section 1)) if and only if H is equicontinuous at $0 \in E$; that is, if and only if for each 0-neighborhood V in F, there exists a 0-neighborhood U in E such that $u(U) \subset V$ whenever $u \in H$.

Of course an equicontinuous set of linear mappings of E into F is a subset of $\mathcal{L}(E, F)$.

As before, we shall denote by $\mathcal{L}(E, F)$ the space of all continuous linear maps of E into F, E and F being Hausdorff t.v.s. over the same field K, and by $\mathcal{L}_{\mathfrak{S}}(E, F)$ the same space under an $\mathfrak{S}$-topology with $\mathfrak{S}$ a (directed) family of bounded subsets of E whose union is total in E. Finally, $L(E, F)$ will denote the vector space of all linear maps (continuous or not) of E into F.

The proof of the following statement is quite similar to the proof of (3.3) and will be omitted.

4.1

Let H be a subset of $\mathcal{L}(E, F)$. The following assertions are equivalent:

(a) *H is equicontinuous.*
(b) *For each 0-neighborhood V in F, $\bigcap_{u \in H} u^{-1}(V)$ is a 0-neighborhood in E.*
(c) *For each 0-neighborhood V in F, there exists a 0-neighborhood U in E such that $\bigcup_{u \in H} u(U) \subset V$.*

(4.1)(b) implies (3.3)(b), hence:

COROLLARY. *Each equicontinuous subset of $\mathcal{L}(E, F)$ is bounded for every $\mathfrak{S}$-topology.*

The converse of this corollary is not valid (Exercise 10), but there are important instances in which even a simply bounded subset of $\mathcal{L}(E, F)$ is necessarily equicontinuous.

4.2

Theorem. *Let E, F be l.c.s. such that E is barreled, or let E, F be t.v.s. such that E is a Baire space. Every simply bounded subset H of $\mathcal{L}(E, F)$ is equicontinuous.*

Proof. We give the proof first for the case where E is barreled and F is any l.c.s. If V is any closed, convex, circled 0-neighborhood in F, $W = \bigcap_{u \in H} u^{-1}(V)$ is a closed, convex, circled subset of E which, by condition (b) of (3.3), absorbs finite subsets in E; thus W is a barrel and hence a 0-neighborhood in E, whence H is equicontinuous by (4.1)(b).

If E is a Baire space, F is any t.v.s., and V is a given 0-neighborhood in F, select a closed, circled 0-neighborhood V_1 such that $V_1 + V_1 \subset V$. By (3.3)(b), $W = \bigcap_{u \in H} u^{-1}(V_1)$ is a closed, circled subset of E which is radial, whence $E = \bigcup_1^{\infty} nW$. Since E is a Baire space, nW must have an interior point for at least one n; hence W must have an interior point, whence $U = W + W$ is a neighborhood of 0 in E. Now $u(W) \subset V_1$ and hence $u(U) \subset V$ for all $u \in H$, which proves H to be equicontinuous.

An immediate consequence is the following classical result due to Banach-Steinhaus [1], and known as the **principle of uniform boundedness**:

COROLLARY. *Let E be a normed space, let F be a normed space, and let H be a subset of $\mathscr{L}(E, F)$ such that $\sup\{\|u(x)\|: u \in H\}$ is finite for every $x \in M$, where M is not meager in E. Then $\sup\{\|u\|: u \in H\}$ is finite.*

Proof. The linear hull E_M of M, which is clearly dense in E and a Baire space, since M is not meager in E, has the property that H_0 is simply bounded in $\mathscr{L}(E_M, F)$, where H_0 is the set obtained by restricting all $u \in H$ to E_M. Hence by (4.2), H_0 is equicontinuous and thus norm bounded in $\mathscr{L}(E_M, F)$. Now since the unit ball of E_M is dense in the unit ball of E, the mapping $u \to u_0$ (u_0 the restriction of $u \in \mathscr{L}(E, F)$ to E_M) is a norm isomorphism of $\mathscr{L}(E, F)$ into $\mathscr{L}(E_M, F)$; hence H is norm bounded as asserted.

Before we can prove the Banach-Steinhaus theorem (see (4.6) below) in appropriate generality, we have to gather further information on equicontinuous sets which will also be needed in Chapter IV. We note first that the subspace $L(E, F)$ of F^E is closed in F^E for the topology of simple convergence (which is the topology of the product of E copies of F (Section 3, Example 1)): Since F is assumed to be Hausdorff and since $f \to f(x)$ is continuous on F^E into F for each $x \in E$, it follows that the set

$$M(x, y, \lambda, \mu) = \{f \in F^E: f(\lambda x + \mu y) - \lambda f(x) - \mu f(y) = 0\}$$

is closed for each fixed quadruple (x, y, λ, μ), and $L(E, F) = \bigcap M(x, y, \lambda, \mu)$ where (x, y, λ, μ) ranges over $E \times E \times K \times K$.

4.3

If $H \subset \mathscr{L}(E, F)$ is equicontinuous and H_1 is the closure of H in F^E for the topology of simple convergence, then $H_1 \subset \mathscr{L}(E, F)$ and H_1 is equicontinuous.

Proof. If $u_1 \in H_1$, then $u_1 \in L(E, F)$ by the preceding remark. Since H is equicontinuous, there exists a 0-neighborhood U in E such that for all $u \in H$, $u(U) \subset V$, where V is a given 0-neighborhood in F which can, without restriction of generality, be assumed closed. From the continuity of $f \to f(x)$ on F^E into F, we conclude that $u_1(x) \in V$ for all $u_1 \in H_1$ and $x \in U$. Thus H_1 is equicontinuous in $\mathscr{L}(E, F)$.

Combining this result with Tychonov's theorem on products of compact spaces, we obtain the following well-known result, known as the theorem of Alaoglu-Bourbaki.

COROLLARY. *Let E be a t.v.s. with dual E'; every equicontinuous subset of E' is relatively compact for $\sigma(E', E)$.*

Proof. The weak topology $\sigma(E', E)$ is the topology of simple convergence on $E' = \mathscr{L}(E, K_0)$ and hence induced by the product topology of K_0^E. By the Tychonov theorem, a subset $H \subset K_0^E$ is relatively compact if (and only if) for each $x \in E$, $\{f(x): f \in H\}$ is relatively compact in K_0. Now if $H \subset E'$ is equicontinuous, there exists a 0-neighborhood U in E such that $|u(x)| \leq 1$ for all $u \in H$ and $x \in U$; thus if $x_0 \in E$ is given, there exists $\lambda > 0$ such that $\lambda x_0 \in U$, whence $|u(x_0)| \leq \lambda^{-1}$ for all $u \in H$. Thus the closure H_1 of H in K_0^E is compact; but since $H_1 \subset E'$ by (4.3), H_1 agrees with the closure $\bar{H}$ of H in $(E', \sigma(E', E))$, so $\bar{H}$ is weakly compact, which proves the assertion.

4.4

If F is quasi-complete and $\mathfrak{S}$ covers E, every closed, equicontinuous set is complete in $\mathscr{L}_{\mathfrak{S}}(E, F)$.

Proof. Let $H \subset \mathscr{L}_{\mathfrak{S}}(E, F)$ be closed and equicontinuous. If $\mathfrak{F}$ is a Cauchy filter on H, it is a fortiori a Cauchy filter on H for the uniformity associated with the topology of simple convergence; hence for each $x \in E$ the sets $\{\Phi(x) : \Phi \in \mathfrak{F}\}$ are bounded and a base of a Cauchy filter in F (for H is bounded and $u \to u(x)$ is linear and continuous). Since F is quasi-complete, this filter base converges to an element $u_1(x) \in F$ and by (4.3), $x \to u_1(x)$ is in $\mathscr{L}(E, F)$. Moreover, $\mathfrak{F}$ being a Cauchy filter for the $\mathfrak{S}$-topology, there exists $\Phi \in \mathfrak{F}$ such that $u(x) - v(x) \in V$ for all $u \in \Phi$, $v \in \Phi$ and $x \in S$, where $S \in \mathfrak{S}$ and the 0-neighborhood V in F can be preassigned. Hence if V is chosen to be closed, it follows that $u(x) - u_1(x) \in V$ for all $u \in \Phi$ and all $x \in S$, implying that $u_1 = \lim \mathfrak{F}$ for the $\mathfrak{S}$-topology.

COROLLARY. *If E, F satisfy the assumptions of* (4.2) *and F is quasi-complete, then $\mathscr{L}_{\mathfrak{S}}(E, F)$ is quasi-complete for every $\mathfrak{S}$-topology such that $\mathfrak{S}$ covers E.*

For another condition guaranteeing quasi-completeness or completeness of $\mathscr{L}_{\mathfrak{S}}(E, F)$ for certain $\mathfrak{S}$-topologies, see Exercise 8.

4.5

Let H be an equicontinuous subset of $\mathscr{L}(E, F)$. The restrictions to H of the following topologies are identical:

1. *The topology of simple convergence on a total subset of E.*
2. *The topology of simple convergence (on E).*
3. *The topology of precompact convergence.*

Proof. Each of the three topologies is finer than the preceding one. The result will be established if we can show that when restricted to H, topology 1 is finer than topology 3. Let A be a total subset of E. We have to show that for each $u_0 \in H$, 0-neighborhood V in F, and precompact set $S \subset E$, there exist a finite subset $S_0 \subset A$ and a 0-neighborhood V_0 in F such that

$$[u_0 + M(S_0, V_0)] \cap H \subset u_0 + M(S, V),$$

where the notation is that employed in Section 3. Select a 0-neighborhood W in F such that $W + W + W + W + W \subset V$, and a circled 0-neighborhood U in E such that $w(U) \subset W$ whenever $w \in H$. S ($\neq \varnothing$) being precompact, there exist elements $y_i \in S$ $(i = 1, ..., m)$ for which $S \subset \bigcup_i (y_i + U)$. Since the linear hull of A is dense in E, there exist (supposing that $E \neq \{0\}$) elements $x_{ij} \in A$ and scalars λ_{ij} $(i = 1, ..., m; j = 1, ..., n)$ such that $y_i \in \sum\limits_{j=1}^{m} \lambda_{ij}x_{ij} + U$. It follows that

$$S \subset \bigcup_{i=1}^{m} \left(\sum_{j=1}^{n} \lambda_{ij}x_{ij} + U + U \right).$$

Choose a circled 0-neighborhood V_0 in F with $\sum\limits_{i,j} (\lambda_{ij}V_0) \subset W$, and denote by S_0 the finite set $\{x_{ij}: i = 1, ..., m; j = 1, ..., n\}$. If $v \in M(S_0, V_0)$, then $v(x_{ij}) \in V_0$ for all i, j and

$$v(S) \subset \bigcup_{i=1}^{m} \sum_{j=1}^{n} (\lambda_{ij}V_0) + v(U) + v(U) \subset W + v(U) + v(U).$$

Now let $u_0 \in H$ and $w \in H \cap [u_0 + M(S_0, V_0)]$; then $w = u_0 + v$, where $v \in M(S_0, V_0)$. Since $v = w - u_0$, $v(U) \subset w(U) + u_0(U) \subset W + W$, since U is circled. Thus $v(S) \subset V$, $w = u_0 + v \in u_0 + M(S, V)$, and the proof is complete.

The preceding results make it possible to prove the following theorem, called the theorem of Banach-Steinhaus, in substantial generality (cf. Bourbaki [8], chap. III). For briefness we call a filter $\mathfrak{F}$ on a t.v.s. E *bounded* if $\mathfrak{F}$ contains a bounded subset of E.

4.6

Theorem. *Let E, F be l.c.s. such that E is barreled; or let E, F be t.v.s. such that E is a Baire space. If $\mathfrak{F}$ is a filter in $\mathscr{L}(E, F)$, bounded for the topology of simple convergence and which converges pointwise to a mapping $u_1 \in F^E$, then $u_1 \in \mathscr{L}(E, F)$ and $\mathfrak{F}$ converges uniformly to u_1 on every precompact subset of E.*

Proof. Let Φ be an element of $\mathfrak{F}$ bounded for the topology of simple convergence; by (4.2), Φ is equicontinuous. If Φ_1 denotes the closure of Φ in F^E, then $u_1 \in \Phi_1$ by hypothesis and by (4.3), Φ_1 is contained in $\mathscr{L}(E, F)$ and equicontinuous. Since by (4.5) the topologies of simple and precompact convergence agree on Φ_1, the theorem is proved.

The theorem applies, in particular, to a sequence $\{u_n\}$ such that for each $x \in E$, $\{u_n(x)\}$ is a Cauchy sequence in F, provided that F is quasi-complete. More generally, it applies when $\mathfrak{F}$ is a filter (not necessarily bounded) with countable base (Exercise 11). The following corollary is an extended version of the classical form of the theorem (cf. Banach [1], chap. V, theor. 3–5.)

COROLLARY. *Let E, F be Banach spaces, and let $M \subset E$ be a subset not meager in E. If $\{u_n\} \subset \mathscr{L}(E, F)$ is a sequence such that $\{u_n(x)\}$ is a Cauchy sequence in F for every $x \in M$, then $\{u_n\}$ converges to an element $u \in \mathscr{L}(E, F)$ uniformly on each compact subset of E.*

Proof. Let E_M be the linear hull of M; E_M is a non-meager subspace of E, and hence a Baire space. Denoting by $\tilde{u}_n$ the restriction of u_n to E_M ($n \in N$), it follows from (4.6) that $\lim \tilde{u}_n(x) = \tilde{u}(x)$ for all $x \in E_M$, where $\tilde{u} \in \mathscr{L}(E_M, F)$. By the corollary of (4.2), the sequence $\{u_n\}$ is norm bounded in $\mathscr{L}(E, F)$ and hence equicontinuous; thus if we denote by u the unique continuous extension of $\tilde{u}$ to E, the set $H = \{u_n: n \in N\} \cup \{u\}$ is still equicontinuous. It follows now from (4.5) that $\lim u_n = u$ uniformly on every precompact (or equivalently, since E is complete, on every compact) subset of E.

We conclude this section by giving conditions under which an equicontinuous set $H \subset \mathscr{L}(E, F)$ is metrizable and separable. Recall that a metric space possesses a countable base of open sets if and only if it is separable.

4.7

If $H \subset \mathscr{L}(E, F)$ is equicontinuous, if E is separable and if F is metrizable, then the restriction to H of the topology of simple convergence is metrizable. If, in addition, F is separable, then H is separable for this topology.

Proof. In view of (4.5) it is sufficient to prove the theorem for the topology, restricted to H, of simple convergence on a total subset of E. Since E is separable, there exists an at most countable subset $A = \{x_n\}$ of E which is linearly independent and total in E. Take $\{V_m\}$ to be a countable 0-neighborhood base in F, and let $S_n = \{x_k: k \le n\}$. Clearly, the sets $M(S_n, V_m)$, $(n, m) \in N \times N$, form a neighborhood base of 0 in $\mathscr{L}(E, F)$ (notation as in Section 3) for the topology of simple convergence on A; hence this topology is metrizable by (I, 6.1) and so is its restriction to H.

In view of the remark preceding (4.7), the second assertion will be proved when we show that the $\mathfrak{S}$-topology, $\mathfrak{S} = \{S_n: n \in N\}$, on $L(E, F)$ (which is, in general, not a Hausdorff topology) possesses a countable base of open sets. To this end, extend A to a vector space basis B of E, let $Y = \{y_n: n \in N\}$ be a dense subset of F, and define Q to be the set of elements $u \in L(E, F)$ such that $u(z) = 0$ for all $z \in B$ except for finitely many $x_\nu \in A$ ($\nu = 1, \ldots, n$) for which $u(x_\nu) = y_{n_\nu}$, where $\{y_{n_\nu}\}$ is any non-empty, finite subset of Y. Q is clearly countable, and dense in $L(E, F)$ for the $\mathfrak{S}$-topology: If $u \in L(E, F)$ and $u + M(S_n, V_m)$ is a given neighborhood of u, then we choose $u_0 \in Q$ such that $u_0(x) \in u(x) + V_m$ for each $x \in S_n$ (which is possible, since Y is dense in F), whence it follows that $u_0 \in u + M(S_n, V_m)$. Thus if we denote by $M(S_n, V_m)^0$ the interior of $M(S_n, V_m)$ in $L(E, F)$, it is immediate that the countable family $\{u + M(S_n, V_m)^0: u \in Q, (n, m) \in N \times N\}$ is a base of open sets for the $\mathfrak{S}$-topology.

5. BILINEAR MAPPINGS

Let E, F, G be vector spaces over K; a mapping f of $E \times F$ into G is called **bilinear** if for each $x \in E$ and each $y \in F$, the partial mappings $f_x: y \to f(x, y)$ and $f_y: x \to f(x, y)$ are linear. If E, F, G are t.v.s., it is not difficult to prove that a bilinear map f is continuous if and only if f is continuous at $(0, 0)$

(Exercise 16); accordingly, a family B of bilinear maps is equicontinuous if and only if B is equicontinuous at $(0, 0)$. A bilinear map f is said to be **separately continuous** if all partial maps f_x and f_y are continuous; that is, if $f_x \in \mathscr{L}(F, G)$ for all $x \in E$ and $f_y \in \mathscr{L}(E, G)$ for all $y \in F$. Accordingly, a family B of bilinear maps of $E \times F$ into G is **separately equicontinuous** if for each $x \in E$ and each $y \in F$ the families $\{f_x : f \in B\}$ and $\{f_y : f \in B\}$ are equicontinuous. Finally, if $G = K_0$, then a bilinear map of $E \times F$ into G is called a **bilinear form** on $E \times F$.

The following important result is a special case of a theorem due to Bourbaki [8] (chap. III, §3, theor. 3):

5.1

Theorem. *Let E, F be metrizable and let G be any t.v.s. If E is a Baire space or if E is barreled and G is locally convex, then every separately equicontinuous family B of bilinear mappings of $E \times F$ into G is equicontinuous.*

Proof. In view of the identity ($f \in B$)

$$f(x, y) - f(x_0, y_0) = f(x - x_0, y - y_0) + f(x - x_0, y_0) + f(x_0, y - y_0)$$

and the separate equicontinuity of B, it is sufficient to prove the equicontinuity of B at $(0, 0)$. Denote by $\{U_n\}$, $\{V_n\}$ decreasing sequences that constitute a 0-neighborhood base in E, F respectively; $\{U_n \times V_n\}$ is a 0-neighborhood base in $E \times F$. Now if B were not equicontinuous at $(0, 0)$, there would exist a 0-neighborhood W_0 in G and sequences $\{x_n\}$, $\{y_n\}$, with $x_n \in U_n$, $y_n \in V_n$ ($n \in N$) such that for all n, $f_n(x_n, y_n) \notin W_0$, where $\{f_n\}$ is a sequence suitably chosen from B. We shall show that this is impossible. Since for each fixed $x \in E$, the family $\{f_x : f \in B\}$ is equicontinuous, by the corollary of (4.1) it is bounded for the topology of compact convergence on $\mathscr{L}(F, G)$; thus $\{f_x(\{y_n\}) : f \in B\}$ is bounded in G by (3.3), since $\{y_n\}$, being a null sequence in F, is relatively compact. Therefore by (3.3)(c) the family $\{x \to f_n(x, y_n) : n \in N\}$ of linear maps is simply bounded in $\mathscr{L}(E, G)$ and hence is equicontinuous by (4.2); it follows that $f_n(U, y_n) \subset W_0$ ($n \in N$) for a suitable 0-neighborhood U in E, which conflicts with the assumption that $f_n(x_n, y_n) \notin W_0$ ($n \in N$), since $\{x_n\}$ is a null sequence in E.

COROLLARY 1. *Under the assumptions made on E, F, G in (5.1), every separately continuous bilinear mapping on $E \times F$ into G is continuous.*

COROLLARY 2. *In addition to the assumptions made on E, F, G in (5.1), suppose that F is a Baire space or (if G is locally convex) that F is barreled. If B is a family of separately continuous bilinear maps of $E \times F$ into G such that $\{f(x, y) : f \in B\}$ is bounded in G for each $(x, y) \in E \times F$, then B is equicontinuous.*

The proof of Corollary 1 is obtained by applying (5.1) to the family B consisting of a single element f; the proof of Corollary 2 is also easy, since by (4.2) the assumptions imply that B is separately equicontinuous.

As simple examples show (Exercise 17), in general, a separately continuous bilinear map is not continuous; it has thus proved fruitful to introduce an intermediate concept that is closely related to the notion of an $\mathfrak{S}$-topology. Let E, F, G be t.v.s., let $\mathfrak{S}$ be a family of bounded subsets of E, and let f be a bilinear map on $E \times F$ into G. f is called $\mathfrak{S}$-**hypocontinuous** if f is separately continuous and if, for each $S \in \mathfrak{S}$ and each 0-neighborhood W in G, there exists a 0-neighborhood V in F such that $f(S \times V) \subset W$. By (4.1) it amounts to the same to require that for each $S \in \mathfrak{S}$ the family $\{f_x : x \in S\}$ be equicontinuous. The $\mathfrak{T}$-hypocontinuity of f is analogously defined if $\mathfrak{T}$ is a family of bounded subsets of F: f is $\mathfrak{T}$-hypocontinuous if, for each $T \in \mathfrak{T}$, $\{f_y : y \in T\}$ is equicontinuous, and if f is separately continuous. Finally, a bilinear map is $(\mathfrak{S}, \mathfrak{T})$-**hypocontinuous** if it is both $\mathfrak{S}$-hypocontinuous and $\mathfrak{T}$-hypocontinuous. Note that separate continuity emerges as a particular case when $\mathfrak{S}$ and $\mathfrak{T}$ are the families of all finite subsets of E and F, respectively.

5.2

If F is barreled and G is locally convex (or if F is a Baire space), every separately continuous bilinear map f of $E \times F$ into G is $\mathfrak{B}$-hypocontinuous, where $\mathfrak{B}$ is the family of all bounded subsets of E.

Proof. The separate continuity of f is obviously equivalent to the assertion that the linear map $x \to f_x$ of E into $L(F, G)$ maps E into $\mathscr{L}(F, G)$ and is continuous for the topology of simple convergence on $\mathscr{L}(F, G)$. Thus if $B \subset E$ is bounded, $\{f_x : x \in B\}$ is simply bounded in $\mathscr{L}(F, G)$ and hence is equicontinuous by (4.2); this establishes the proposition.

5.3

Let $\mathfrak{S}$, $\mathfrak{T}$ be families of bounded subsets of E, F, respectively, and let f be a bilinear map of $E \times F$ into G, where E, F, G are t.v.s. If f is $\mathfrak{S}$-hypocontinuous, then f is continuous on $S \times F$ for each $S \in \mathfrak{S}$; if f is $(\mathfrak{S}, \mathfrak{T})$-hypocontinuous, then f is uniformly continuous on $S \times T$ for each $S \in \mathfrak{S}$ and $T \in \mathfrak{T}$.

Proof. The first assertion is an immediate consequence of the $\mathfrak{S}$-hypocontinuity of f and the identity

$$f(x, y) - f(x_0, y_0) = f(x, y - y_0) + f(x - x_0, y_0)$$

to be applied for $x, x_0 \in S$ and $y, y_0 \in F$. To prove the second assertion, allow $x, \bar{x}$ to be variable in S and $y, \bar{y}$ to be variable in T. Since f is $(\mathfrak{S}, \mathfrak{T})$-hypocontinuous, for a given 0-neighborhood W in G, there exist 0-neighborhoods U, V in E, F, respectively, such that $f(S \times V) \subset W$ and $f(U \times T) \subset W$. If $x - \bar{x} \in U$, $y - \bar{y} \in V$, it follows that

$$f(x, y) - f(\bar{x}, \bar{y}) = f(x, y - \bar{y}) + f(x - \bar{x}, \bar{y}) \in W + W;$$

hence f is uniformly continuous on $S \times T$.

The preceding result is useful for the extension of $(\mathfrak{S}, \mathfrak{T})$-hypocontinuous bilinear maps.

Let E, E_1, F, F_1 be t.v.s. such that E is a dense subspace of E_1 and F is a dense subspace of F_1. Suppose that $\mathfrak{S}$ is a family of bounded subsets of E with the property that $\mathfrak{S}_1$ covers E_1, where $\mathfrak{S}_1$ denotes the family of the closures, taken in E_1, of all $S \in \mathfrak{S}$; suppose further that $\mathfrak{T}$ is a family of bounded subsets of F, such that the corresponding family $\mathfrak{T}_1$ of closures covers F_1; finally, let G be a quasi-complete Hausdorff t.v.s. Under these assumptions, the following extension theorem holds:

5.4

Every $(\mathfrak{S}, \mathfrak{T})$-hypocontinuous bilinear mapping of $E \times F$ into G has a unique extension to $E_1 \times F_1$ (and into G) which is bilinear and $(\mathfrak{S}_1, \mathfrak{T}_1)$-hypocontinuous.

Proof. As before (Section 3), we suppose $\mathfrak{S}, \mathfrak{T}$ to be directed under " $\subset$ " (which is, incidentally, no restriction of generality); then so are the families $\{S \times T\}$ and $\{S_1 \times T_1\}$. Since $S \times T$ is dense in the uniform space $S_1 \times T_1$, the restriction $f_{S,T}$ of the bilinear map f to $S \times T$ has by (5.3) a unique (uniformly) continuous extension $\bar{f}_{S_1, T_1}$ to $S_1 \times T_1$ with values in G (since $f(S \times T)$ is bounded and G quasi-complete). Since the family $\{S_1 \times T_1 : S \in \mathfrak{S}, T \in \mathfrak{T}\}$ is directed and covers $E_1 \times F_1$, it follows that in their totality, the extensions $\bar{f}_{S_1, T_1}$ define an extension $\bar{f}$ of f to $E_1 \times F_1$. This argument also shows that a possible extension of f with the desired properties is necessarily unique; it remains to show that $\bar{f}$ is bilinear and $(\mathfrak{S}_1, \mathfrak{T}_1)$-hypocontinuous.

Let $\bar{x} \in E_1$ be given; there exists S_1 with $\bar{x} \in S_1$. The map $\phi_{\bar{x}}: y \to \bar{f}(\bar{x}, y)$ ($y \in F$) is an element of $\mathscr{L}(F, G)$ by (4.3) since, f being $\mathfrak{S}$-hypocontinuous, $\{f_x : x \in S\}$ is equicontinuous in $\mathscr{L}(F, G)$. Now $\phi_{\bar{x}}$ has a unique continuous extension to F_1 with values in G, which must necessarily agree with $\bar{f}_{\bar{x}}$: $\bar{y} \to \bar{f}(\bar{x}, \bar{y})$, since G is separated (uniqueness of limits). Hence each $\bar{f}_{\bar{x}}$ (and by symmetry, each $\bar{f}_{\bar{y}}$) is linear and continuous, which shows $\bar{f}$ to be bilinear and separately continuous.

Since f is $\mathfrak{S}$-hypocontinuous, for each $S \in \mathfrak{S}$ there exists a 0-neighborhood V in F such that $f(S \times V) \subset W$, W being a given 0-neighborhood in G which can be assumed closed. Denoting by V_1 the closure of V in F_1 (V_1 is a 0-neighborhood in F_1, cf. (I, 1.5)), it follows from the separate continuity of $\bar{f}$ that $\bar{f}(S \times V_1) \subset W$ and, repeating the argument, that $\bar{f}(S_1 \times V_1) \subset W$. Thus $\bar{f}$ is $\mathfrak{S}_1$-hypocontinuous and (by symmetry) $\mathfrak{T}_1$-hypocontinuous, which completes the proof.

We remark that if E and G are locally convex, an $\mathfrak{S}$-hypocontinuous bilinear map of $E \times F$ into G is also $\bar{\mathfrak{S}}$-hypocontinuous, where $\bar{\mathfrak{S}}$ denotes the saturated hull of $\mathfrak{S}$ (Section 3), with a corresponding statement holding under $(\mathfrak{S}, \mathfrak{T})$-hypocontinuity.

The set of all bilinear mappings of $E \times F$ into G is a vector space $B(E, F; G)$ which is a subspace of $G^{E \times F}$; the subspaces of $B(E, F; G)$ (supposing E, F, G

to be t.v.s.) consisting of all separately continuous and all continuous bilinear maps, respectively, will be denoted by $\mathfrak{B}(E, F; G)$ and by $\mathscr{B}(E, F; G)$. The corresponding spaces of bilinear forms will be denoted by $B(E, F)$, $\mathfrak{B}(E, F)$, and $\mathscr{B}(E, F)$.

If $\mathfrak{S}$ and $\mathfrak{T}$ are families of bounded subsets of E and F, respectively, and if D is a subspace of $B(E, F; G)$, we consider the topology of $\mathfrak{S} \times \mathfrak{T}$-convergence on D (Section 3), that is, the topology of uniform convergence on the sets $S \times T$, where $S \in \mathfrak{S}$ and $T \in \mathfrak{T}$. We recall that D is a t.v.s. under this topology if (and only if) for all $S \in \mathfrak{S}$, $T \in \mathfrak{T}$ and $f \in D, f(S \times T)$ is bounded in G; this is, in particular, always the case when $D \subset \mathscr{B}(E, F; G)$ (cf. Exercise 16). If the preceding condition is satisfied and G is locally convex, then the $\mathfrak{S} \times \mathfrak{T}$-topology is locally convex. We leave it to the reader to verify that if G is separated and $D \subset \mathfrak{B}(E, F; G)$, the $\mathfrak{S} \times \mathfrak{T}$-topology is a Hausdorff topology whenever $\mathfrak{S}$ and $\mathfrak{T}$ are total families (that is, families whose union is a total subset of E or F, respectively).

The following is a general condition under which $\mathfrak{B}(E, F; G)$ is a l.c.s. for an $\mathfrak{S} \times \mathfrak{T}$-topology.

5.5

Let E, F, G be locally convex spaces; denote by $\mathfrak{S}$ a total, saturated family of bounded subsets of E such that the closure of each $S \in \mathfrak{S}$ is complete, and denote by $\mathfrak{T}$ a total family of bounded subsets of F. Then $\mathfrak{B}(E, F; G)$ is a locally convex space under the $\mathfrak{S} \times \mathfrak{T}$-topology.

Proof. We have to show that for each $f \in \mathfrak{B}(E, F; G)$ and all sets $S \in \mathfrak{S}$, $T \in \mathfrak{T}, f(S \times T)$ is bounded in G; since $\mathfrak{S}$ is saturated, we can suppose S to be closed, convex, and circled. Now since T is bounded in F and since (by the separate continuity of f) the linear map $y \to f_y$ is continuous on F into $\mathscr{L}(E, G)$ when $\mathscr{L}(E, G)$ carries the topology of simple convergence, the set $\{f_y: y \in T\}$ is simply bounded in $\mathscr{L}(E, G)$. Thus if W is a closed, convex, circled 0-neighborhood in G, the set $U = \bigcap \{f_y^{-1}(W): y \in T\}$ is closed, convex, circled, and by (3.3) radial; hence U is a barrel in E. It follows from (II, 8.5) that U absorbs S, whence we have $f(S \times T) \subset \lambda W$ for a suitable scalar λ. Since W was an arbitrary element of a 0-neighborhoood base in $G, f(S \times T)$ is bounded.

The conditions of the preceding proposition are, in particular, satisfied if E and F are replaced by the weak duals (Chapter II, Section 5) $E'_\sigma = (E', \sigma(E', E))$ and F'_σ of two arbitrary l.c.s. E and F, and if $\mathfrak{S}$ and $\mathfrak{T}$ are taken to be the families of all equicontinuous subsets of E' and F', respectively; for $\mathfrak{S}$ and $\mathfrak{T}$ are saturated families of bounded sets whose closed members are compact (hence complete) in E'_σ and F'_σ, respectively, by the corollary of (4.3). This $\mathfrak{S} \times \mathfrak{T}$-topology is called the topology of **bi-equicontinuous convergence** (Grothendieck [13]), and under this topology, $\mathfrak{B}(E'_\sigma, F'_\sigma; G)$ is a locally convex space which will be denoted by $\mathfrak{B}_e(E'_\sigma, F'_\sigma; G)$.

6. TOPOLOGICAL TENSOR PRODUCTS

Let E, F be vector spaces over K and let $B(E, F)$ be the vector space of all bilinear forms on $E \times F$. For each pair $(x, y) \in E \times F$, the mapping $f \to f(x, y)$ is a linear form on $B(E, F)$, and hence an element $u_{x,y}$ of the algebraic dual $B(E, F)^*$. It is easily seen that the mapping $\chi: (x, y) \to u_{x,y}$ of $E \times F$ into $B(E, F)^*$ is bilinear. The linear hull of $\chi (E \times F)$ in $B(E,F)^*$ is denoted by $E \otimes F$ and is called the **tensor product** of E and F; χ is called the **canonical bilinear map** of $E \times F$ into $E \otimes F$. The element $u_{x,y}$ of $E \otimes F$ will be denoted by $x \otimes y$ so that each element of $E \otimes F$ is a finite sum $\sum \lambda_i (x_i \otimes y_i)$ (the sum over the empty set being 0). We shall also find it convenient to write $A \otimes B = \chi(A \times B)$ for arbitrary subsets $A \subset E$, $B \subset F$, although this usage is inconsistent with the notation $E \otimes F$. Ambiguity can be avoided if, only for subspaces $M \subset E$, $N \subset F$, the symbol $M \otimes N$ denotes the linear hull of $\chi(M \times N)$ rather than the set $\chi(M \times N)$ itself.

One verifies without difficulty the rules $\lambda(x \otimes y) = (\lambda x) \otimes y = x \otimes (\lambda y)(\lambda \in K)$, $(x_1 + x_2) \otimes y = x_1 \otimes y + x_2 \otimes y$, and $x \otimes (y_1 + y_2) = x \otimes y_1 + x \otimes y_2$. Hence each element $u \in E \otimes F$ is of the form $u = \sum x_i \otimes y_i$. Obviously, the representation of u is not unique, but it can be assumed that both sets $\{x_i\}$ and $\{y_i\}$ are linearly independent sets of $r(\geq 0)$ elements. The number r is uniquely determined by u and called the **rank** of u; it is the minimal number of summands by means of which u can be represented (Exercise 18).

One of the principal advantages of tensor products lies in the fact that they permit us to consider vector spaces of bilinear (more generally, of multilinear) maps as vector spaces of linear mappings. We recall this more precisely:

6.1

Let E, F be vector spaces over K and let χ be the canonical bilinear map of $E \times F$ into $E \otimes F$. For any vector space G over K, the mapping $u \to u \circ \chi$ is an isomorphism of $L(E \otimes F, G)$ onto $B(E, F; G)$.

Proof. It is clear that $u \to u \circ \chi = f$ is a linear map of $L(E \otimes F, G)$ into $B(E, F; G)$, which is one-to-one, since $f = 0$ implies $u(x \otimes y) = f(x, y) = 0$ for all $x \in E$ and $y \in F$, hence $u = 0$. It remains to show that the map is onto $B(E, F; G)$. For any $f \in B(E, F; G)$ define $u(\sum x_i \otimes y_i) = \sum f(x_i, y_i)$; it is clear that the definition is consistent, that u is linear on $E \otimes F$ into G, and that $f = u \circ \chi$.

COROLLARY. *The algebraic dual of $E \otimes F$ can be identified with $B(E, F)$; under this identification, each vector space of linear forms on $E \otimes F$ is a vector space of bilinear forms on $E \times F$, and conversely.*

In particular, $E^* \otimes F^*$ can be identified with a space of bilinear forms on $E \times F$ by means of $(x^* \otimes y^*)(x, y) = x^*(x)y^*(y)$, and hence with a subspace of $(E \otimes F)^*$; it is readily seen that $E^* \otimes F^*$ separates points in $E \otimes F$.

In order to define useful topologies on $E \otimes F$ when E, F are t.v.s., we restrict our attention to locally convex spaces E, F and locally convex topologies on $E \otimes F$. Consider the family $\mathcal{T}$ of all locally convex topologies on $E \otimes F$ for which the canonical bilinear map $(E, F$ being l.c.s.) on $E \times F$ into $E \otimes F$ is continuous: The upper bound $\mathfrak{T}_p$ of $\mathcal{T}$ (Chapter II, Section 5) is a locally convex topology, called the **projective** (tensor product) **topology** on $E \otimes F$. It is immediate that when $\mathfrak{U}, \mathfrak{V}$ are 0-neighborhood bases in E, F, respectively, the family of convex, circled hulls $\{\Gamma(U \otimes V) : U \in \mathfrak{U}, V \in \mathfrak{V}\}$ is a neighborhood base of 0 for $\mathfrak{T}_p$; thus the projective topology is the finest locally convex topology on $E \otimes F$ for which the canonical bilinear map is continuous. We shall see at once that $\mathfrak{T}_p$ is always a Hausdorff topology.

6.2

Let E, F, G be locally convex spaces and provide $E \otimes F$ with the projective topology. Then the isomorphism $u \to u \circ \chi$ of (6.1) maps the space of continuous linear mappings $\mathscr{L}(E \otimes F, G)$ onto the space of continuous bilinear mappings $\mathscr{B}(E, F; G)$.

Proof. It is clear that the continuity of u implies that of $u \circ \chi$, since χ is continuous. Conversely, if W is a convex, circled 0-neighborhood in G and $f = u \circ \chi$ is continuous, then $f^{-1}(W)$ contains a 0-neighborhood $U \times V$ in $E \times F$. It follows that $u^{-1}(W)$ contains $U \otimes V$. Since $u^{-1}(W)$ is convex and circled, it contains $\Gamma(U \otimes V)$, which proves the continuity of u.

COROLLARY. *The dual of $E \otimes F$ for the projective topology can be identified with the space $\mathscr{B}(E, F)$ of all continuous bilinear forms on $E \times F$. Under this identification, the equicontinuous subsets of $(E \otimes F)'$ are the equicontinuous sets of bilinear forms on $E \times F$.*

This corollary implies that $\mathfrak{T}_p$ is necessarily Hausdorff, for evidently $E' \otimes F' \subset \mathscr{B}(E, F)$; thus if we show that $E' \otimes F'$ separates points in $E \otimes F$, it follows that $\sigma(E \otimes F, \mathscr{B}(E, F))$ and a fortiori $\mathfrak{T}_p$ is a Hausdorff topology. Now if $u \in E \otimes F$ is of rank $r \geq 1$, say, $u = \sum_{i=1}^{r} x_i \otimes y_i$, then $\{x_i\}$ and $\{y_i\}$ are linearly independent, whence by (II, 4.2), Corollary 1, there exist linear forms $f_1 \in E'$ and $g_1 \in F'$ such that $f_1(x_i) = \delta_{1i}$ and $g_1(y_i) = \delta_{1i}$ ($i = 1, ..., r$), and it follows that $f_1 \otimes g_1(u) = \sum f_1(x_i)g_1(y_i) = 1$. For a description of $\mathfrak{T}_p$ by semi-norms, we need the following result:

6.3

Let p, q be semi-norms on E, F respectively, such that p is the gauge of $U \subset E$, and q is the gauge of $V \subset F$. Then the semi-norm on $E \otimes F$,

$$u \to r(u) = \inf \{\textstyle\sum_i p(x_i)q(y_i) : u = \sum x_i \otimes y_i\},$$

is the gauge of $\Gamma(U \otimes V)$ and has the property that $r(x \otimes y) = p(x)q(y)$ for all $x \in E$, $y \in F$.

Proof. It is immediate that r is a semi-norm on $E \otimes F$; let $M_0 = \{u: r(u) < 1\}$, $M_1 = \{u: r(u) \leqq 1\}$. To prove that r is the gauge of $\Gamma(U \otimes V)$, it suffices to show that $M_0 \subset \Gamma(U \otimes V) \subset M_1$. If $u \in \Gamma(U \otimes V)$, then $u = \sum \lambda_i(x_i \otimes y_i)$, where $x_i \in U$, $y_i \in V$ for all i and $\sum |\lambda_i| \leqq 1$. Now $u = \sum \bar{x}_i \otimes y_i$, where $\bar{x}_i = \lambda_i x_i$, whence $r(u) \leqq \sum p(\bar{x}_i) q(y_i) = \sum |\lambda_i| p(x_i) q(y_i) \leqq 1$. On the other hand, if $u \in M_0$, then $u = \sum x_i \otimes y_i$, where $\sum p(x_i) q(y_i) < 1$. Thus there exist real numbers $\varepsilon_i > 0$ such that $\sum \mu_i < 1$, where $\mu_i = [p(x_i) + \varepsilon_i][q(y_i) + \varepsilon_i]$ for all i. Set $\bar{x}_i = x_i/[p(x_i) + \varepsilon_i]$ and $\bar{y}_i = y_i/[q(y_i) + \varepsilon_i]$; then $\bar{x}_i \in U$, $\bar{y}_i \in V$, and hence $u = \sum \mu_i(\bar{x}_i \otimes \bar{y}_i) \in \Gamma(U \otimes V)$.

To prove the second assertion, let $x_0 \in E$, $y_0 \in F$ be given; we conclude from (II, 3.2) that there exist linear forms $f \in E^*$, $g \in F^*$ such that $f(x_0) = p(x_0)$, $g(y_0) = q(y_0)$, and $|f(x)| \leqq p(x)$, $|g(y)| \leqq q(y)$ for all $x \in E$, $y \in F$. [Define f on the subspace generated by x_0 by $f(\lambda x_0) = \lambda p(x_0)$ and extend to E.] It is immediate that for the linear form $f \otimes g$ on $E \otimes F$ and $u = \sum x_i \otimes y_i$, we have $|f \otimes g(u)| \leqq \sum p(x_i) q(y_i)$, whence $|f \otimes g(u)| \leqq r(u)$; hence $p(x_0) q(y_0) \leqq r(x_0 \otimes y_0)$. Since clearly $r(x_0 \otimes y_0) \leqq p(x_0) q(y_0)$, the proof is complete.

The semi-norm r is called the tensor product of the semi-norms p and q, and is denoted by $p \otimes q$. It is not difficult to prove that $p \otimes q$ is a norm on $E \otimes F$ if and only if p and q are norms on E and F, respectively (Exercise 20). A family P of semi-norms on E is directed if, for each pair $p_1, p_2 \in P$, there exists $p_3 \in P$ such that sup $(p_1, p_2) \leqq p_3$; if P is directed, the sets $U_{p,n} = \{x \in E: p(x) \leqq n^{-1}\}$ $(p \in P, n \in N)$ form a neighborhood base of 0 for a locally convex topology on E (Chapter II, Exercise 8). Thus we obtain this corollary of (6.3):

COROLLARY. *Let E, F be locally convex spaces and let P and Q be directed families of semi-norms generating the topologies of E and F, respectively. The projective topology on $E \otimes F$ is generated by the directed family $\{p \otimes q: (p, q) \in P \times Q\}$.*

In particular, if E and F are normed spaces, then the tensor product of the respective norms generates the projective topology on $E \otimes F$ (Exercise 21).

If E, F are any l.c.s., then $(E \otimes F, \mathfrak{T}_p)$ is a l.c.s. as we have seen above; hence by (I, 1.5) it can be imbedded in a complete l.c.s. which is unique (to within isomorphism) and will be denoted by $E \tilde{\otimes} F$. It results from the corollary of (6.3) that if E, F are metrizable, then $E \tilde{\otimes} F$ is an (F)-space. It is one of the fundamental results (also due to Grothendieck [13]) of the theory to have an explicit representation of $E \tilde{\otimes} F$, when E, F are metrizable l.c.s. (For the definition of an absolutely convergent series in a t.v.s. E, see Exercise 23.)

6.4

Theorem. *Let E, F be metrizable l.c.s.; each element $u \in E \tilde{\otimes} F$ is the sum of an absolutely convergent series,*

$$u = \sum_{i=1}^{\infty} \lambda_i x_i \otimes y_i,$$

where $\sum_i |\lambda_i| < + \infty$, and $\{x_i\}$, $\{y_i\}$ are null sequences in E, F, respectively.

Proof. (The following simple proof is due to A. Pietsch.) Let $\{p_n\}$, $\{q_n\}$ be increasing sequences of semi-norms generating the topologies of E, F respectively, and denote by r_n the semi-norm $p_n \otimes q_n$ $(n \in N)$; the sequence $\{r_n\}$ generates the projective topology of $E \otimes F$ and is increasing. Denote by $\bar{r}_n (n \in N)$ the continuous extension of r_n to $E \tilde{\otimes} F$.

Given $u \in E \tilde{\otimes} F$, there exists a sequence $\{u_n\}$ in $E \otimes F$ such that $\bar{r}_n(u - u_n) < n^{-2} 2^{-(n+1)}$. Let $\sum\limits_{i=1}^{i_1} \lambda_i x_i \otimes y_i$ be any representation of u_1, and set $v_n = u_{n+1} - u_n$ for all $n \in N$.

We have

$$r_n(v_n) \leq \bar{r}_n(u - u_n) + \bar{r}_n(u - u_{n+1})$$

$$\leq \bar{r}_n(u - u_n) + \bar{r}_{n+1}(u - u_{n+1}) < n^{-2}2^{-n}.$$

We conclude from (6.3) that there exists a representation

$$v_n = \sum_{i=i_n+1}^{i_{n+1}} \lambda_i x_i \otimes y_i$$

such that $p_n(x_i) \leq n^{-1}$, $q_n(x_i) \leq n^{-1}$ whenever $i_n < i \leq i_{n+1}$, and such that $\sum\limits_{i=i_n+1}^{i_{n+1}} |\lambda_i| \leq 2^{-n}$.

Therefore, we have $u = u_1 + \sum\limits_1^\infty v_n = \sum\limits_1^\infty \lambda_i x_i \otimes y_i$, where the sequences $\{x_i\}$, $\{y_i\}$, and $\{\lambda_i\}$ have the desired properties, and the proof is complete.

We shall now consider a general example of a projective tensor product. Let (X, Σ, μ) be a measure space (Chapter II, Section 2, Example 2) so that μ is a positive measure on X, and $L^1(\mu)$ the Banach space of (equivalence classes modulo μ-null functions of) real-valued μ-summable functions on X, with $\|f\| = \int |f| \, d\mu$. Let E be any Banach space over R and let S_E be the vector space over R of all E-valued simple functions; that is, functions ϕ of the form $t \to \sum\limits_{i=1}^n \psi_i(t) x_i$, where ψ_i are the characteristic functions of n sets $S_i \in \Sigma$ such that $\mu(S_i) < +\infty$, and x_i are arbitrary elements of E. It is clear that $\phi \to \int \|\phi\| \, d\mu$ is a semi-norm p on S_E; now $L_E^1(\mu)$ is defined to be the completion of the Hausdorff (hence normed) space $(S_E, p)/p^{-1}(0)$. The space $L_E^1(\mu)$ is called the space of (classes of) E-valued μ-summable functions.

We show that the Banach space $L_E^1(\mu)$ is norm-isomorphic with $L^1(\mu) \tilde{\otimes} E$. There exists a natural imbedding $u \to \tilde{u}$ of $L^1(\mu) \otimes E$ into $L_E^1(\mu)$ such that for $u = \sum f_i \otimes x_i$, $\tilde{u}$ is the class containing the function $t \to \sum f_i(t) x_i$; evidently $u \to \tilde{u}$ is linear and maps $S \otimes E$ onto $\hat{S}_E$, where S denotes the subspace of $L^1(\mu)$ whose elements contain a simple function, and where $\hat{S}_E = S_E/p^{-1}(0)$. Denote by r the tensor product of the norms of $L^1(\mu)$ and E, respectively; since

$$p(\tilde{u}) = \int \| \sum g_j(t) y_j \| \, d\mu(t) \leq \sum \|g_j\| \, \|y_j\|$$

holds for all representations $\sum g_j \otimes y_j$ of a fixed element u, it follows that $p(\tilde{u}) \leqq r(u)$. On the other hand, if $u \in S \otimes E$, we can choose a representation $u = \sum \psi_i \otimes x_i$ such that the characteristic functions ψ_i have disjoint carriers S_i, which implies $r(u) \leqq \sum \|\psi_i\| \|x_i\| = p(\tilde{u})$. Thus $u \to \tilde{u}$ is a norm isomorphism of $S \otimes E$ onto $\hat{S}_E$ and the assertion follows from the fact that $S \otimes E$ is dense in $L^1(\mu) \hat{\otimes} E$, since S is dense in $L^1(\mu)$.

In the preceding considerations, it is not essential that E be a Banach (or even normable) space. If E is any l.c.s. with P a family of semi-norms generating the topology of E, we define S_E as before and a locally convex topology on S_E by means of the semi-norms $\phi \to \int p[\phi(t)] \, d\mu(t) \, (p \in P)$; the completion of the associated Hausdorff t.v.s. then serves to define $L^1_E(\mu)$, and we prove, as before, that $u \to \tilde{u}$ is an ismorphism of $S \otimes E$ (under the projective topology) onto $\hat{S}_E$. Hence:

6.5

The natural imbedding of $L^1(\mu) \otimes E$ into $L^1_E(\mu)$ induces an isomorphism of $L^1(\mu) \hat{\otimes} E$ onto $L^1_E(\mu)$ which is norm-preserving if E is a Banach space.

We have seen above that the projective topology $\mathfrak{T}_p$ on $E \otimes F$ is the finest l.c. topology for which the canonical bilinear map is continuous. Another topology of importance on $E \otimes F$ is the **inductive** (tensor product) **topology** $\mathfrak{T}_i$, defined to be the finest l.c. topology on $E \otimes F$ for which the canonical bilinear map is separately continuous. $\mathfrak{T}_i$ is an inductive topology in the sense of Chapter II, Section 6; in analogy to (6.2) above, we show that for every l.c.s. G, the isomorphism of (6.1) carries the space of $\mathfrak{T}_i$-continuous linear maps into G to the space of all separately continuous bilinear maps on $E \times F$ into G. In particular, the dual of $(E \otimes F, \mathfrak{T}_i)$ is the space $\mathfrak{B}(E, F)$ (Exercise 22). We shall not be further concerned with $\mathfrak{T}_i$, for which we refer the reader to Grothendieck [13] as well as for other topologies on $E \otimes F$ whose definition is based on the $(\mathfrak{S}, \mathfrak{T})$-hypocontinuity (Section 5) of the canonical bilinear map χ. Let us point out that under the assumptions of (5.1), $\mathfrak{T}_p$ and $\mathfrak{T}_i$ agree on $E \otimes F$.

A topology on $E \otimes F$ of considerably greater importance than $\mathfrak{T}_i$ is the topology $\mathfrak{T}_e$ of **bi-equicontinuous convergence**; viewing $E \otimes F$ as a space of linear maps on $E' \otimes F'$ by virtue of $x \otimes y(x' \otimes y') = x'(x)y'(y)$, $\mathfrak{T}_e$ is the topology of uniform convergence on the sets $S \otimes T$, where S, T are arbitrary equicontinuous subsets of E', F', respectively. $\mathfrak{T}_e$ can be equally characterized as the topology induced on (the subspace) $E \otimes F$ by $\mathfrak{B}_e(E'_\sigma, F'_\sigma)$ which is a l.c.s. (see end of Section 5). The completion of $(E \otimes F, \mathfrak{T}_e)$ will be denoted by $E \tilde{\otimes} F$. It is not difficult to see that $\mathfrak{T}_e$ is coarser than $\mathfrak{T}_p$ on $E \otimes F$; for a successful study of this topology, we need a number of results on duality (Chapter IV). For the moment, we mention only that if E, F are complete l.c.s., then $\mathfrak{B}_e(E'_\sigma, F'_\sigma)$ is complete, whence in this case, $E \tilde{\otimes} F$ can be identified with the closure of $E \otimes F$ in $\mathfrak{B}_e(E'_\sigma, F'_\sigma)$.

7. NUCLEAR MAPPINGS AND SPACES

If E is a vector space over K and V is a convex, circled, and radial subset of E, then $\{n^{-1}V: n \in N\}$ is a 0-neighborhood base for a locally convex topology $\mathfrak{T}_V$ on E. The Hausdorff t.v.s. associated with $(E, \mathfrak{T}_V)$ is the quotient space $(E, \mathfrak{T}_V)/p^{-1}(0)$, where p is the gauge of V; this quotient space is normable by the norm $\hat{x} \to \|\hat{x}\| = p(x)$, where $x \in \hat{x}$. We shall denote by E_V the normed space $(E/p^{-1}(0), \| \ \|)$ just introduced, and by $\tilde{E}_V$ its completion, which is a Banach space. If E is a i.c.s. and V is a convex, circled neighborhood of 0, the topology of the quotient space $E/p^{-1}(0)$ is (in general, strictly) finer than the topology of E_V. Thus the quotient map (called the **canonical** map) is continuous on E into $\tilde{E}_V$; this map will be denoted by ϕ_V.

Dually, if E is a l.c.s. and $B \neq \varnothing$ a convex, circled, and bounded subset of E, then $E_1 = \bigcup\limits_{n=1}^{\infty} nB$ is a (not necessarily closed) subspace of E. The gauge function p_B of B in E_1 is quickly seen to be a norm on E_1; the normed space (E_1, p_B) will henceforth be denoted by E_B. It is immediate that the imbedding map $\psi_B: E_B \to E$ (again called **canonical**) is continuous. Moreover, if B is complete in E, then E_B is a Banach space by (I, 1.6). We finally note that no confusion can arise if $V = B$ is a convex, circled subset of E which is radial and bounded, for in this case the spaces E_V and E_B are identical.

If U, V are convex, circled, and radial subsets of E with respective gauge functions p, q and such that $U \subset V$, then $p^{-1}(0) \subset q^{-1}(0)$ and each equivalence class $\hat{x} \bmod p^{-1}(0)$ is contained in a unique equivalence class $\hat{y} \bmod q^{-1}(0)$; $\hat{x} \to \hat{y}$ is a linear map $\phi_{V,U}$, which is called the **canonical** map of E_U onto E_V. Since $\phi_{V,U}$ is clearly continuous (in fact, of norm ≤ 1), it has a unique continuous extension on $\tilde{E}_U$ into $\tilde{E}_V$, which is again called *canonical*, and also denoted by $\phi_{V,U}$.

Likewise, if B and C are convex, circled, and bounded sets of a l.c.s. E such that $\varnothing \neq B \subset C$, then $E_B \subset E_C$ and the **canonical imbedding** $\psi_{C,B}: E_B \to E_C$ is continuous. Finally, if U, V, B, C are as before and $\phi_U, \phi_V, \psi_B, \psi_C$ are the canonical maps $E \to \tilde{E}_U$, $E \to \tilde{E}_V$, $E_B \to E$ and $E_C \to E$, we have the relations $\phi_V = \phi_{V,U} \circ \phi_U$ and $\psi_C = \psi_{C,B} \circ \psi_B$.

The two methods of constructing auxiliary normed spaces were systematically employed by Grothendieck [13] and will be extremely useful in what follows. We have used these methods before in Chapter II (proof of (II, 5.4) and the discussion preceding (II, 8.4)).

Let E, F be l.c.s. and let E' be the dual of E. Each element $v \in E' \otimes F$ defines a linear map $u \in \mathscr{L}(E, F)$ by virtue of

$$x \to u(x) = \sum_{i=1}^{r} f_i(x) y_i$$

if $v = \sum\limits_{i=1}^{r} f_i \otimes y_i$, and $v \to u$ is even an (algebraic) isomorphism of $E' \otimes F$ into

$\mathcal{L}(E, F)$ (Exercise 18). The mappings $u \in \mathcal{L}(E, F)$, which originate in this fashion from an element $v \in E' \otimes F$, are called continuous maps of **finite rank**; the rank r of u is defined to be the rank of v (Section 6). The mappings of finite rank are very special cases of compact linear maps on E into F: A linear map u on E into F is **compact** if, for a suitable 0-neighborhood U in E, $u(U)$ is a relatively compact subset of F.

Suppose now that E, F are Banach spaces, and let E' be the Banach space which is the strong dual of E (Chapter II, Section 2). Then the imbedding $v \rightarrow u$ is continuous for the projective topology on $E' \otimes F$ and the norm topology (the topology of bounded convergence) on $\mathcal{L}(E, F)$: If $v \in E' \otimes F$ then

$$\|u\| = \sup_{\|x\| \leq 1} \|u(x)\| \leq \sup_{\|x\| \leq 1} \sum_{i=1}^{r} |f_i(x)| \, \|y_i\| \leq \sum_{i=1}^{r} \|f_i\| \, \|y_i\|$$

for all representations $v = \sum_{i=1}^{r} f_i \otimes y_i$; hence $\|u\| \leq r(v)$, where the norm r is the tensor product of the respective norms of E and F (cf. (6.3) and its corollary). Since $\mathcal{L}(E, F)$ is complete under the norm topology by the corollary of (4.4), the imbedding $v \rightarrow u$ has a continuous extension τ to $E' \tilde{\otimes} F$, with values in $\mathcal{L}(E, F)$. The linear maps contained in the range of τ are called **nuclear**; that is, $u \in \mathcal{L}(E, F)$ is nuclear if $u = \tau(v)$ for some $v \in E' \tilde{\otimes} F$. (It is not known whether τ is necessarily biunivocal (cf. Chapter IV, Exercise 30).)

The definition of a nuclear map generalizes to arbitrary l.c.s. E, F as follows. A linear map u on E into F is **bounded** if for a suitable 0-neighborhood U in E, $u(U)$ is a bounded subset of F (for example, every compact map is bounded); every bounded map is continuous. A bounded map can be decomposed as follows: Let U be a convex, circled, 0-neighborhood in E such that $u(U) \subset B$, where B is convex, circled, and bounded in F; then $u = \psi_B \circ u_0 \circ \phi_U$, where u_0 is the map in $\mathcal{L}(E_U, F_B)$ induced by u. If, in addition, F_B is complete, then u_0 has a continuous extension $\bar{u}_0 \in \mathcal{L}(\tilde{E}_U, F_B)$ for which $u = \psi_B \circ \bar{u}_0 \circ \phi_U$. The definition is now this:

A linear map u of a l.c.s. E into another l.c.s. F is **nuclear** if there exists a convex, circled 0-neighborhood U in E such that $u(U) \subset B$, where B is bounded with F_B complete, and such that the induced mapping $\bar{u}_0$ is nuclear on $\tilde{E}_U$ into F_B.

It follows at once that every continuous linear map of finite rank is nuclear; moreover, if u is nuclear in $\mathcal{L}(E, F)$, there exists a 0-neighborhood U in E and a bounded, convex, circled subset of F for which F_B is complete, such that u is the uniform limit on U of a sequence of maps of finite rank in $\mathcal{L}(E, F_B)$. Hence for every $\mathfrak{S}$-topology on $\mathcal{L}(E, F)$, the nuclear maps are contained in the closure of $E' \otimes F$ (the latter being viewed as a subspace of $\mathcal{L}(E, F)$). With the aid of Theorem (6.4), we obtain the following explicit characterization of nuclear maps.

7.1

A linear map $u \in \mathcal{L}(E, F)$ is nuclear if and only if it is of the form

$$x \to u(x) = \sum_{n=1}^{\infty} \lambda_n f_n(x) y_n,$$

where $\sum_{n=1}^{\infty} |\lambda_n| < + \infty$, $\{f_n\}$ is an equicontinuous sequence in E', and $\{y_n\}$ is a sequence contained in a convex, circled, and bounded subset B of F for which F_B is complete.

Proof. The condition is necessary. For, if u is nuclear, then $u = \psi_B \circ \bar{u}_0 \circ \phi_U$, where $\bar{u}_0$ is nuclear in $\mathcal{L}(\tilde{E}_U, F_B)$, U being a suitable 0-neighborhood in E, and B being a suitable bounded subset of F for which F_B is complete. Hence $\bar{u}_0$ originates from an element v of $[E_U]' \hat{\otimes} F_B$, which is, by (6.4), of the form $v = \sum_{n=1}^{\infty} \lambda_n h_n \otimes y_n$ with $\sum_{1}^{\infty} |\lambda_n| < + \infty$ and where $\{h_n\}$ and $\{y_n\}$ are null sequences in $[E_U]'$ and F_B, respectively. Define a sequence $\{f_n\}$ of linear forms on E by setting $f_n = h_n \circ \phi_U$. Since $\{h_n\}$ is a bounded sequence in $[E_U]'$, the sequence $\{f_n\}$ is uniformly bounded on U and hence is equicontinuous. It is clear now that the mapping $u = \psi_B \circ \tau(v) \circ \phi_U$ is of the form indicated above.

The condition is sufficient. For if u is as indicated in the proposition, let $U = \{x \in E: |f_n(x)| \le 1, n \in N\}$; U is convex and circled and is a 0-neighborhood in E by the equicontinuity of $\{f_n\}$. Defining h_n $(n \in N)$ by $f_n = h_n \circ \phi_U$ on E_U and subsequent extension to $\tilde{E}_U$, we obtain $\|h_n\| \le 1$ for all n; evidently, $\bar{u}_0$ is the map $\hat{x} \to \sum_{n=1}^{\infty} \lambda_n h_n(\hat{x}) y_n$. Since $\sum_{1}^{\infty} |\lambda_n| \, \|h_n\| \, \|y_n\|$ converges, the series $\sum_{1}^{\infty} \lambda_n h_n \otimes y_n$ is absolutely convergent in $[E_U]' \hat{\otimes} F_B$ by (6.3) and its corollary, and hence defines an element $v \in [E_U]' \hat{\otimes} F_B$; clearly, $\bar{u}_0 = \tau(v)$, whence u is nuclear.

REMARK. If u is of the form indicated in (7.1), we shall find it convenient to write $u = \sum_{1}^{\infty} \lambda_n f_n \otimes y_n$, keeping in mind that u is not, properly speaking, an element of a topological tensor product. It follows then from the first part of the proof that for nuclear u, there exists a representation $u = \sum_{1}^{\infty} \lambda_n f_n \otimes y_n$ such that $\{f_n\}$ is a sequence converging to 0 uniformly on a suitable 0-neighborhood U of E, and $\{y_n\}$ converges to 0 in a suitable Banach space F_B; finally, $(\lambda_n) \in l^1$.

COROLLARY 1. *Every nuclear map is compact.*

Proof. Let $u = \sum_{1}^{\infty} \lambda_n f_n \otimes y_n$ and let $U = \{x \in E: |f_n(x)| \le 1, n \in N\}$. In view of the preceding remark, it can be assumed that $\{y_n\}$ is a null sequence in a suitable space F_B and, in addition, that $\sum_{1}^{\infty} |\lambda_n| \le 1$. It follows that the image

$u(U)$ of the 0-neighborhood U is contained in the closed, convex, circled hull C of the null sequence $\{y_n\}$ in F_B; since $\{y_n\}$ is relatively compact in F_B and F_B is complete, C is compact in F_B (cf. (I, 5.2) and (II, 4.3)), and hence a fortiori compact in F by the continuity of $F_B \to F$.

COROLLARY 2. *Let E, F, G, H be l.c.s., let $u \in \mathscr{L}(E, F)$, let $w \in \mathscr{L}(G, H)$, and let v be a nuclear map on F into G. Then $v \circ u$ and $w \circ v$ (and hence $w \circ v \circ u$) are nuclear maps.*

Proof. It is evident from (7.1) that $v \circ u$ is nuclear. By Corollary 1, there exists a convex, circled 0-neighborhood V in F such that $\overline{v(V)} = B$ is compact in G. Thus $B_1 = w(B)$ is compact in H, hence H_{B_1} is complete. It is now clear that $w \circ v$ is nuclear in $\mathscr{L}(F, H)$.

COROLLARY 3. *If $u \in \mathscr{L}(E, F)$ is nuclear, then u has a unique extension $\bar{u} \in \mathscr{L}(\tilde{E}, F)$, where $\tilde{E}$ is the completion of E, and $\bar{u}$ is nuclear.*

Proof. The first of the stated properties is shared by u with all compact maps on E into F. In fact, if U is a 0-neighborhood in E such that $u(U) \subset C$, where C is compact, then since u is uniformly continuous, its restriction to U has a unique continuous extension to $\overline{U}$ (the closure of U in $\tilde{E}$) with values in C, since C is complete. It is immediately clear that this extension is the restriction to $\overline{U}$ of a linear map $\bar{u}$ of $\tilde{E}$ into F which is compact, hence continuous; that $\bar{u}$ is nuclear is a direct consequence of the definition of a nuclear map, or of (7.1).

We are now ready to define a nuclear space. A locally convex space E is **nuclear** if there exists a base $\mathfrak{B}$ of convex, circled 0-neighborhoods in E such that for each $V \in \mathfrak{B}$, the canonical mapping $E \to \tilde{E}_V$ is nuclear.

It is at once clear from this definition and (7.1) that a l.c.s. E is nuclear if and only if its completion $\tilde{E}$ is nuclear. The space K_0^d (d any cardinal) is a first example of a nuclear space; in fact, for any convex, circled 0-neighborhood V, the space $E_V = \tilde{E}_V$ is of finite dimension; thus $E \to \tilde{E}_V$ is of finite rank and hence nuclear. Further and more interesting examples will be given below and in Section 9, Chapter IV. Let us note, however, that a normed space E cannot be nuclear unless it is of finite dimension; for if V is a convex, circled 0-neighborhood which is bounded, then $E \to E_V$ is a topological automorphism; hence if $E \to \tilde{E}_V$ is a nuclear map, it is compact by Corollary 1 above. Thus (I, 3.6) implies that E is finite-dimensional. We shall have use for the following alternative characterizations of nuclear spaces.

7.2

Let E be a l.c.s. The following assertions are equivalent:

(a) *E is nuclear.*
(b) *Every continuous linear map of E into any Banach space is nuclear.*

(c) *Each convex, circled 0-neighborhood U in E contains another, V, such that the canonical map $\tilde{E}_V \to \tilde{E}_U$ is nuclear.*

Proof. (a) $\Rightarrow$ (b): Let F be any Banach space and $u \in \mathcal{L}(E, F)$. There exists a convex, circled 0-neighborhood V in E such that $\phi_V: E \to \tilde{E}_V$ is nuclear, and such that $u(V)$ is bounded in F. Since $\phi_V(E) = E_V$, u determines a unique $v \in \mathcal{L}(\tilde{E}_V, F)$ such that $u = v \circ \phi_V$, and it follows from Corollary 2 of (7.1) that u is nuclear. (b) $\Rightarrow$ (c): Let U be any convex, circled 0-neighborhood in E. By assumption, the canonical map $E \to \tilde{E}_U$ is nuclear, and hence of the form $\phi_U = \sum \lambda_n f_n \otimes y_n$ as described in (7.1). Set $V = U \cap \{x: |f_n(x)| \leq 1, n \in N\}$, then $V \subset U$ is convex, circled, and a 0-neighborhood by the equicontinuity of the sequence $\{f_n\}$. Now each f_n induces a continuous linear form (of norm ≤ 1) on E_V. Denote by h_n its continuous extension to $\tilde{E}_V$: It is now trivial that the canonical map $\phi_{U,V}: \tilde{E}_V \to \tilde{E}_U$ is given by $\sum \lambda_n h_n \otimes y_n$, and hence nuclear by (7.1).

(c) $\Rightarrow$ (a): If U is a given convex, circled 0-neighborhood in E, there exists another, V, such that $\phi_{U,V}$ is nuclear. Since $\phi_U = \phi_{U,V} \circ \phi_V$, it follows from Corollary 2 of (7.1) that $E \to \tilde{E}_U$ is nuclear, whence E is a nuclear space by definition.

COROLLARY 1. *If E is a nuclear space, then $E \to \tilde{E}_V$ is a nuclear map for every convex, circled neighborhood V of 0 in E.*

For $E \to \tilde{E}_V$ is continuous and $\tilde{E}_V$ is a Banach space.

COROLLARY 2. *Every bounded subset of a nuclear space is precompact.*

Proof. If $\mathfrak{B}$ is a neighborhood base of 0 in E consisting of convex, circled sets, then by Corollary 2 of (II, 5.4) E is isomorphic with a subspace of $\prod_{V \in \mathfrak{B}} \tilde{E}_V$ by virtue of the mapping $x \to \{\phi_V(x): V \in \mathfrak{B}\}$. This isomorphism carries a bounded set $B \subset E$ into the set $\prod \phi_V(B)$. Now if E is nuclear, each $\phi_V(B)$ is precompact in E_V by Corollary 1 of (7.1). Thus the product $\prod \phi_V(B)$ is precompact, which proves the assertion.

We recall the common usage to understand by l^p ($1 \leq p < +\infty$) the Banach space of all (real or complex) sequences $x = (x_1, x_2, \ldots)$ whose pth powers are (absolutely) summable, under the norm $\|x\|_p = (\sum |x_n|^p)^{1/p}$; l^∞ is the Banach space of bounded sequences with $\|x\|_\infty = \sup_n |x_n|$.

The following result reveals the special structure of nuclear spaces.

7.3

Let E be a nuclear space, let U be a given 0-neighborhood in E, and let p be a number such that $1 \leq p \leq \infty$. There exists a convex, circled 0-neighborhood $V \subset U$ for which $\tilde{E}_V$ is (norm) isomorphic with a subspace of l^p.

Proof. We show that there exists a continuous linear map $v \in (\mathcal{L}E, l^p)$ such that $v^{-1}(B) \subset U$, where B is the open unit ball of l^p; $V = v^{-1}(B)$ will be the neighborhood in question. Assume without loss of generality that U is convex and circled. The canonical map ϕ_U is nuclear by (7.2), Corollary 1, hence of the form $\phi_U = \sum \lambda_n f_n \otimes y_n$, where we can assume that $\lambda_n > 0$ $(n \in N)$, $\sum_{n=1}^{\infty} \lambda_n = 1$, $\|y_n\| = 1$ in $\tilde{E}_U$ $(n \in N)$ and that the sequence $\{f_n\}$ is equicontinuous. Define v by

$$v(x) = (\sqrt[p]{\lambda_1} f_1(x), \sqrt[p]{\lambda_2} f_2(x), \ldots)$$

for all $x \in E$ (set $\sqrt[p]{\lambda_n} = 1$ for all n if $p = \infty$). By the equicontinuity of the sequence $\{f_n\}$, we have $v(x) \in l^p$ and evidently $v \in \mathscr{L}(E, l^p)$. Now let $p^{-1} + q^{-1} = 1$ $(q = 1$ if $p = \infty$ and $q = \infty$ if $p = 1$) and apply Hölder's inequality to $\sum_{n=1}^{\infty} \alpha_n \beta_n$ with $\alpha_n = \sqrt[p]{\lambda_n} f_n(x)$, $\beta_n = \sqrt[q]{\lambda_n}$. Denoting by $\| \ \|$ the norm in E_U, we obtain

$$\|\phi_U(x)\| = \| \sum_{n=1}^{\infty} \lambda_n f_n(x) y_n \| \leq \sum_{1}^{\infty} \lambda_n |f_n(x)| \leq \|v(x)\|_p,$$

whence $v^{-1}(B) \subset U$. Letting $V = v^{-1}(B)$, the definition of v implies that E_V is norm isomorphic with $v(E)$; hence $\tilde{E}_V$ is norm isomorphic with the closed subspace $\overline{v(E)}$ of l^p.

In the three corollaries that follow, denote by A a set whose cardinality is the minimal cardinality of a neighborhood base of 0 in E.

COROLLARY 1. *Let E be nuclear, and let $\{E_\alpha: \alpha \in A\}$ be a family of Banach spaces, each of which is isomorphic with a space $l^p(1 \leq p \leq \infty)$. There exist linear maps f_α of E into E_α $(\alpha \in A)$ such that the topology of E is the coarsest topology for which all mappings f_α are continuous.*

In other words, the topology of E is the projective topology with respect to the family $\{(E_\alpha, f_\alpha): \alpha \in A\}$ (Chapter II, Section 5). If we apply (7.3), with $p = 2$, to each element $U_\alpha(\alpha \in A)$ of a 0-neighborhood base in E, we obtain a base $\{V_\alpha: \alpha \in A\}$ of 0-neighborhoods such that for each $\alpha \in A$, $\tilde{E}_\alpha = \tilde{E}_{V_\alpha}$ is a Hilbert space (not necessarily of infinite dimension (cf. Chapter II, Section 2, Example 5)). Now if $\tilde{E}_\alpha$ is a Hilbert space, the norm of $\tilde{E}_\alpha$ originates from a positive definite Hermitian form $(\hat{x}, \hat{y}) \to [\hat{x}, \hat{y}]_\alpha$ on $\tilde{E}_\alpha \times \tilde{E}_\alpha$; hence if ϕ_α denotes the canonical map $E \to \tilde{E}_\alpha$, then $(x, y) \to [\phi_\alpha(x), \phi_\alpha(y)]_\alpha$ is a positive semi-definite Hermitian form on $E \times E$ such that $x \to [\phi_\alpha(x), \phi_\alpha(x)]_\alpha^{\frac{1}{2}}$ is the gauge function p_α of V_α.

COROLLARY 2. *In every nuclear space E there exists a 0-neighborhood base $\{V_\alpha: \alpha \in A\}$ such that for each $\alpha \in A$, $\tilde{E}_{V_\alpha}$ is a Hilbert space; hence the topology of E can be generated by a family of semi-norms, each of which originates from a positive semi-definite Hermitian form on $E \times E$.*

Combining this result with the construction used in the proof of (II, 5.4), we obtain a representation of nuclear spaces as dense subspaces of projective limits of Hilbert spaces. Thus the completion of a nuclear space E is isomorphic with a projective limit of Hilbert spaces, and obviously nuclear by Corollary 3 of (7.1).

COROLLARY 3. *Every complete nuclear space is isomorphic with the projective limit of a suitable family (of cardinality card* A*) of Hilbert spaces. A Fréchet space E is nuclear if and only if it is the projective limit of a sequence of Hilbert spaces,* $E = \lim_{\leftarrow} g_{mn} H_n$ *such that* g_{mn} *is a nuclear map whenever* $m < n$.

Proof. We have only to prove the second assertion. If E is a nuclear (F)-space, by Corollary 2 there exists a base $\{V_n : n \in N\}$ at 0 which can be supposed decreasing, and such that each $\tilde{E}_n$ is a Hilbert space. By (7.2)(c) we can even suppose that each of the canonical maps $\phi_{V_n, V_{n+1}} : \tilde{E}_{n+1} \to \tilde{E}_n$ is nuclear. The desired representation is then obtained with $H_n = \tilde{E}_n$ and $g_{mn} = \phi_{V_m, V_n}$ $(m \leq n)$. Conversely, if E is of the form indicated and V is a convex, circled 0-neighborhood chosen from a suitable base in E, then $E \to \tilde{E}_V$ can be identified with the projection p of E into a finite product of spaces H_n, say $\prod_{k=1}^{m} H_k$. Denoting by p_n the projection of E into H_n $(n \in N)$ we have $p = (p_1, ..., p_m)$; hence $p = (g_{1n} \circ p_n, ..., g_{mn} \circ p_n)$ for any $n > m$, which implies that p is nuclear.

The following important theorem of permanence is also due to Grothendieck (cf. [13], chap. II, theor. 9).

7.4

Theorem. *Every subspace and every separated quotient space of a nuclear space is nuclear. The product of an arbitrary family of nuclear spaces is nuclear, and the locally convex direct sum of a countable family of nuclear spaces is a nuclear space.*

Before proving the theorem, we note the following immediate consequence:

COROLLARY. *The projective limit of any family of nuclear spaces, and the inductive limit of a countable family of nuclear spaces, are nuclear.*

Proof of (7.4)

1. The proof for countable direct sums and arbitrary products will be based on property (b) of (7.2). Let $E = \bigoplus_{i=1}^{\infty} E_i$, E_i $(i \in N)$ be nuclear spaces, and let u be a continuous linear map of E into a given Banach space F. If u_i is the restriction of u to the subspace E_i of E, u_i is continuous and hence nuclear, and thus of the form

$$u_i = \sum_{n=1}^{\infty} \mu_n^{(i)} h_n^{(i)} \otimes y_{n,i} \qquad (i \in N).$$

Here we can assume that $\|y_{n,i}\| \leq 1$ in F for all $(n, i) \in N \times N$, that $\sum_{n=1}^{\infty} |\mu_n^{(i)}| \leq i^{-2}$ $(i \in N)$, and that each of the sequences $\{h_n^{(i)} : n \in N\}$ is equicontinuous on E_i. Let U_i be a 0-neighborhood in E_i such that $|h_n^{(i)}(x_i)| \leq 1$ for all $x_i \in U_i$ and all $n \in N$, and define $f_{n,i}$ to be the continuous linear form on E which is the extension of $h_n^{(i)}$ to E that vanishes on the complementary subspace

$\bigoplus\limits_{j \neq i} E_j$ of E_i. The family $\{f_{n,i}: (n, i) \in N \times N\}$ is equicontinuous, for if U is the

0-neighborhood $\Gamma_i U_i$ in E, then $x \in U$ implies $|f_{n,i}(x)| \leq 1$ for all n and i. Since u can be written as

$$u = \sum_{n,i=1}^{\infty} \mu_n^{(i)} f_{n,i} \otimes y_{n,i},$$

it follows from (7.1) that u is nuclear.

Let $\{E_\alpha: \alpha \in A\}$ be any family of nuclear spaces, $E = \prod_\alpha E_\alpha$ and let u be a continuous linear map of E into a given Banach space F. There exists a 0-neighborhood V in E such that $u(V)$ is bounded in F, and by definition of the product topology, V contains a 0-neighborhood of the form $V_{\alpha_1} \times \cdots \times V_{\alpha_n} \times \prod\limits_{\beta \neq \alpha_i} E_\beta$. It follows that u vanishes on the subspace $G = \prod\limits_{\beta \neq \alpha_i} E_\beta$ of E. Since $E = \prod\limits_{i=1}^{n} E_{\alpha_i} \oplus G$, it remains to show that the restriction of u to $\prod\limits_i E_{\alpha_i}$ is nuclear. But this is clear from the preceding proof, since $\bigoplus\limits_{i=1}^{n} E_{\alpha_i}$ is identical with $\prod\limits_{i=1}^{n} E_{\alpha_i}$ (Chapter II, Section 6).

2. The proof of nuclearity for subspaces and quotient spaces will be based on property (c) of (7.2) and Corollary 1 of (7.3). Let E be a nuclear space and let M be a subspace of E. For each convex, circled 0-neighborhood U in E, set $V = M \cap U$. We show that for each V, there exists another such neighborhood, $V_1 \subset V$, such that the canonical map $\tilde{M}_{V_1} \to \tilde{M}_V$ is nuclear. We can assume without loss of generality that $V = M \cap U$, where U is such that $\tilde{E}_U$ is a Hilbert space. There exists a 0-neighborhood $U_1 \subset U$ such that the canonical map $\phi_{U,U_1}: \tilde{E}_{U_1} \to \tilde{E}_U$ is nuclear; let $V_1 = M \cap U_1$. Now it is not difficult to see that $\tilde{M}_{V_1}$ and $\tilde{M}_V$ can be identified with closed subspaces of $\tilde{E}_{U_1}$ and $\tilde{E}_U$, respectively, so that the canonical map ϕ_{V,V_1} is the restriction of ϕ_{U,U_1} to $\tilde{M}_{V_1}$ (Exercise 3). But ϕ_{U,U_1} is of the form $\sum\limits_{i=1}^{\infty} \lambda_i f_i \otimes y_i$ with $\{\lambda_i\}$ summable, $\{f_i\}$ equicontinuous in $[\tilde{E}_{U_1}]'$, and $\{y_i\}$ bounded in $\tilde{E}_U$. Denote by p the orthogonal projection of $\tilde{E}_U$ onto $\tilde{M}_V$, let $w_i = py_i$, and denote by g_i the restriction of f_i to $\tilde{M}_{V_1}$ ($i \in N$). Then $\{g_i\}$ is equicontinuous, $\{w_i\}$ is bounded in $\tilde{M}_V$, and ϕ_{V,V_1} necessarily of the form $\sum\limits_{i=1}^{\infty} \lambda_i g_i \otimes w_i$, and hence nuclear by (7.1).

We employ the same pattern of proof for quotient spaces: Let E be nuclear, let M be a closed subspace of E, $F = E/M$ (topological), and let ϕ be the canonical map $E \to F$. For a given convex, circled 0-neighborhood V in F, we show the existence of another, $V_1 \subset V$, such that $\phi_{V,V_1}: \tilde{F}_{V_1} \to \tilde{F}_V$ is nuclear. For this we can suppose that $V = \phi(U)$, $\tilde{E}_U$ is a Hilbert space, and $U_1 \subset U$ is such that $\tilde{E}_{U_1}$ is a Hilbert space and $\phi_{U,U_1}: \tilde{E}_{U_1} \to \tilde{E}_U$ is nuclear. The point of the proof consists now in recognizing that $\tilde{F}_V$ can be identified with a quotient space of $\tilde{E}_U$. In fact, $\tilde{F}_V$ is isomorphic with the space $\tilde{E}_U/L$, where L is the closure of $\phi_U(M)$ in $\tilde{E}_U$. Similarly, letting $V_1 = \phi(U_1)$, $\tilde{F}_{V_1}$ can be identified with $\tilde{E}_{U_1}/L_1$, where L_1 is the closure of $\phi_{U_1}(M)$ in $\tilde{E}_{U_1}$ (Exercise 3).

We note further that ϕ_{U,U_1} maps L_1 into L, and ϕ_{V,V_1} is nothing else but the map of $\tilde{E}_{U_1}/L_1$ into $\tilde{E}_U/L$ induced by ϕ_{U,U_1} under the identification just made.

Since ϕ_{U,U_1} is nuclear, it is of the form $\sum_{i=1}^{\infty} \lambda_i f_i \otimes y_i$ as described in (7.1). We decompose $\tilde{E}_{U_1} = L_1 \oplus L_1^{\perp}$, $\tilde{E}_U = L \oplus L^{\perp}$ (orthogonal complements). Let $f_i = f_i' + f_i''$ and $y_i = y_i' + y_i''$ ($i \in N$) be the corresponding decompositions (so that, for f_i, we have $f_i'(L_1^{\perp}) = f_i''(L_1) = \{0\}$). Since ϕ_{U,U_1} maps L_1 into L, it follows that $\sum_i \lambda_i f_i' \otimes y_i''$ vanishes on $\tilde{E}_{U_1}$ whence,

$$\phi_{U,U_1} = \sum_{i=1}^{\infty} \lambda_i f_i \otimes y_i' + \sum_{i=1}^{\infty} \lambda_i f_i'' \otimes y_i''.$$

If now g_i denotes the linear form on $\tilde{E}_{U_1}/L_1$ determined by f_i'', and w_i denotes the equivalence class of y_i'' mod L ($i \in N$), then ϕ_{V,V_1} (being the map induced by ϕ_{U,U_1}) is of the form $\sum \lambda_i g_i \otimes w_i$, and hence is nuclear by (7.1).

The proof of the theorem is complete.

We supplement theorem (7.4) by showing that the projective tensor product of two nuclear spaces is nuclear. To this end, we need the concept of the tensor product of two linear mappings: Let E, F, G, H be vector spaces over K and $u \in L(E, G)$, $v \in L(F, H)$. The mapping $(x, y) \to u(x) \otimes v(y)$ is bilinear on $E \times F$ into $G \otimes H$; the linear mapping of $E \otimes F$ into $G \otimes H$, which corresponds to the former, is denoted by $u \otimes v$, and is called the **tensor product** of u and v. It is obvious that $(u, v) \to u \otimes v$ is bilinear on $L(E, G) \times L(F, H)$ into $L(E \otimes F, G \otimes H)$. Thus again by (6.1), to this map there corresponds a linear map of $L(E, G) \otimes L(F, H)$ into $L(E \otimes F, G \otimes H)$ (called the canonical imbedding), which is an isomorphism. If $G = H = K_0$, that is, if $u = f, v = g$ are linear forms, then tensor multiplication in $K_0 \otimes K_0$ can be identified with ordinary multiplication in K (proof!) and we have $f \otimes g(x \otimes y) = f(x)g(y)$ so that the tensor products $f \otimes g$ and $E^* \otimes F^*$ considered earlier are special cases of the present definition.

7.5

If E and F are nuclear spaces, the projective tensor product of E and F, as well as its completion $E \,\tilde{\otimes}\, F$, are nuclear.

Proof. Let U, V be convex, circled 0-neighborhoods in E, F respectively; set $G = E \otimes F$ and $W = \Gamma U \otimes V$ in G. It is clear from (6.3) that G_W is identical with the normed space $(E_U \otimes F_V, r)$, where r is the tensor product of the respective norms of E_U and F_V. Hence if ϕ_U, ϕ_V, ϕ_W denote the respective canonical maps $E \to \tilde{E}_U$, $F \to \tilde{F}_V$, $G \to \tilde{G}_W$, we have $\phi_W = \phi_U \otimes \phi_V$. Since E, F are nuclear, (7.1) implies that $\phi_U = \sum \lambda_i f_i \otimes \hat{x}_i$, $\phi_V = \sum \mu_j g_j \otimes \hat{y}_j$, where $\{\lambda_i\}$, $\{\mu_j\}$, etc., have the properties enumerated in (7.1). For $x \in E, y \in F$ we have by definition

$$\phi_U \otimes \phi_V(x \otimes y) = \left(\sum_{i=1}^{\infty} \lambda_i f_i(x)\hat{x}_i \right) \otimes \left(\sum_{j=1}^{\infty} \mu_j g_j(y)\hat{y}_j \right),$$

which, as an element of $\tilde{G}_W = E_U \tilde{\otimes} F_V$, can be written

$$\phi_W(x \otimes y) = \sum_{i,j=1}^{\infty} \lambda_i \mu_j f_i(x) g_j(y)(\hat{x}_i \otimes \hat{y}_j),$$

so that ϕ_W is represented by $\sum_{i,j} \lambda_i \mu_j (f_i \otimes g_j) \otimes (\hat{x}_i \otimes \hat{y}_j)$. Now $\{\lambda_i \mu_j : (i,j)$ $\in N \times N\}$ is a summable family, $\{f_i \otimes g_j\}$ is an equicontinuous family (namely, uniformly bounded on $\Gamma U_1 \otimes V_1$ for suitable 0-neighborhoods U_1, V_1 in E, F, respectively), and, clearly, the family $\{\hat{x}_i \otimes \hat{y}_j\}$ is bounded in $\tilde{G}_W$ because of $\|\hat{x}_i \otimes \hat{y}_j\| = \|\hat{x}_i\| \|\hat{y}_j\|$. Hence ϕ_W is nuclear for each element W of a 0-neighborhood base of the projective topology on $E \otimes F$. The nuclearity of $E \tilde{\otimes} F$ is immediate from (7.1), Corollary 3.

8. EXAMPLES OF NUCLEAR SPACES

1. Let T be the k-dimensional torus. The space $\mathscr{D}_T$ of (real- or complex-valued) infinitely differentiable functions on T, endowed with the topology of uniform convergence in all derivatives, is a nuclear (F)-space. By (7.4) this implies that the space $\mathscr{D}_I$ of infinitely differentiable functions on R^k, whose support is contained in the k-dimensional interval I, is nuclear (notation as in Chapter II, Section 6, Example 2, except that the domain is sometimes written in parentheses). For $\mathscr{D}_I$ can be identified with a subspace of $\mathscr{D}_T$ by considering each $f \in \mathscr{D}_I$ as a k-fold periodic function on I. The method of proof will be sufficiently exhibited by considering the case $k = 1$.

Denote by $g_k (k = 0, \pm 1, \pm 2, \ldots)$ the normalized trigonometric functions $g_k(t) = (2\pi)^{-\frac{1}{2}} e^{ikt}$, and set $[f, g_k] = \int_{-\pi}^{\pi} f(t) g_k^*(t) dt$ for $f \in \mathscr{D}_T$. It is well known from elementary analysis that f has a Fourier expansion $f = \sum_k a_k g_k$ that converges to f uniformly in all derivatives; that is, $f = \sum_k a_k g_k$ is an expression valid in $\mathscr{D}_T$. The coefficients are given by $a_k = [f, g_k]$, and by repeated partial integration it follows that

$$a_k = [f, g_k] = (ik)^{-m}[f^{(m)}, g_k]$$

for all $k = \pm 1, \pm 2, \ldots$ and all integers $m \geq 0$, $f^{(m)}$ denoting the mth derivative of f. The family of norms $f \to p_n(f) = \sup\{|f^{(m)}(t)| : t \in T,$ $m \leq n\}$ generates the topology of $\mathscr{D}_T$, and the 0-neighborhoods $V_n = \{f : p_n(f) \leq n^{-1}\}$ where $n \in N$ form a base at 0. Note also that each of the spaces E_{V_n} is algebraically isomorphic with $E = \mathscr{D}_T$, since the p_n are norms. For fixed n, say $n = m$, the expansion of f can be written

$$f = [f, g_0] g_0 - \sum_{k \neq 0} k^{-2} [f^{(m+2)}, g_k](ik)^{-m} g_k.$$

Now the linear forms $f \to h_k(f) = [f^{(m+2)}, g_k]$ $(k = \pm 1, \pm 2, \ldots)$ are uniformly bounded on V_{m+2} and hence are equicontinuous, and the functions $y_k = (ik)^{-m} g_k$ $(k = \pm 1, \pm 2, \ldots)$ can be arranged to form a bounded sequence in $\tilde{E}_{V_m}$. Hence the canonical map $E \to \tilde{E}_{V_m}$, being of the form

$$f \to [f, g_0]g_0 - \sum_{k \neq 0} k^{-2}h_k(f)y_k,$$

is nuclear. Since $m \in N$ was arbitrary, $\mathscr{D}_T$ is nuclear as asserted.

The same conclusion holds if C is any compact subset of R (and more generally, of R^k), for then C is contained in an interval I as considered above, and $\mathscr{D}_C$ is a closed subspace of $\mathscr{D}_I$, hence nuclear by (7.4).

2. The space $\mathscr{D}$ of L. Schwartz (Chapter II, Section 6, Example 2), being the inductive limit of a sequence of spaces $\mathscr{D}_C$, is nuclear by the corollary of (7.4).

3. Let $\mathscr{E}$ be the space of infinitely differentiable complex functions on R^k (with no restrictions on their supports), under the topology of compact convergence in all derivatives. Let $\{C_n\}$ be an increasing sequence of convex compact sets with non-empty interior in R^k such that $\bigcup_n C_n = R^k$. Then each compact set $C \subset R^k$ is contained in some C_n, and the topology of $\mathscr{E}$ is generated by the semi-norms

$$f \to p_n(f) = \sup\{|D^m f(t)|: t \in C_n, m \leq n\} \qquad (n \in N)$$

where $|D^m f(t)|$ stands for the sum of the absolute values (at t) of all derivatives of f that have order m (≥ 0). As $\mathscr{D}_C$, $\mathscr{E}$ is an (F)-space which is nuclear. We shall not verify the nuclearity of $\mathscr{E}$ directly, since it will be a consequence of (IV, 9.7).

4. Let $\mathscr{H}(C)$ be the space of all entire functions of one complex variable under the topology of compact convergence. It is clear from the elements of complex function theory that $\mathscr{H}(C)$ is an (F)-space; moreover, by a classical theorem of Weierstrass, $\mathscr{H}(C)$ is not only algebraically, but also topologically, a closed subspace of $\mathscr{E}(R^2)$. Hence by Example 3 and (7.4), $\mathscr{H}(C)$ is nuclear.

5. The space $\mathscr{S}$ (cf. L. Schwartz [2], chap. VII, §3), or $\mathscr{S}(R^k)$, is (algebraically) defined to be the subspace of $\mathscr{E}(R^k)$ such that $|t|^m D^n f(t) = 0$ for any derivative of f of any order n and any integer $m \in N$, with $|t| = [\sum_{i=1}^k t_i^2]^{\frac{1}{2}}$ denoting the Euclidean norm of $t = (t_1, ..., t_k)$ $\in R^k$. The topology of $\mathscr{S}$ is defined by the sequence of semi-norms

$$f \to p_n(f) = \sup\{(1 + |t|^n)|D^m f(t)|: |t| \leq n, m \leq n\}$$

$(n \in N)$. $\mathscr{S}$ is an (F)-space, called the space of rapidly decreasing, infinitely differentiable functions on R^k. The space $\mathscr{S}$ is nuclear. This can be proved directly by applying the method used in Example 1 to the expansion of $f \in \mathscr{S}$ by the functions of Hermite (cf. L. Schwartz [2], vol. II, p. 117). Another proof uses the fact that $\mathscr{S}(R^k)$ is isomorphic with a closed subspace of $\mathscr{D}(S^k)$(l.c. p. 91), where $\mathscr{D}(S^k)$ is the space of infinitely differentiable functions on the k-sphere S^k. For $k = 1$, the nuclearity of $\mathscr{D}(S^1)$ was shown in Example 1; for $k > 1$, one can use expansions by spherical functions.

In Examples 1, 2, and 3 R^k can be replaced by an open subset of R^k or, more generally, by an infinitely differentiable manifold, and C by an open subset in Example 4. Further examples will be obtained in Chapter IV, Section 9. For examples of nuclear sequence spaces, see Exercise 25.

9. THE APPROXIMATION PROBLEM. COMPACT MAPS

Let H be a Hilbert space. If $\{x_\alpha : \alpha \in A\}$ is an orthonormal basis of H (Chapter II, Section 2, Example 5), if $[,]$ is the inner product of H, and if $f_\alpha(\alpha \in A)$ is the continuous linear form on H for which $f_\alpha(x) = [x, x_\alpha]$, then it is known and easy to prove that for each $x \in H$ the family $\{f_\alpha(x)x_\alpha : \alpha \in A\}$ is summable to x (for the definition of summability in H, see Exercise 23):

$$x = \sum_{\alpha \in A} f_\alpha(x)x_\alpha.$$

The convergence of this sum can be interpreted in the following way: If for each finite subset $\Phi \subset A$, we denote by u_Φ the linear map $x \to \sum_{\alpha \in \Phi} f_\alpha(x)x_\alpha$ (that is, if $u_\Phi = \sum_{\alpha \in \Phi} f_\alpha \otimes x_\alpha$), then $\{u_\Phi\}$ converges pointwise to the identity map e of H, the convergence being along the family of finite subsets of A directed by inclusion $\subset$. Now since each u_Φ, being an orthogonal projection, is (if $\Phi \neq \varnothing$) of norm 1 in $\mathscr{L}(H)$, it follows from (4.6) that the convergence of $\{u_\Phi\}$ is uniform on every compact subset of H. [We write $\mathscr{L}(E) = \mathscr{L}(E, E)$ and endow, for any pair of normed spaces E, F, the space $\mathscr{L}(E, F)$ with the standard norm $u \to \|u\| = \sup\{\|u(x)\| : \|x\| \leqq 1\}$.]

This implies that for every Hilbert space H, the identity map e is in the closure of $H' \otimes H \subset \mathscr{L}(H)$ for the topology of precompact convergence. It can be shown (Karlin [2]) that even a separable Banach space does not, in general, contain an unconditional basis (see below), i.e., a sequence $\{x_n\}$ for which there exists a sequence $\{f_n\} \subset E'$ satisfying $f_m(x_n) = \delta_{mn}$ $(m, n \in N)$, and such that $\{f_n(x)x_n : n \in N\}$ is summable to x for all $x \in E$. Karlin shows the non-existence of such a basis for $\mathscr{C}[0, 1]$. But it is not known whether in every l.c.s. E the identity map e can be approximated, uniformly on every precompact set in E, by continuous linear maps of finite rank. The question whether or not this is true constitutes the **approximation problem**. Any l.c.s. for which the answer is positive (and which consequently has all four of the equivalent properties stated in (9.1)) is said to have the **approximation property** (a.p.). It appears that all known locally convex spaces have the a.p. (cf. Grothendieck [13], chap. I, §5). In the following we denote the topology of precompact convergence by a subscript "c".

9.1

Let E be any locally convex space with dual E'. The following properties of E are equivalent:

(a) *The closure of $E' \otimes E$ in $\mathscr{L}_c(E)$ contains the identity map e.*

(b) $E' \otimes E$ *is dense in* $\mathscr{L}_c(E)$.

(c) *For every l.c.s.* F, $E' \otimes F$ *is dense in* $\mathscr{L}_c(E, F)$.

(d) *For every l.c.s.* F, $F' \otimes E$ *is dense in* $\mathscr{L}_c(F, E)$.

Proof. (a) $\Rightarrow$ (b): Given a $u \in \mathscr{L}(E)$, a precompact set A, and a 0-neighborhood V in E, we have to show the existence of $u_0 \in E' \otimes E$ such that $u(x) - u_0(x) \in V$ for all $x \in A$. Let U be a 0-neighborhood in E for which $u(U) \subset V$. By (a) there exists an $e_0 \in E' \otimes E$ such that $x - e_0(x) \in U$ for all $x \in A$. Clearly, if $e_0 = \sum_1^n x_i' \otimes x_i$, then the map $u_0 = u \circ e_0 = \sum_1^n x_i' \otimes u(x_i)$ satisfies the requirement.

(b) $\Rightarrow$ (c): For each fixed $v \in \mathscr{L}(E, F)$ the mapping $u \to v \circ u$ is continuous on $\mathscr{L}_c(E)$ into $\mathscr{L}_c(E, F)$. Thus since $E' \otimes E$ is dense in $\mathscr{L}_c(E)$, $E' \otimes v(E)$ is dense in a subspace of $\mathscr{L}_c(E, F)$ containing $v \circ e = v$, which establishes the assertion.

(b) $\Rightarrow$ (d): Likewise, for each fixed $w \in \mathscr{L}(F, E)$, $u \to u \circ w$ is continuous on $\mathscr{L}_c(E)$ into $\mathscr{L}_c(F, E)$, since for each precompact set $B \subset F$, $w(B)$ is precompact in E (for w is uniformly continuous). It follows that w is in the closure of the subspace $(E' \otimes E) \circ w$ of $\mathscr{L}_c(F, E)$, and it is quickly seen that $(E' \otimes E) \circ w \subset F' \otimes E$.

Finally, the implications (d) $\Rightarrow$ (b), (c) $\Rightarrow$ (b), and (b) $\Rightarrow$ (a) are trivial.

The following result reduces the approximation problem entirely to Banach spaces. Besides, it gives a positive answer for a large class of locally convex spaces (including all nuclear spaces).

9.2

Let E be a l.c.s. with a 0-neighborhood base $\mathfrak{B}$ of convex, circled sets such that for each $V \in \mathfrak{B}$, $\tilde{E}_V$ has the approximation property. Then E possesses the approximation property.

Proof. Let $V \in \mathfrak{B}$ be given, and denote by ϕ the canonical map $E \to \tilde{E}_V = F$. Let $W = \frac{1}{4}\overline{\phi(V)}$, where the closure is taken in F. It follows that $\phi^{-1}(W) \subset \frac{1}{2}V + V_0 \subset V$ if $V_0 = \bigcap_{\lambda > 0} \lambda V$ denotes the null space of V. Note further that $E_V = \phi(E)$ is dense in F. Since F has the a.p. by assumption, (9.1)(d) implies that $E' \otimes F$ is dense in $\mathscr{L}_c(E, F)$, whence $E' \otimes \phi(E)$ is also dense in $\mathscr{L}_c(E, F)$. Hence, ϕ being in $\mathscr{L}(E, F)$, for a given precompact set $A \subset E$ there exists $w \in E' \otimes \phi(E)$ such that $w(x) - \phi(x) \in W$ for all $x \in A$. Let $w = \sum_1^n x_i' \otimes \phi(x_i)$. It follows that $\phi[\sum_i x_i'(x)x_i - x] \in W$ hence (since $\phi^{-1}(W) \subset V$) $\sum_1^n x_i'(x)x_i - x \in V$ for all $x \in A$. This proves the assertion, since V was any member of a neighborhood base of 0 in E.

COROLLARY 1. *Every subspace of an arbitrary product of Hilbert spaces possesses the approximation property.*

Proof. Let $E = \prod_{\alpha \in A} H_\alpha$ be a product of Hilbert spaces. There exists a 0-neighborhood base $\mathfrak{B}$ in E such that for each $V \in \mathfrak{B}$, $\tilde{E}_V$ is isomorphic with the Hilbert space $\prod_{\alpha \in \Phi} H_\alpha$, where Φ is a suitable finite subset of A; thus $\tilde{E}_V$ has the a.p. If M is a subspace of E and $W = M \cap V$ ($V \in \mathfrak{B}$), then $\tilde{M}_W$ can be identified with the closure of $\phi_V(M)$ in $\tilde{E}_V$ (Exercise 3); hence $\tilde{M}_W$ is a Hilbert space if $\tilde{E}_V$ is.

COROLLARY 2. *Every projective limit of Hilbert spaces and every subspace of such a projective limit (in particular, every nuclear space) has the approximation property.*

This is an immediate consequence of Corollary 1 and (7.3), Corollary 2.

COROLLARY 3. *If there exists a locally convex space not having the a.p., then there exists a Banach space not having the a.p.*

In view of the last corollary, we shall analyze the approximation problem for Banach spaces somewhat further. For this we need several results on compact maps and sets. Since these results are also of independent interest, they will be proved in detail. We denote by a subscript "b" the topology of bounded convergence on $\mathscr{L}(E, F)$; recall (Section 3) that when E, F are normed spaces, this is the topology of the normed space $\mathscr{L}(E, F)$.

9.3

Let E be normed and let F be a quasi-complete l.c.s.; the set of all compact linear maps of E into F is a closed subspace of $\mathscr{L}_b(E, F)$.

Proof. The subset of $\mathscr{L}(E, F)$ consisting of all compact maps is evidently a subspace M. Let us show that M is closed. Let $v \in \bar{M} \subset \mathscr{L}_b(E, F)$ and a 0-neighborhood V in F be given, let W be a circled 0-neighborhood in F such that $W + W + W \subset V$, and denote by U the unit ball of E. There exists $u \in M$ such that $v(x) - u(x) \in W$ for all $x \in U$, and since u is compact, $u(U)$ is relatively compact, so that $u(U) \subset \bigcup_i (b_i + W)$ for a suitable finite subset $\{b_i\}$ of $u(U)$. Since W is circled, it follows that $u(U) \subset v(U) + W$, whence $b_i \in a_i + W$ ($i = 1, ..., n$) for a suitable subset $\{a_i\}$ of $v(U)$. Now

$$v(U) \subset u(U) + W \subset \bigcup_1^n (b_i + W + W) \subset \bigcup_1^n (a_i + V),$$

which shows $v(U)$ to be precompact, and hence relatively compact, since F is quasi-complete.

REMARK. The preceding proof shows that if E, F are t.v.s., F is separated and quasi-complete, and v is the limit of compact maps, uniformly on some non-empty open subset of E (or even on a non-meager subset of E if E is a Baire space), then v is compact.

Although the study of adjoint maps is deferred to Chapter IV (Sections 2 and 7), since it can be handled successfully only with the aid of duality, we shall make use in what follows of a few elementary facts with which the reader is likely to be familiar. Recall that if E, F are normed spaces and if E', F' are the Banach spaces that are their respective strong duals, then every $u \in \mathscr{L}(E, F)$ induces a $v \in \mathscr{L}(F', E')$ by means of $y' \to v(y') = y' \circ u$. It is immediate that for all $x \in E$, $y' \in F'$ $|v(y')[x]| \leq \|y'\| \|u\| \|x\|$, whence $\|v\| \leq \|u\|$. An application of the Hahn-Banach theorem (in its analytical form (II, 3.2)) then shows that $\|v\| = \|u\|$. v is called the adjoint of u, and denoted by u'. The following result is due to Schauder.

9.4

Let E, F be normed spaces and let F be complete. A linear map $u \in \mathscr{L}(E, F)$ is compact if and only if its adjoint $u' \in \mathscr{L}(F', E')$ is compact.

Proof. Let U, V be the respective unit balls in E, F and let U^0, V^0 be the corresponding dual unit balls in E', F'. We show that $u'(V^0) = B$ is relatively compact in E'. For this it suffices (since E' is metric) to show that each sequence $\{x'_n\} \subset B$ has a cluster point. Let $x'_n = u'(y'_n)(n \in N)$, where $\{y'_n\} \subset V^0$. Since V^0 is equicontinuous and closed for $\sigma(F', F)$, it is $\sigma(F', F)$-compact by (4.3), Corollary. Hence $\{y'_n\}$ has a weak cluster point y' which, by (4.5), is also a cluster point for the topology of compact convergence. Thus if $A = u(U)$ is relatively compact (i.e., if u is compact), there exist infinitely many $k \in N$ such that $|y'_k(ux) - y'(ux)| \leq \varepsilon$ for all $x \in U$ and a given $\varepsilon > 0$. Let $z' = u'(y')$. It follows that $\|u'(y'_k) - u'(y')\| = \|x'_k - z\| \leq \varepsilon$ for the same k, which shows z' to be a cluster point of the sequence $\{x'_n\}$. The converse is clear, since if u' is compact, then u'' is compact in $\mathscr{L}(E'', F'')$ by the preceding, and u is the restriction of u'' to the subspace E of E''.

We further need the following lemma which, as an inspection of the proof shows, is actually valid in (F)-spaces. We shall confine ourselves to (B)-spaces for convenience, in particular since the first assertion will be obtained for (F)-spaces in a different context (IV, 6.3, Corollary 1).

LEMMA 1. *Let A be a compact subset of the Banach space E. There exists a null sequence $\{x_n\}$ in E whose closed, convex, circled hull contains A; and there exists a compact, convex, circled subset B of E such that A is compact as a subset of E_B.*

Proof. Let $\{\lambda_n\}$ be a sequence of positive numbers such that $\sum_1^\infty \lambda_n = 1$, and put $\varepsilon_n = \lambda_{n+1}^2$. Denote by $P_i (i \in N)$ a finite subset of $A(\neq \varnothing)$ such that for $x \in A$, there exists $y \in P_i$ satisfying $\|x - y\| < \varepsilon_i$; clearly, $\bigcup_1^\infty P_i$ is dense in A. We define a sequence $\{Q_i\}$ of finite subsets of E, as follows: Let Q_1 be the set $\lambda_1^{-1} P_1$. If $i > 1$, select for each $y \in P_i$ an element $z \in P_{i-1}$ satisfying $\|y - z\| < \varepsilon_{i-1}$, and form $x = (y - z)/\lambda_i$; the resulting set (which has the same number of elements as P_i) is Q_i. Since each element of Q_i is of norm $< \lambda_i$ when $i > 1$, the sequence $\{Q_1, Q_2, \ldots\}$ defines a null sequence $\{x_n : n \in N\}$ such that, say, $x_n \in Q_i$ exactly when $n_i \leqq n < n_{i+1}$. It is now readily seen that each $y \in P_k$ is of the form $y = \lambda_1 x_{i_1} + \cdots + \lambda_k x_{i_k}$ and hence in the convex, circled hull of $\bigcup_1^\infty Q_i$, which proves the first assertion.

[1] To prove the second assertion, it is certainly legitimate to assume A is convex (since E is complete), and even that A is the closed, convex hull of the range Q of a suitable null sequence $\{x_n : n \in N\}$ in E. Notice that there exists a sequence $\{\alpha_n : n \in N\}$ of positive numbers such that $\alpha_n \to +\infty$ and $\{\alpha_n x_n : n \in N\}$ is still a null sequence in E; it suffices, for example, to take $\alpha_n = 1/\sqrt{\|x_n\|}$ if $x_n \neq 0$, $\alpha_n = n$ if $x_n = 0$ $(n \in N)$. Let B be the closed, convex, circled hull of the range of $\{\alpha_n x_n : n \in N\}$; B is compact. It is clear that $\{x_n\}$ is a null sequence in E_B; for if p_B is the gauge of B, then $p_B(\alpha_n x_n) \leqq 1$, whence $p_B(x_n) \leqq \alpha_n^{-1}$ for all $n \in N$. To show that A is compact in E_B denote by A_1 the closed, convex hull of Q in E_B; A_1 is compact in E_B and, as a subset of E, dense in A. But since $E_B \to E$ is continuous, A_1 is a fortiori compact in E, and hence identical with A. The lemma is proved.

The results established so far are not quite sufficient to prove (9.5) below; we shall have to use Proposition (IV, 1.2) in two places, and Lemma 2, below. To be sure, Lemma 2 is an easy consequence of (IV, 2.3), Corollary, and (IV, 3.3), but we shall give a direct proof which involves only the geometrical form (II, 3.1) of the Hahn-Banach theorem. For a full understanding of (9.5) the reader is advised to defer reading its proof until he is familiar with the material contained in the first three sections of Chapter IV.

LEMMA 2. *Let* $(E, \mathfrak{T})$ *be a l.c.s.,* $B \neq \varnothing$ *a compact, convex, circled subset of* E, *let* Q *be the subspace of* $[E_B]'$ *whose elements are continuous for the topology induced by* $\mathfrak{T}$, *and let* B^0 *be the unit ball of* $[E_B]'$. *Then* $Q \cap B^0$ *is dense in* B^0 *for the topology of uniform convergence on all compact subsets of* E_B.

Proof. By (4.5) it suffices to show that $Q \cap B^0$ is dense in B^0 for the topology of simple convergence. Let $y' \in B^0$, $\varepsilon > 0$ and $y_i \in B$ $(i = 1, \ldots, n)$ be given. We can assume that $y'(y_i) \neq 0$ for some i. Denote by M the finite dimensional subspace of E_B (and of E) generated by the y_i $(i = 1, \ldots, n)$; $H_1 = \{x \in M : y'(x) = 1 + \varepsilon\}$ is a hyperplane in M and a linear manifold in E, not intersecting B. Since B is compact, there exists a convex, open set V in E containing B and not intersecting H_1. By (II, 3.1) there exists a closed hyperplane H in E containing H_1 and not intersecting V, say, $H = \{x \in E : x'(x) = 1\}$. Since $H \cap B = \varnothing$, the element of Q which x' defines is in B^0. Moreover, $H \cap M = H_1$, whence for $x \in M$, $y'(x) = (1 + \varepsilon)x'(x)$, which implies $|y'(y_i) - x'(y_i)| \leqq \varepsilon$ for all i, since $y_i \in B$.

We can now establish the following theorem (Grothendieck [13]) on the a.p. of Banach spaces and their strong duals.

9.5

Theorem. *Let E be a Banach space and let E′ be its strong dual. Consider the following assertions:*

(a) *E has the approximation property.*

(b) *For every Banach space F, the closure of $F′ \otimes E$ in $\mathscr{L}_b(F, E)$ is identical with the space of compact maps in $\mathscr{L}(F, E)$.*

(c) *E′ has the approximation property.*

(d) *For every Banach space F, the closure of $E′ \otimes F$ in $\mathscr{L}_b(E, F)$ is identical with the space of compact maps in $\mathscr{L}(E, F)$.*

Then (a) ⇔ (b) and (c) ⇔ (d).

REMARK. Assertion (a) is also equivalent to the following: (a′) The canonical map of $E′ \widetilde{\otimes} E$ into $\mathscr{L}(E)$ is one-to-one. With the aid of the equivalence (a) ⇔ (a′), it can also be shown that (c) ⇒ (a); but the proofs require further results on duality. The interested reader is referred to Chapter IV, Exercise 30.

Proof of (9.5). (a) ⇒ (b): If $w \in \mathscr{L}(F, E)$ is compact and V is the unit ball of F, then $A = w(V)$ is precompact in E. Thus there exists $e_0 \in E′ \otimes E$ such that, given $\varepsilon > 0$, $\|e_0(x) - x\| < \varepsilon$ for all $x \in A$. Thus $\|w_0 - w\| \leq \varepsilon$, where $w_0 = e_0 \circ w$. Since w_0 is an element of $F′ \otimes E$ (namely, $\sum w′(x_i′) \otimes x_i$ if $e_0 = \sum x_i′ \otimes x_i$), the implication is proved.

(b) ⇒ (a): Given $\varepsilon > 0$ and a compact subset $A \subset E$, we show the existence of $e_0 \in E′ \otimes E$ satisfying $\|e_0(x) - x\| < \varepsilon$ for $x \in A$. By Lemma 1 above, there exists a compact, convex, circled subset $B \subset E$ such that $A \subset B$ and A is compact in E_B. Letting $F = E_B$, (b) implies that there exists $w_0 \in [E_B]′ \otimes E$ such that $\|w_0 - \psi_B\| < \varepsilon/2$, for the canonical map ψ_B is a compact map of E_B into E. Since by Lemma 2 each $y′ \in [E_B]′$ can be approximated, uniformly on A, by elements $x′ \in E′$, it follows that $\|e_0(x) - w_0(x)\| < \varepsilon/2$ for all $x \in A$ and a suitable $e_0 \in E′ \otimes E$, which implies, by (9.1)(a), that E has the a.p.

(c) ⇒ (d): Let $u \in \mathscr{L}(E, F)$ be a compact map. If we imbed, as usual, E and F as subspaces of their strong biduals $E″$ and $F″$ respectively, then the second adjoint $u″ \in \mathscr{L}(E″, F″)$ is an extension of u. Now the unit ball U of E is $\sigma(E″, E′)$-dense in the unit ball $\overline{U}$ of $E″$. We obtain this result by applying Lemma 2 to the weak dual $E_\sigma′ = (E′, \sigma(E′, E))$ of E, substituting for B the unit ball U^0 of the strong dual $E′$ and using the fact that, by virtue of (IV, 1.2), E is to be substituted for Q. It is an easy matter to verify that $u″$ is continuous for $\sigma(E″, E′)$ and $\sigma(F″, F′)$. On the other hand, since u is compact, $u(U)$ is contained in a compact (hence a fortiori $\sigma(F, F′)$-compact) subset C of F; hence it follows that $u″(\overline{U}) \subset C$, which implies that $u″(E″) \subset F$. (This is also a special case of Chapter IV, Section 9, Lemma 1.)

Now by (9.4), $u′ \in \mathscr{L}(F′, E′)$ is compact; that is, the image under $u′$ of the unit ball of $F′$ is contained in a compact subset A of $E′$. Since $E′$ has the a.p. by hypothesis, there exists a mapping $v_0 = \sum f_i \otimes x_i′ \in E″ \otimes E′$ such that, $\varepsilon > 0$ being preassigned, we have $\|x′ - \sum f_i(x′)x_i′\| < \varepsilon$ for all $x′ \in A$. Now $v_0 \circ u′ =$

$\sum (f_i \circ u') \otimes x_i'$ and we have $f_i \circ u' = u''(f_i) = y_i \in E$ by the preceding. This implies $\|u' - \sum y_i \otimes x_i'\| < \varepsilon$ and, therefore, $\|u - \sum x_i' \otimes y_i\| < \varepsilon$. In view of (9.3) the implication is proved.

(d) $\Rightarrow$ (c): Let A be any compact, convex, circled subset of E'; A is norm bounded, hence equicontinuous, and it follows that $U = \{x \in E: |x'(x)| \leqq 1$ for all $x' \in A\}$ is a convex, circled 0-neighborhood in E. Using the fact that the dual of $(E', \sigma(E', E))$ can be identified with E, (IV, 1.2), it is quickly verified by an application of the Hahn-Banach theorem that $[E']_A$ can be identified with the strong dual of E_U and that under this identification ψ_A is the adjoint map of ϕ_U (Exercise 3). Since ψ_A is compact, so is ϕ_U by (9.4). Hence (d) implies that, given $\varepsilon > 0$, there exists an element $w \in E' \otimes \tilde{E}_U$ satisfying $\|w - \phi_U\| < \varepsilon/2$. Since $\phi_U(E)$ is dense in $\tilde{E}_U$, there exists $w_0 \in E' \otimes \phi_U(E)$ such that $\|w_0 - \phi_U\| < \varepsilon$. Let $w_0 = \sum x_i' \otimes \phi_U(x_i)$. It follows that $\|w_0' - \psi_A\| < \varepsilon$ or, equivalently, that

$$\|x' - \sum x'(x_i)x_i'\| < \varepsilon$$

whenever $x' \in A$; this shows that E' has the a.p.

This completes the proof of (9.5).

When E, F are normed spaces and W, V^0 are the respective unit balls of the Banach spaces E'', F', then the topology of bi-equicontinuous convergence on $E' \otimes F$ is the topology of uniform convergence on $W \otimes V^0$, and hence normable. It is not hard to see (Exercise 24) that the natural norm for this topology is identical with the norm induced by $\mathcal{L}(E, F)$. Hence the following corollary:

COROLLARY. *Let E be a Banach space whose strong dual E' has the approximation property. Then for every Banach space F, the canonical imbedding of the Banach space $E' \tilde{\otimes} F$ into $\mathcal{L}(E, F)$ is a norm isomorphism onto the subspace of compact linear maps of E into F.*

For separable (B)-spaces, a stronger form of the approximation property is obtained as follows: A sequence $\{x_n\} \subset E$ is called a **Schauder basis** if each $x \in E$ has a unique representation

$$x = \sum_{n=1}^{\infty} \alpha_n x_n,$$

where the series converges in E (in the ordinary sense that its partial sums $\sum_1^n \alpha_k x_k$ converge to x as $n \to \infty$; cf. Exercise 23). For example, the sequence $\{e_n\}$ (e_n being the vector $x = (x_1, x_2, ...)$ for which $x_n = 1$, $x_m = 0$ when $m \neq n$) constitutes a Schauder basis in each of the spaces $l^p (1 \leqq p < \infty)$ and c_0 (the space of null sequences under the sup-norm); bases for most standard (B)-spaces were constructed by Schauder [1]. We call a Schauder basis **normalized** if each of its members has norm 1. Clearly, every Schauder basis can be normalized. It is immediate from the postulated unicity of the representation of x by a Schauder basis that the maps $x \to \alpha_n$ ($n \in N$) are linear forms. The following stronger result, a beautiful application of the Banach-Steinhaus and the homomorphism theorem, was essentially known to Banach [1].

9.6

If $\{x_n\}$ is a normalized Schauder basis of the Banach space E, then the coefficient forms $x \to \alpha_n$ ($n \in N$) are equicontinuous linear forms f_n on E, and the expansion $x = \sum_n f_n(x)x_n$ converges uniformly on every compact subset of E.

Proof. Letting $\|x\|_1 = \sup_n \|\sum_{k=1}^{n} \alpha_k x_k\|$, $x \to \|x\|_1$ is a new norm on E under which E is complete. Since $\|x\| \leqq \|x\|_1$, the new norm also generates the topology of E by Corollary 2 of (2.1). It follows that there exists a number $C \geqq 1$ such that $\|x\|_1 \leqq C\|x\|$ for all $x \in E$ (cf. Chapter II, Exercise 5). Now for each $n \in N$,

$$|\alpha_n| = \|\alpha_n x_n\| = \|\sum_{k=1}^{n+1} \alpha_k x_k - \sum_{k=1}^{n} \alpha_k x_k\| \leqq 2\|x\|_1 \leqq 2C\|x\|,$$

which implies that $x \to \alpha_n = f_n(x)$ are equicontinuous linear forms. The remainder is immediate from (4.6).

COROLLARY. *Every (separable) (B)-space that contains a Schauder basis possesses the approximation property.*

This is immediate since (9.6) implies that $e = \sum_1^{\infty} f_n \otimes x_n$, where the series converges in $\mathscr{L}_c(E)$. The preceding considerations can be carried over to separable (F)-spaces without difficulty. For an enlightening discussion of bases in the framework of separable barreled spaces, see Dieudonné [8]. It is not known whether every separable (B)-space has a Schauder basis; this constitutes the **basis problem**. As we have remarked earlier, not every separable (B)-space has an **unconditional basis**, that is, a Schauder basis such that for each x, $\{f_n(x)x_n: n \in N\}$ is summable to x (cf. Exercise 23 for notation); the result is due to Karlin [2]. For the many ramifications of the basis problem in (B)-spaces, we refer to Day [2].

EXERCISES

1. Let L, M be Hausdorff t.v.s., let L_0 be a dense subspace of L, and let u be a continuous linear map of L into M whose restriction u_0 to L_0 is a topological homomorphism; then u is a topological homomorphism. In particular, if u is a topological homomorphism of L into M, and $\bar{u}$ is its continuous extension to the completion $\tilde{L}$ with values in $\tilde{M}$, then $\bar{u}$ is a topological homomorphism. [Note that, in general, $\bar{u}(\tilde{L}) \neq \tilde{M}$ even if $u(L) = M$; cf. Exercise 2, below, and Chapter IV, Exercise 11].

2. Let L be a metrizable t.v.s. and let N be a closed subspace of L; Show that the completion $(L/N)^{\sim}$ is isomorphic (norm isomorphic if L is normed) with $\tilde{L}/\bar{N}$, where $\bar{N}$ is the closure of N in $\tilde{L}$. (By the

method used in the proof of (I, 6.3), show that $\bar{u}(\tilde{L}) = (L/N)^{\sim}$ and that $\bar{N} = \bar{u}^{-1}(0)$, where u denotes the quotient map $L \to L/N$.)

3. Let E be a vector space, let M be a subspace of E, and let U be a radial, convex, circled subset of E. Denote by ϕ_U the canonical map of E into $\tilde{E}_U$. (For notation see the beginning of Section 7.)

(a) If $V = U \cap M$, there exists a natural norm isomorphism of M_V onto the subspace $\phi_U(M)$ of E_U.

(b) Let ϕ be the quotient map $E \to E/M$ and set $W = \phi(U)$. There exists a natural norm isomorphism of $(E/M)_W$ onto E_U/N, where N is the closure of $\phi_U(M)$ in the normed space E_U.

(c) In addition to the hypotheses above, suppose that E is a l.c.s. and that U is a 0-neighborhood in E. Let B be the set of all $x' \in E'$ such that $|x'(x)| \leq 1$ whenever $x \in U$. There exists a natural norm isomorphism of the strong dual of E_U (or of the strong dual of $\tilde{E}_U$) onto $[E']_B$.

4. Let $E(\mathfrak{T}_1)$ be a l.c.s. such that $\mathfrak{T}_1$ is the inductive topology with respect to a family $\{(E_\alpha, g_\alpha): \alpha \in A\}$, where all E_α are Banach spaces and such that $E = \bigcup_\alpha g_\alpha(E_\alpha)$ (e.g., let $E(\mathfrak{T}_1)$ be a quasi-complete bornological space, (II, 8.4)). Let $F(\mathfrak{T}_2)$ be a l.c.s. such that $\mathfrak{T}_2$ is the inductive topology with respect to a sequence $\{(F_n, h_n): n \in N\}$, where all F_n are Fréchet spaces and such that $F = \bigcup_n h_n(F_n)$. Generalize (2.2) as follows:

(a) If v is a linear map of F onto E which is continuous, then v is a topological homomorphism.

(b) If u is a linear map of E into F with closed graph, then u is continuous.

(For a proof, see Grothendieck [13], Intro., theor. B.)

5. Let E be a t.v.s. over K and let F be a Hausdorff t.v.s. over K.

(a) If E_0 denotes the Hausdorff t.v.s. associated with E, and ϕ the canonical map $E \to E_0$, show that $\chi: u \to u \circ \phi$ is an (algebraic) isomorphism of $\mathscr{L}(E_0, F)$ onto $\mathscr{L}(E, F)$; if $\mathfrak{S}$ is a family of bounded subsets of E and $\mathfrak{S}_0 = \phi(\mathfrak{S})$, then χ is a (topological) isomorphism of $\mathscr{L}_{\mathfrak{S}_0}(E_0, F)$ onto $\mathscr{L}_{\mathfrak{S}}(E, F)$.

(b) Suppose that E is Hausdorff, F is complete, and $\tilde{E}$ is the completion of E. For each $u \in \mathscr{L}(E, F)$ denote by $\bar{u}$ the continuous extension of u to $\tilde{E}$. $u \to \bar{u}$ is an isomorphism of $\mathscr{L}(E, F)$ onto $\mathscr{L}(\tilde{E}, F)$, which is topological for the $\mathfrak{S}$-topology and the $\tilde{\mathfrak{S}}$-topology, respectively, if for a family $\mathfrak{S}$ of bounded subsets of E, $\tilde{\mathfrak{S}}$ denotes the family of their closures in $\tilde{E}$.

(c) The isomorphism $u \to \bar{u}$ maps the respective families of equicontinuous subsets onto each other. If $\mathfrak{S}$ and $\mathfrak{S}_1$ are total families of precompact subsets of E and $\tilde{E}$ respectively, this correspondence $H \to H_1$ induces a set of homeomorphisms (and even uniform isomorphisms). (Use (4.5).)

6. Let E be a vector space over K, let $\{E_\alpha: \alpha \in A\}$ be a family of l.c.s. over K, and let $\{g_\alpha: \alpha \in A\}$ be a family of linear maps of E_α into E, respectively, such that $E = \bigcup_\alpha g_\alpha(E_\alpha)$. Denote by $\mathfrak{S}_\alpha$ ($\alpha \in A$) a total family of bounded subsets of E_α. Let $\mathfrak{S} = \bigcup_\alpha g_\alpha(\mathfrak{S}_\alpha)$, provide E with the inductive topology with respect to the class $\{(E_\alpha, g_\alpha): \alpha \in A\}$, and let F be any l.c.s. The $\mathfrak{S}$-topology on $\mathscr{L}(E, F)$ is the projective topology with respect to the mappings $u \to u \circ g_\alpha$ of $\mathscr{L}(E, F)$ into $\mathscr{L}_{\mathfrak{S}_\alpha}(E_\alpha, F)$ ($\alpha \in A$).

Deduce from this that if $E = \oplus_\alpha E_\alpha$, then $\mathscr{L}_{\mathfrak{S}}(E, F)$ is isomorphic with $\prod_\alpha \mathscr{L}_{\mathfrak{S}_\alpha}(E_\alpha, F)$.

7. Let E, F be l.c.s. such that $F \neq \{0\}$, and let $\mathfrak{S}_1$ and $\mathfrak{S}_2$ be saturated families of bounded subsets of E with $\mathfrak{S}_1 \subset \mathfrak{S}_2$. If $\mathfrak{S}_1 \neq \mathfrak{S}_2$, the $\mathfrak{S}_2$-topology is strictly finer than the $\mathfrak{S}_1$-topology on $\mathscr{L}(E, F)$. Deduce from this that, under the conditions stated, the family of $\mathfrak{S}$-topologies on $\mathscr{L}(E, F)$ is in biunivocal correspondence with the saturated families of bounded subsets of E.

8. Let E be a bornological space, and let $\mathfrak{S}$ be a family of bounded subsets of E such that the range of each null sequence in E is contained in some $S \in \mathfrak{S}$. Show that if F is a quasi-complete (respectively, complete) l.c.s., then $\mathscr{L}_{\mathfrak{S}}(E, F)$ is quasi-complete (respectively, complete). (Use (8.3) and Exercise 17, Chapter II.)

9. (Theorem of Osgood). Let X be a non-empty topological space which is a Baire space.

(a) Let $\{f_n\}$ be a pointwise convergent sequence of continuous functions on X with values in a metric space Y; the set of points where the sequence is equicontinuous is not meager in X.

(b) Let $\{f_n\}$ be a simply bounded sequence of continuous functions with values in F, where F is a t.v.s. possessing a fundamental sequence of bounded sets; there exists a subset X_0 of X with non-empty interior such that the sequence is uniformly bounded on X_0.

Deduce from this the classical versions of the principle of uniform boundedness and the Banach-Steinhaus theorem.

10. Show that under the conditions of Chapter II, Exercise 14, the sequence of linear forms $f \to nf(n^{-1})(n \in N)$ on E is simply bounded (and, in fact, uniformly bounded on every convex, circled subset of E which is complete) but not equicontinuous.

11. Let T be a set. A filter $\mathfrak{F}$ on T is *elementary* if it is the section filter of a sequence $\{t_n\}$ in T, that is, if the sets $F_n = \{t_k: k \geq n\}(n \in N)$ form a base of $\mathfrak{F}$. Show that every filter $\mathfrak{G}$ on T which possesses a countable base is the intersection of all elementary filters $\mathfrak{F} \supset \mathfrak{G}$. Use this to extend (4.6) to filters with countable base.

12. (Principle of the Condensation of Singularities. Banach-Steinhaus [1]). Let E, F be t.v.s. such that E is a Baire space. If $H \subset \mathscr{L}(E, F)$ is not equicontinuous, show that $M = \{x: H(x) \text{ is bounded in } F\}$ is a meager subspace of E. Thus if $\{H_n\}$ is a sequence of subsets of $\mathscr{L}(E, F)$ each of which is not simply bounded, there exists $x_0 \in E$ such that $H_n(x_0)$ is unbounded in F for all $n \in N$, and the set of these x_0 is not meager in E.

13. Let $E = \mathscr{C}(I)$ be the Banach space of continuous complex-valued functions on the real unit interval $I = [0, 1]$ and let H be the Hilbert space $L^2(\mu)$, where μ denotes Lebesgue measure on I. Identify E algebraically with a subspace of H, and denote by $\{g_k\}$ any orthonormal basis of H which is also a total family in E; for example, the normalized trigonometric functions. The formal series $\sum_k [f, g_k] g_k$, where $[,]$ denotes the inner product in H, is called the Fourier expansion of f (with respect to $\{g_k\}$). Show that for each given $t_0 \in I$

such that lim sup $L_n(t_0) = \infty$, where $L_n(s) = \int |\sum_1^n g_k(s)\, g_k^*(t)|\, d\mu(t)$, there exists a continuous function $f \in E$ whose Fourier expansion is unbounded (hence not convergent) at $t = t_0$. (Establish the result by applying (4.2).)

14. With the notation of Exercise 13, let P be a countable subset of I and let $\mathfrak{S}$ be a countable family of orthonormal bases in $L^2(\mu)$, each of which satisfies lim sup $L_n(t) = \infty$ for all $t \in P$.

(a) Show that there exists an $f \in E$ whose Fourier expansion with respect to any member of $\mathfrak{S}$ is unbounded at each $t \in P$. (Use Exercises 12, 13.)

(b) Generalize the foregoing result to the case where X is a metrizable, locally compact space, μ a bounded positive measure on X with $\mu(G) > 0$ for each open $G \neq \varnothing$, with $E = \mathscr{C}(X)$, $H = L^2(\mu)$ (Chapter II, Section 2, Examples) and P a countable subset of X such that $\mu(\{t\}) = 0$ for every $t \in P$.

15. Let $\{a_{mn}\}$ be a numerical double sequence such that for each $m \in N$ there exists a summable sequence $\{x_n^{(m)}: n \in N\} \in l^1$ for which $\sum\limits_{n=1}^{\infty} a_{mn} x_n^{(m)}$ is not convergent. There exists a sequence $\{x_n\} \in l^1$ such that $\sum\limits_{n=1}^{\infty} a_{mn} x_n$ is divergent for all $m \in N$. (Use Exercise 12.)

16. Let E, F, G be t.v.s. over K.

(a) Show that a bilinear map on $E \times F$ into G which is continuous at $(0, 0)$, is continuous (everywhere).

(b) Let H be a vector space of $(\mathfrak{S}, \mathfrak{T})$-hypocontinuous bilinear maps of $E \times F$ into G. Show that H is a t.v.s. under the topology of $\mathfrak{S} \times \mathfrak{T}$-convergence (it suffices that each $f \in H$ be either $\mathfrak{S}$- or $\mathfrak{T}$-hypocontinuous).

(c) A family B of bilinear maps of $E \times F$ into G is $\mathfrak{S}$-equihypocontinuous if, for each $S \in \mathfrak{S}$, the family $\{f_x: x \in S, f \in B\}$ is equicontinuous in $\mathscr{L}(F, G)$. Define the corresponding notions of $\mathfrak{T}$- and $(\mathfrak{S}, \mathfrak{T})$-equihypocontinuity, and prove two propositions analogous to (5.2) and (5.3).

17. Let E be an infinite-dimensional normable space and let F be its weak dual $(E', \sigma(E', E))$. The bilinear form $(x, f) \to f(x)$ is $\mathfrak{T}$-hypocontinuous where $\mathfrak{T}$ is the family of all equicontinuous subsets of F; but it is not $\mathfrak{S}$-hypocontinuous for any saturated family $\mathfrak{S}$ of bounded subsets of E other than the one generated by the finite subsets of E, and a fortiori not continuous on $E \times F$. (Let S be a bounded subset of E not contained in a finite-dimensional subspace; whatever the 0-neighborhood $U = \{f: |f(x_i)| \leq 1\}$, there exists $y \in S$ and, by the Hahn-Banach theorem, $f \in U$ such that $f(y)$ is a given number.)

18. Let E, F be vector spaces of respective dimensions d_1, d_2 over K.

(a) The map $x \otimes y \to [x^* \to x^*(x)y]$ induces an isomorphism of $E \otimes F$ into $L(E^*, F)$; similarly, $E \otimes F$ is isomorphic with a subspace of $L(F^*, E)$.

(b) For each $u \in E \otimes F$, the minimal number $k(\geq 0)$ of summands $x_i \otimes y_i$ such that $u = \sum_1^n x_i \otimes y_i$, is identical with the dimension r of the range of v: $x^* \to v(x^*) = \sum_1^n x^*(x_i)y_i$. In each representation of u by r summands, both sets $\{x_i\}$ and $\{y_i\}$ are linearly independent.

(c) The isomorphism of (a) is onto $L(E^*, F)$ if and only if at least one of the cardinals d_1, d_2 is finite.

(d) The dimension of $E \otimes F$ is $d_1 d_2$.

19. Let $E_i(i = 1, \ldots n)$, G be vector spaces over K. A mapping $(x_1, \ldots, x_n) \to f(x_1, \ldots, x_n)$ of $\prod_i E_i$ into G is **multilinear** (*n*-linear) if each of the partial maps $E_i \to G$, obtained by fixing the coordinates $x_j (j \neq i)$ of x_i is linear. Define the tensor product $F = \otimes_i E_i$, and if E_i are l.c.s. define the projective tensor product topology on F. Formulate and prove for this case results analogous to those in Section 6.

20. If E, F are vector spaces and p, q are semi-norms on E, F, respectively, the tensor product $p \otimes q$ is a norm on $E \otimes F$ if and only if both p and q are norms.

21. Let E, F, G be normed spaces with respective norms p, q, r. A norm s on $E \otimes F$ is called a **cross-norm** of p and q if for all $(x, y) \in E \times F$, $s(x \otimes y) = p(x)q(y)$.

(a) The tensor product $p \otimes q$ can be characterized as the unique norm on $E \otimes F$ such that for each normed space (G, r), the canonical isomorphism of $\mathscr{L}(E \otimes F, G)$ onto $\mathscr{B}(E, F; G)$ (cf. (6.2)) is a norm isomorphism, $\mathscr{B}(E, F; G)$ being provided with the norm $f \to \|f\| = \sup\{r[f(x, y)]: p(x) \leq 1, q(y) \leq 1\}$.

(b) If s is any cross-norm of p and q, then $s \leq p \otimes q$.

(c) Denote by E' (respectively, F') the strong dual of E (respectively, of F) with their standard norms p', q'. Show that

$$u \to s(u) = \sup\{\sum x'(x_i)y'(y_i): u = \sum x_i \otimes y_i, p'(x') \leq 1, q'(y') \leq 1\}$$

is a cross-norm on $E \otimes F$ that generates the topology of bi-equicontinuous convergence (Section 6).

(d) If t is a cross-norm of p and q, the following assertions are equivalent:

(1) $s \leq t \leq p \otimes q$, where s is the norm introduced in (c).

(2) The functional $z \to t'(z) = \sup\{|z(w)|: t(w) \leq 1\}$, where $w \in E \otimes F$ and $z \in E' \otimes F'$ is a cross-norm of p' and q'. (Cf. Schatten [1].)

22. Let E, F be l.c.s., χ the canonical bilinear map of $E \times F$ into $E \otimes F$, $\mathfrak{T}_i$ the inductive tensor product topology, and $E \overline{\otimes} F$ the completion of $(E \otimes F, \mathfrak{T}_i)$.

(a) For every l.c.s. G, the isomorphism of (6.1) maps $\mathscr{L}((E \otimes F, \mathfrak{T}_i), G)$ onto $\mathscr{B}(E, F; G)$.

(b) $\mathfrak{T}_i$ is the inductive topology (Chapter II, Section 6) with respect to the family of mappings $\{\chi_x, \chi_y: x \in E, y \in F\}$ of F (respectively, E) into $E \otimes F$. Deduce from this that if E, F are both bornological or both barreled, $E \overline{\otimes} F$ is barreled. (Use Exercise 15, Chapter II).

23. Let E be a Hausdorff t.v.s.

(i) A formal series $\sum_1^\infty x_n$, where $x_n \in E \, (n \in N)$ is **convergent** to $x \in E$ if the sequence of partial sums $s_n = \sum_{k=1}^n x_k \, (n \in N)$ converges to x in E. This is expressed by writing $x = \sum_1^\infty x_n$.

(ii) A family $\{x_\alpha : \alpha \in A\} \subset E$ is **summable** to $x \in E$ if for each 0-neighborhood U in E, there exists a finite subset $\Phi_U \subset A$ such that for each finite set Φ satisfying $\Phi_U \subset \Phi \subset A$, it is true that $\sum_{\alpha \in \Phi} x_\alpha \in x + U$. This is expressed by writing $x = \sum_\alpha x_\alpha$. If $A = N$ and $\{x_n\}$ is summable (a summable sequence), the series $\sum_1^\infty x_n$ is called **unconditionally convergent**.

(iii) Suppose E to be locally convex. A family $\{x_\alpha : \alpha \in A\}$ E is **absolutely summable** if it is summable in E and if for each continuous semi-norm p on E, the family $\{p(x_\alpha) : \alpha \in A\}$ is summable (in $\mathbf{R}$). If $A = N$ and $\{x_n\}$ is absolutely summable, the series $\sum_1^\infty x_n$ is called **absolutely convergent**.

(a) If E is complete, $\{x_\alpha : \alpha \in A\}$ is summable if and only if for each 0-neighborhood U in E, there exists a finite subset $\Phi_U \subset A$ such that $\sum_{\alpha \in \Phi} x_\alpha \in U$ whenever $\Phi \subset A$ is finite and $\Phi \cap \Phi_U = \varnothing$. If E is a complete l.c.s., $\{x_\alpha : \alpha \in A\}$ is absolutely summable if for each member p of a generating family of semi-norms the family $\{p(x_\alpha) : \alpha \in A\}$ is summable.

(b) (ii) and (iii) are equivalent if E is finite dimensional (hence locally convex).

(c) If E is a l.c.s. on which there exists a continuous norm, an absolutely summable family in E cannot contain more than countably many non-zero members.

(d) Let E be a complete l.c.s., A an index set of cardinality $d > 0$, and denote by S_a the subspace of E^A whose elements constitute absolutely summable families in E. Let $\mathscr{P}$ be a generating family of semi-norms on E; under the topology generated by the semi-norms $\mathbf{x} \to \bar{p}(\mathbf{x}) = \sum_\alpha p(x_\alpha) \, (p \in \mathscr{P})$, S_a is a l.c.s. isomorphic with $l_d^1 \, \tilde\otimes \, E$. (Use (6.5).)

(e) Let $\{x_\alpha : \alpha \in A\}$ be a summable family in the l.c.s. E. Show that for each equicontinuous set $B \subset E'$, $\sum_\alpha |x'(x_\alpha)|$ converges uniformly with respect to $x' \in B$.

24. Let E, F be normed spaces and identify $E' \otimes F$ with a subspace of $\mathscr{L}(E, F)$ (Section 7). The topology of bounded convergence on $\mathscr{L}(E, F)$ induces the topology of bi-equicontinuous convergence on $E' \otimes F$ (E' being the strong dual of E). (By an application of the Hahn-Banach theorem (II, 3.2), Corollary, show that for any normed space G and $z \in G$, $\|z\| = \sup\{|z'(z)| : z' \in G', \|z'\| \leqq 1\}$, and use Exercise 21(c).)

25. Let E be a gestufter Raum (Köthe [1]). Algebraically, E is defined as follows: Let $\{\sigma_n : n \in N\}$ be a family of sequences $\sigma_n = (s_1^{(n)}, s_2^{(n)}, \dots)$ of real numbers, such that $0 \leqq s_m^{(n)} \leqq s_m^{(n+1)}$ for all $m, n \in N$ and such that for each m, there exists n satisfying $s_m^{(n)} > 0$. Consider the subspace of K_0^N for whose elements $x = (x_1, x_2, \dots)$ each of the sequences

$\{x_m s_m^{(n)}: m \in N\}(n \in N)$ is summable (Exercise 23). Provided with the topology generated by the semi-norms

$$x \rightarrow p_n(x) = \sum_{m=1}^{\infty} |x_m s_m^{(n)}| \qquad (n \in N)$$

E is an (F)-space. This space is nuclear if and only if for each n, there exists $p \in N$ such that $\sum_{m=1}^{\infty} s_m^{(n)}/s_m^{(n+p)} < +\infty$. (Replace any quotients 0/0 by 0.) (For the proof, use (7.2)(c). Concerning the necessity of the condition, note that the dual E' of E can be identified with the space of sequences, each of which is absolutely majorized by some $n\sigma_n$. If B_n denotes the set of sequences $\{y_m\}$ such that $|y_m| \leq s_m^{(n)}$ for all m, the family $\{nB_n\}$ is a fundamental family of equicontinuous subsets of E'.)

26. A family $\{x_\alpha: \alpha \in A\}$ in a t.v.s. E is **topologically free** if for each $\alpha \in A$, x_α is not contained in the smallest closed subspace of E containing the subfamily $\{x_\beta: \beta \neq \alpha\}$. Let E be a separable (B)-space (more generally, a separable barreled space), and let $\{x_n\}$ be a maximal, topologically free sequence in E. Show that there exists a unique sequence $\{f_n\} \subset E'$ biorthogonal to $\{x_n\}$, and show that $\{x_n\}$ is a Schauder basis of E if for each $x \in E$ and each $g \in E'$, the numerical sequence $\{\sum_{k=1}^{n} f_k(x)g(x_k): n \in N\}$ is bounded.

27. Let S be a set, F a l.c.s. and let G be a vector space of functions on S into F that are bounded on S. Let V be a fixed 0-neighborhood in F and let Z be a subset of G such that for each finite subset $H \subset Z$, there exists $x \in S$ satisfying $f(x) \notin V$ whenever $f \in H$. The complements of all these sets Z (as V runs through a base of 0-neighborhoods in F) form a base at 0 for a locally convex topology on G, called the topology of **almost uniform convergence** on S (Brace [1]). If E, F are Banach spaces, a map $u \in \mathcal{L}(E, F)$ is compact if and only if it is a cluster point, for the topology of almost uniform convergence on the unit ball of E, of a sequence in $E' \otimes F$. (Brace [2].)

Chapter IV

DUALITY

The study of a locally convex space in terms of its dual is the central part of the modern theory of topological vector spaces, for it provides the setting for the deepest and most beautiful results of the subject; the present elaborate form of duality theory is largely due to Bourbaki [8] (cf. also Dieudonné [1] and Dieudonné-Schwartz [1]). The first five sections of this chapter contain the basic information, the remaining six being concerned with more refined and advanced results; as in the other chapters of the book, supplementary information can be found in the exercises. We proceed to survey the chapter briefly.

Section 1 is concerned with weak topologies, the bipolar theorem and its first consequences. Section 2 follows with a brief discussion of the adjoint of a weakly continuous linear map. Section 3 presents the Mackey-Arens theorem characterizing the locally convex topologies consistent with a given duality, and Mackey's theorem on the identity of the respective families of bounded sets for these topologies. The duality of subspaces and quotients and the duality of products and direct sums are discussed in detail in Section 4; one obtains the permanence properties of the weak and Mackey topologies, and some applications are made to the duality of projective and inductive limits. Section 5 introduces the strong topology on the dual of a locally convex space, then turns to the discussion of the strong dual and the bidual. This includes a detailed study of semi-reflexive and reflexive spaces, and the section concludes with a short discussion of Montel spaces (for these, see also Exercise 19).

Section 6 presents Grothendieck's completeness theorem and some related further going results on metrizable l.c.s., in particular, the theorems of Banach-Dieudonné and Krein-Šmulian; it ends with a brief discussion of Grothendieck's (DF)-spaces (see also Exercises 24, 32, 33). Section 7 continues the study of adjoints, now for densely defined closed linear maps of one l.c.s. into another. There are dual characterizations of continuous and open linear maps, followed by several results relating to Fréchet and Banach spaces. The section also paves the way for the general open mapping and closed graph theorems derived in Section 8. These theorems, essentially due to Pták, show a rather unexpected relationship between Banach's homomorphism theorem and the theorem of Krein-Šmulian; they provide an excellent example of the power of duality theory. Section 9 continues the study of topological tensor products and nuclear spaces from Chapter III, presenting several fundamental results on nuclear spaces. Section 10 is devoted to a study of the relationship between the concepts of absolute summability and nuclear space, and opens an approach (due to Pietsch) to nuclear spaces independently of the theory of topological tensor products. As a by-result one obtains the theorem of Dvoretzky-Rogers (for a sharpened form of the theorem see Exercise 36). The chapter concludes with a section on weak compactness, a subject that has received a great deal of attention in the literature; included are the theorems of Eberlein and Krein in their general versions due to Dieudonné and Grothendieck.

1. DUAL SYSTEMS AND WEAK TOPOLOGIES

Let F, G be a pair of vector spaces over K, and let f be a bilinear form on $F \times G$ satisfying the separation axioms:

(S_1) $f(x_0, y) = 0$ *for all* $y \in G$ *implies* $x_0 = 0$.
(S_2) $f(x, y_0) = 0$ *for all* $x \in F$ *implies* $y_0 = 0$.

The triple (F, G, f) is called a **dual system** or **duality** (over K). It is also customary to say that f places F and G in duality, or separated duality if the validity of (S_1) and (S_2) is to be stressed. To distinguish f from other bilinear forms on $F \times G$, f is called the **canonical bilinear form** of the duality, and is usually denoted by $(x, y) \to \langle x, y \rangle$. The triple $(F, G, \langle\ ,\ \rangle)$ is more conveniently denoted by $\langle F, G \rangle$.

Examples

1. Let E be a vector space and let E^* be its algebraic dual; the bilinear form $(x, x^*) \to x^*(x) = \langle x, x^* \rangle$ defines the duality $\langle E, E^* \rangle$.
2. If E is a locally convex space with (topological) dual E', then E' is a subspace of E^* separating points in E by Corollary 1 of (II, 4.2); hence the duality $\langle E, E^* \rangle$ of Example 1 induces a duality $\langle E, E' \rangle$ on the subspace $E \times E'$ of $E \times E^*$.
3. Let E, F be l.c.s. with respective duals E', F'. The algebraic tensor products $E \otimes F$ and $E^* \otimes F^*$ are in duality with respect to the bilinear form determined by $\langle x \otimes y, x^* \otimes y^* \rangle = \langle x, x^* \rangle \langle y, y^* \rangle$. It has been shown in Chapter III (Section 6) that this duality induces a duality between $E \otimes F$ and $E' \otimes F'$ (cf. Exercise 2).

4. Denote by ϕ_d and ω_d, respectively, the direct sum and product of d copies of the scalar field K. For any vector space λ over K such that $\phi_d \subset \lambda \subset \omega_d$, let $\lambda^\times$ be the subspace of ω_d such that $y = (y_\alpha) \in \lambda^\times$ whenever the family $\{x_\alpha y_\alpha\}$ is summable for every $x = (x_\alpha) \in \lambda$. The bilinear form $(x, y) \to \sum_\alpha x_\alpha y_\alpha$ places λ and $\lambda^\times$ in duality (Exercise 5).

If $\langle F, G \rangle$ is a duality, the mapping $x \to \langle x, y \rangle$ is, for each $y \in G$, a linear form f_y on F. Since $y \to f_y$ is linear and, by virtue of (S_2), biunivocal, it is an isomorphism of G into the algebraic dual F^* of F; thus G can be identified with a subspace of F^*. This identification will be made in the following unless the contrary is explicitly stated. Note that under this identification, the canonical bilinear form of $\langle F, G \rangle$ is induced by the canonical bilinear form of $\langle F, F^* \rangle$ (Example 1, above).

In this and the following sections, any proposition on F can also be made on G by simply interchanging the roles of F and G; this is immediate from the symmetry of $\langle F, G \rangle$ with respect to F and G, and will not be repeated. We begin our investigation with a simple algebraic observation; δ_{ij} is, as usual, the Kronecker symbol.

1.1

Let $\langle F, G \rangle$ be a duality and let $\{y_i: i = 1, ..., n\}$ $(n \in \mathbf{N})$ be a linearly indepen-dent subset of G. There exist n (necessarily linearly independent) elements $x_i \in F$ such that $\langle x_i, y_j \rangle = \delta_{ij}$ $(i, j = 1, ..., n)$.

Proof. The proof is by induction with respect to n. By (S_2) the assertion holds for $n = 1$; if $n > 1$ there exists by assumption a set $\{\bar{x}_i: i = 1, ..., n - 1\}$ for which $\langle \bar{x}_i, y_j \rangle = \delta_{ij}$ $(i, j = 1, ..., n - 1)$. Let M_n be the subspace generated by the elements $\bar{x}_i$ $(1 \leq i \leq n - 1)$ and let $F_n = \{x \in F: \langle x, y_j \rangle = 0, j = 1, ..., n - 1\}$. Clearly, $F = F_n + M_n$ is an algebraic direct sum. Now y_n cannot vanish on F_n, or else it would be a linear combination of $\{y_i: i = 1, ..., n - 1\}$. Hence there exists $x_n \in F_n$ such that $\langle x_n, y_n \rangle = 1$ and, defining x_i $(i = 1, ..., n - 1)$ by $x_i = \bar{x}_i - \langle \bar{x}_i, y_n \rangle x_n$, we obtain the desired set $\{x_i: i = 1, ..., n\}$.

COROLLARY. *Let F be a vector space, and let f_i $(i = 1, ..., n)$ and g be linear forms on F such that the relations $f_i(x) = 0$, $i = 1, ..., n$ imply $g(x) = 0$ (equiva-lently, such that $\bigcap_{i=1}^{n} f_i^{-1}(0) \subset g^{-1}(0)$). Then g is a linear combination of the forms f_i $(i = 1, ..., n)$.*

We recall (Chapter II, Section 5) that the weak topology $\sigma(F, G)$ is the coarsest topology on F for which the linear forms $x \to \langle x, y \rangle$, $y \in G$ are continuous; by (S_1) F is a l.c.s. under $\sigma(F, G)$. If B is any Hamel basis of G, the topology $\sigma(F, G)$ is generated by the semi-norms $x \to |\langle x, y \rangle|$, $y \in B$.

1.2

The dual of $(F, \sigma(F, G))$ is G; that is, a linear form f on F is $\sigma(F, G)$-con-tinuous if and only if it is of the form $f(x) = \langle x, y \rangle$ for a (unique) $y \in G$.

Proof. In view of the definition of $\sigma(F, G)$, we have to show only that a given continuous f can be written as indicated. By (III, 1.1) there exist elements $y_i \in G$ ($i = 1, ..., n$) such that $|f(x)| \leq c \sup_i |\langle x, y_i \rangle|$ for all $x \in F$ and a suitable constant c. Viewing the y_i as linear forms on F, the corollary of (1.1) shows f to be a linear combination of the y_i, whence the proposition follows.

COROLLARY. *Let $\langle F, G \rangle$ and $\langle F, G_1 \rangle$ be dual systems such that $G_1 \subset G$. Unless $G_1 = G$, $\sigma(F, G_1)$ is strictly coarser than $\sigma(F, G)$.*

1.3

Let $\langle F, G \rangle$ be a duality and let G_1 be a subspace of G; the canonical bilinear form of $\langle F, G \rangle$ places F and G_1 in duality if and only if G_1 is $\sigma(G, F)$-dense in G.

Proof. To prove the sufficiency of the condition, we have to show that the canonical bilinear form satisfies (S_1) on $F \times G_1$. If G_1 is weakly dense in G and $\langle x_0, y \rangle = 0$ for all $y \in G_1$, then the $\sigma(G, F)$-continuity of $y \to \langle x_0, y \rangle$ implies that $\langle x_0, y \rangle = 0$ for all $y \in G$, whence $x_0 = 0$.

For the necessity of the condition, suppose that $\langle F, G \rangle$ induces a duality between F and G_1. If G_1 were not dense in $(G, \sigma(G, F))$, there would exist a $y_0 \in G$ not contained in the closure $\bar{G}_1$ of G_1. Define a linear form f on $\bar{G}_1 + [y_0]$ (where $[y_0]$ is the one-dimensional subspace of G generated by y_0) by $f(y) = 0$ when $y \in \bar{G}_1$ and $f(y_0) = 1$; f is $\sigma(G, F)$-continuous on its domain by (I, 4.2), hence by (II, 4.2) has a continuous extension $\bar{f}$ to G. By (1.2) $\bar{f}(y) = \langle x_0, y \rangle$ for all $y \in G$ and an $x_0 \in F$. Since (S_1) holds on $F \times G_1$ by assumption, it follows that $x_0 = 0$, which conflicts with $f(y_0) = 1$.

COROLLARY. *Let F be a vector space and let G be a subspace of F^*. $\langle F, F^* \rangle$ induces a duality between F and G if and only if G is $\sigma(F^*, F)$-dense in F^*.*

Let $\langle F, G \rangle$ be a duality. For any subset M of F,

$$M^\circ = \{ y \in G : \operatorname{Re}\langle x, y \rangle \leq 1 \quad \text{if } x \in M \},$$

where $\operatorname{Re}\langle x, y \rangle$ denotes the real part of $\langle x, y \rangle$, is a subset of G, called the **polar set** (or **polar**) of M. The **absolute polar** of M is the polar of the circled hull of M; it is the subset $\{ y : |\langle x, y \rangle| \leq 1 \text{ if } x \in M \}$ of G. The following facts are immediate consequences of this definition:

1. $\varnothing^\circ = G$ and $F^\circ = \{0\}$.
2. If $\lambda \neq 0$ and $\lambda M \subset N$ then $N^\circ \subset \lambda^{-1} M^\circ$.
3. For any family $\{M_\alpha\}$ of subsets of F, $[\bigcup M_\alpha]^\circ = \bigcap M_\alpha^\circ$.
4. If $\mathfrak{S}$ is any saturated family of $\sigma(F, G)$-bounded subsets of F, the family of polars $\{S^\circ : S \in \mathfrak{S}\}$ is a 0-neighborhood base for the $\mathfrak{S}$-topology on G. ($G = \mathscr{L}((F, \sigma(F, G)), K_0).$)
5. If L is a t.v.s., a subset M of its dual L' is equicontinuous if and only if the polar M° (with respect to the duality $\langle L, L^* \rangle$) is a 0-neighborhood in L.

The proof of these statements as well as of the following result is left to the reader.

1.4

For any subset $M \subset F$, $M°$ is a $\sigma(G, F)$-closed, convex subset of G containing 0. If M is circled, then so is $M°$; if M is a subspace of F, $M°$ is a subspace of G.

If $M \subset F$, the polar of $M°$ is a subset of F, called the **bipolar** of M, and is denoted by $M°°$; accordingly, the polar of $M°°$ is denoted by $M°°°$. The following result, called the bipolar theorem, is a consequence of the Hahn-Banach theorem and is an indispensable tool in working with dualities.

1.5

Theorem. *Let $\langle F, G \rangle$ be a duality. For any subset $M \subset F$, the bipolar $M°°$ is the $\sigma(F, G)$-closed, convex hull of $M \cup \{0\}$.*

Proof. It follows from (1.4) that $M°°$ is $\sigma(F, G)$-closed, convex, and contains 0; obviously it also contains M. Thus $M_1 \subset M°°$ if M_1 is the closed, convex hull of $M \cup \{0\}$; the assertion will be proved when we show that $x \notin M_1$ implies $x \notin M°°$. Suppose that $x_0 \notin M_1$. By the second separation theorem (II, 9.2), there exists a closed real hyperplane separating M_1 and $\{x_0\}$ strictly. Since $0 \in M_1$, H is of the form $H = \{x \in F: f(x) = 1\}$ for a suitable $\sigma(F, G)$-continuous real linear form f on F. It follows from (1.2) and (I, 7.2) that $f(x) = \text{Re}\langle x, y_0 \rangle$ for all $x \in F$ and some $y_0 \in G$. Now since $0 \in M_1$, we have $\text{Re}\langle x, y_0 \rangle < 1$ if $x \in M_1$, hence $\text{Re}\langle x_0, y_0 \rangle > 1$; it follows that $y_0 \in M_1° \subset M°$, whence $x_0 \notin M°°$.

COROLLARY 1. *For any $M \subset F$, $M°°° = M°$.*

COROLLARY 2. *Let $\{M_\alpha: \alpha \in A\}$ be a family of $\sigma(F, G)$-closed, convex subsets of F, each containing 0, and let $M = \bigcap_\alpha M_\alpha$; then the polar of M is the $\sigma(G, F)$-closed, convex hull of $\bigcup_\alpha M_\alpha°$.*

Proof. Let N be the closed, convex hull of $\bigcup M_\alpha°$. Since $M_\alpha°° = M_\alpha$ ($\alpha \in A$) it follows that $N° = [\bigcup M_\alpha°]° = \bigcap M_\alpha°° = \bigcap M_\alpha = M$ (the first of these equalities holding by (1.5) and Corollary 1, the second by Remark 3 preceding (1.4)), hence $M° = N°° = N$ as was to be shown.

It is clear that for any t.v.s. L, the polars (taken in L^*) of a 0-neighborhood base form a fundamental family of equicontinuous sets (cf. Remark 5 above). For locally convex spaces, the converse is also true:

COROLLARY 3. *If E is a l.c.s., then the polars (taken with respect to $\langle E, E' \rangle$) of any fundamental family of equicontinuous sets in E' form a neighborhood base of 0 in E.*

Proof. Let $\mathfrak{S}$ be a fundamental family of equicontinuous subsets of E' and let U be a given 0-neighborhood in E; since E is locally convex, U can be assumed closed and convex, hence $\sigma(E, E')$-closed by (II, 9.2), Corollary 1.

Since U° is equicontinuous, there exists $S \in \mathfrak{S}$ with $U^\circ \subset S$; it follows that $S^\circ \subset U^{\circ\circ} = U$, which proves the assertion.

In slightly greater generality, the last corollary can be restated as follows:

COROLLARY 4. *If (and only if) E is a t.v.s. whose topology $\mathfrak{T}$ is locally convex, $\mathfrak{T}$ is the topology of uniform convergence on the equicontinuous subsets of E^*.*

It follows from the bipolar theorem that for subspaces $M \subset F$, $M = M^{\circ\circ}$ if and only if M is closed for $\sigma(F, G)$. Hence the mapping $M \to M^\circ$ is one-to-one from the family of all $\sigma(F, G)$-closed subspaces of F to the family of all $\sigma(G, F)$-closed subspaces of G. More precisely, $M \to M^\circ$ is an anti-isomorphism of the lattice of closed subspaces of F onto the lattice of closed subspaces of G, the lattice operations being defined by $\inf(M_1, M_2) = M_1 \cap M_2$ and $\sup(M_1, M_2) = (M_1 + M_2)^-$. For it is immediate from (1.5) and its corollaries that $(\inf(M_1, M_2))^\circ = \sup(M_1^\circ, M_2^\circ)$ and $(\sup(M_1, M_2))^\circ = \inf(M_1^\circ, M_2^\circ)$. It is customary to call the polar M° of a subspace $M \subset F$ the subspace of G **orthogonal** to M (with respect to the duality $\langle F, G \rangle$). If $F = M_1 + M_2$ is the algebraic direct sum of the closed subspaces M_1 and M_2, then $G = (M_1^\circ + M_2^\circ)^-$ by the preceding; it will be seen below (Section 2) that the sum is $\sigma(F, G)$-topological, $F = M_1 \oplus M_2$, if and only if $G = M_1^\circ \oplus M_2^\circ$ for $\sigma(G, F)$.

The most important and most frequent dualities are the systems $\langle E, E' \rangle$, where E is a given l.c.s. (Example 2 above). Note that every dual system $\langle F, G \rangle$ can be interpreted in this way; by (1.2) it suffices to endow F with $\sigma(F, G)$ and consider G as the dual F' of F. Recall that the weak dual of a l.c.s. (more generally, of a t.v.s.) E is the l.c.s. $(E', \sigma(E', E))$; we shall find it convenient to denote this space by E'_σ. The section is concluded with two useful results on E'_σ.

1.6

If E is any l.c.s., the family of all barrels in E and the family of all bounded subsets of E'_σ that are closed, convex, and circled, correspond to each other by polarity (with respect to $\langle E, E' \rangle$).

Proof. Let D be a barrel in E, then D° is bounded in E'_σ. Since by (II, 9.2), Corollary 2, D is closed for $\sigma(E, E')$, it follows from (1.5) that $D = D^{\circ\circ}$ with respect to $\langle E, E' \rangle$; it is hence sufficient to show that for each bounded, closed, convex, circled subset B of E'_σ, B° is a barrel in E. In view of (1.4) there remains to prove only that B° is radial. Let $x \in E$; then $\{x\}^\circ$ is a 0-neighborhood in E'_σ, hence there exists $\lambda > 0$ such that $B \subset \lambda^{-1}\{x\}^\circ = \{\lambda x\}^\circ$. This implies $\lambda x \in B^\circ$, completing the proof.

In view of Corollary 3 of (1.5) there results the following dual characterization of barreled spaces (for a strengthened form, cf. Section 5).

COROLLARY. *A l.c.s. E is barreled if and only if each bounded subset of E'_σ is equicontinuous.*

1.7

Let E be a separable t.v.s. Every closed equicontinuous subset of E'_σ is a compact metrizable space (under the induced topology); if, in addition, E is metrizable then E'_σ is separable.

Proof. The first assertion is a special case of (III, 4.7) for $F = K_0$. If E is metrizable, denote by $\{U_n : n \in N\}$ a neighborhood base of 0. Since each polar U°_n is equicontinuous and closed in E'_σ. U°_n is a compact metrizable, hence separable space for the topology induced by $\sigma(E', E)$. Let A_n be a countable, dense subset of U°_n $(n \in N)$. Since, clearly, each $x' \in E'$ is contained in some U°_n (i.e., since $\{U^\circ_n\}$, being fundamental, covers E'), it follows that the countable set $A = \bigcup_1^\infty A_n$ is dense in E'_σ.

2. ELEMENTARY PROPERTIES OF ADJOINT MAPS

Let F and F_1 be vector spaces over K and let u be a linear map of F into F_1. For each element y^* of F^*_1, the mapping $y^* \circ u : x \rightarrow \langle ux, y^* \rangle$ is a linear form $x^* \in F^*$ (we write ux in place of $u(x)$) commonly denoted by u^*y^*. Thus we have the identity

$$\langle ux, y^* \rangle = \langle x, u^*y^* \rangle$$

on $F \times F^*_1$; the map $y^* \rightarrow u^*y^*$, evidently linear on F^*_1 into F^*, is called the **algebraic adjoint** u^* of u.

2.1

Let $\langle F, G \rangle$ and $\langle F_1, G_1 \rangle$ be dualities over K. A linear map u of F into F_1 is continuous for $\sigma(F, G)$ and $\sigma(F_1, G_1)$ if and only if $u^*(G_1) \subset G$. In this case the restriction u' of u^* to G_1 is continuous for $\sigma(G_1, F_1)$ and $\sigma(G, F)$, and $u'' = (u')' = u$.

Proof. If $u^*(G_1) \subset G$, then $x \rightarrow \langle ux, y' \rangle = \langle x, u'y' \rangle (x \in F, y' \in G_1)$ is continuous for $\sigma(F, G)$, which implies, by definition of $\sigma(F_1, G_1)$, the continuity of u for these two topologies. Conversely, if u is continuous for $\sigma(F, G)$ and $\sigma(F_1, G_1)$, then $x \rightarrow \langle ux, y' \rangle = \langle x, u^*y' \rangle$ is continuous for $\sigma(F, G)$, whence $u^*y' \in G$ by (1.2). By virtue of the identity

$$\langle ux, y' \rangle = \langle x, u'y' \rangle \qquad (x \in F, y' \in G_1)$$

it is now obviously true that u' is continuous for $\sigma(G_1, F_1)$ and $\sigma(G, F)$; the final assertion follows by symmetry.

If $u^*(G_1) \subset G$, the mapping u' is called the **adjoint** (transpose, dual map) of u with respect to the dualities $\langle F, G \rangle$ and $\langle F_1, G_1 \rangle$. Sometimes the relation $u^*(G_1) \subset G$, equivalent with the weak continuity of u, is expressed by saying that the adjoint of u "exists". Notice also that every linear map u of F into F_1 is continuous for $\sigma(F, F^*)$ and $\sigma(F_1, F^*_1)$, these topologies being the weak topologies associated with the finest locally convex topology on F and F_1, respectively, (Chapter II, Section 6 and Exercise 7).

We consider an example. Suppose that $\langle F, G \rangle$ is a duality and $F = M_1 + M_2$ an algebraic direct sum; denote polars with respect to $\langle F, G \rangle$ by $\circ$, and polars with respect to $\langle F, F^* \rangle$ by $\bullet$. Let p be the projection of F onto M_1 that vanishes on M_2. It follows readily from the identity $\langle px, x^* \rangle = \langle x, p^*x^* \rangle$ on $F \times F^*$ that p^* is a projection with range $M_2^\bullet$ and null space $M_1^\bullet$; hence $F^* = M_1^\bullet + M_1^\bullet$ is an algebraic direct sum; if M_1, M_2 are $\sigma(F, G)$-closed, $M_1^\circ + M_2^\circ$ is $\sigma(G, F)$-dense in G by (1.5), Corollary 2. In order that $G = M_1^\circ + M_2^\circ$ it is necessary and sufficient that $p^*(G) = M_2^\circ$ which, by virtue of $M_2^\circ = M_2^\bullet \cap G$, is equivalent with $p^*(G) \subset G$; this, in turn, is equivalent with the $\sigma(F, G)$-continuity of p. Hence an algebraic direct sum $F = M_1 + M_2$ of two closed subspaces is $\sigma(F, G)$-topological if and only if $G = M_1^\circ + M_2^\circ$; in that case, this decomposition of G is $\sigma(G, F)$-topological, and the projections $p: F \to M_1$ and $p': G \to M_2^\circ$ are mutually adjoint.

The proof of the following simple result, recorded for reference, is omitted.

2.2

Let $\langle F_i, G_i \rangle$ $(i = 1, 2, 3)$ *be dualities over* K, *and denote by* u_i *a weakly continuous linear map of* F_i *into* F_{i+1} $(i = 1, 2)$. *The adjoint of* $w = u_2 \circ u_1$ *is* $w' = u_1' \circ u_2'$.

2.3

Let $\langle F, G \rangle$ *and* $\langle F_1, G_1 \rangle$ *be dualities, let* u *be a weakly continuous linear map of* F *into* F_1 *with adjoint* u', *and let* A, B *be subsets of* F, F_1 *respectively. Then the following relations hold:*

(a) $[u(A)]^\circ = (u')^{-1}(A^\circ)$.

(b) $u(A) \subset B$ *implies* $u'(B^\circ) \subset A^\circ$.

(c) *If* A *and* B *are weakly closed, convex sets containing* 0, *then* $u'(B^\circ) \subset A^\circ$ *implies* $u(A) \subset B$.

Proof. (a) $[u(A)]^\circ = \{ y' \in G_1 : \text{Re}\langle ux, y' \rangle \leqq 1$ if $x \in A \} = \{ y' \in G_1 : \text{Re}\langle x, u'y' \rangle \leqq 1$ if $x \in A \} = (u')^{-1}(A^\circ)$.

(b) $u(A) \subset B$ implies $B^\circ \subset [u(A)]^\circ = (u')^{-1}(A^\circ)$, which implies $u'(B^\circ) \subset A^\circ$.

(c) By (b), $u''(A^{\circ\circ}) \subset B^{\circ\circ}$, hence $u(A) \subset B$ in view of $u'' = u$ and the bipolar theorem (1.5) by which $A = A^{\circ\circ}$, $B = B^{\circ\circ}$.

COROLLARY. *The null space* $(u')^{-1}(0)$ *is the subspace of* G_1 *orthogonal to the range* $u(F)$ *of* u. *In particular,* u' *is one–to–one if and only if* $u(F)$ *is* $\sigma(F_1, G_1)$-*dense in* F_1.

The corollary is immediate in view of (2.3)(a) and the bipolar theorem. If E, F are l.c.s. (more generally, t.v.s.) and u is a continuous linear map of E

into F, then u is continuous for the associated weak topologies $\sigma(E, E')$ and $\sigma(F, F')$; in fact, for any $y' \in F'$ the linear form $x \to \langle ux, y' \rangle$ is continuous on E, hence continuous for $\sigma(E, E')$ by the definition of $\sigma(E, E')$, which implies the weak continuity of u. Thus each $u \in \mathcal{L}(E, F)$ has an adjoint u' which is a continuous linear map of the weak dual F'_σ into E'_σ. More generally we prove the following result on the continuity of an adjoint map.

2.4

Let $\langle F, G \rangle$ and $\langle F_1, G_1 \rangle$ be dualities and let u be a weakly continuous linear map of F into F_1 with adjoint u'. Let $\mathfrak{S}$ and $\mathfrak{S}_1$ be saturated families of $\sigma(F, G)$-bounded (respectively, $\sigma(F_1, G_1)$-bounded) subsets of F and F_1, and denote by $\mathfrak{T}$ the $\mathfrak{S}$-topology on G, by $\mathfrak{T}_1$ the $\mathfrak{S}_1$-topology on G_1. Then u' is continuous on $(G_1, \mathfrak{T}_1)$ into $(G, \mathfrak{T})$ if and only if $u(\mathfrak{S}) \subset \mathfrak{S}_1$.

Proof. Let $\hat{\mathfrak{S}}$ and $\hat{\mathfrak{S}}_1$, respectively, be the families of all weakly closed, convex, circled sets contained in $\mathfrak{S}$ and $\mathfrak{S}_1$. Since $\hat{\mathfrak{S}}$ and $\hat{\mathfrak{S}}_1$ are fundamental subfamilies of $\mathfrak{S}$ and $\mathfrak{S}_1$, respectively, the families $\mathfrak{U} = \{S^\circ \colon S \in \hat{\mathfrak{S}}\}$ and $\mathfrak{U}_1 = \{S_1^\circ \colon S_1 \in \hat{\mathfrak{S}}_1\}$ are 0-neighborhood bases for $\mathfrak{T}$ and $\mathfrak{T}_1$. By virtue of (b) and (c) of (2.3), the relations $u(S) \subset S_1$ and $u'(S_1^\circ) \subset S^\circ$ are equivalent for all non-empty $S \in \hat{\mathfrak{S}}$ and $S_1 \in \hat{\mathfrak{S}}_1$, which proves the assertion.

3. LOCALLY CONVEX TOPOLOGIES CONSISTENT WITH A GIVEN DUALITY. THE MACKEY-ARENS THEOREM

Let $\langle F, G \rangle$ denote a given duality over K and let $\mathfrak{S}$ be a family of $\sigma(G, F)$-bounded subsets of G. The topology of uniform convergence on the sets $S \in \mathfrak{S}$ (Chapter III, Section 3) is a locally convex topology on the space $F = \mathcal{L}((G, \sigma(G, F)), K_0)$, where K_0, as always, denotes the one-dimensional t.v.s. associated with the scalar field K. Recall that the $\mathfrak{S}$-topology and the $\bar{\mathfrak{S}}$-topology are identical if $\bar{\mathfrak{S}}$ denotes the saturated hull of $\mathfrak{S}$ (Chapter III, Section 3).

A locally convex topology $\mathfrak{T}$ on F is called **consistent** with the duality $\langle F, G \rangle$ if the dual of $(F, \mathfrak{T})$ is identical with G (G being viewed as a subspace of F^*, Section 1). It follows that a topology on F consistent with $\langle F, G \rangle$ is finer than $\sigma(F, G)$, hence Hausdorff; $\sigma(F, G)$ is by its definition the coarsest consistent topology on F. An immediate consequence of this definition and (II, 9.2), Corollary 2 is the following frequently used fact.

3.1

The closure of a convex subset $C \subset F$ is the same for all (locally convex) topologies on F consistent with $\langle F, G \rangle$ (and hence identical with the $\sigma(F, G)$-closure of C).

It follows from Corollary 3 of the bipolar theorem (1.5) that every topology $\mathfrak{T}$ on F consistent with $\langle F, G \rangle$ is an $\mathfrak{S}$-topology, namely the topology of uniform convergence on the (saturated) class of $\mathfrak{T}$-equicontinuous subsets of G; by the theorem of Alaoglu-Bourbaki, these classes consist entirely of sets relatively compact for $\sigma(G, F)$. The following result, due to G. W. Mackey [5] and R. Arens [1], asserts that conversely, every saturated family, covering G, of $\sigma(G, F)$-relatively compact sets is eligible to be the class of equicontinuous sets for a consistent topology on F. These classes of subsets of G, being saturated, are thus in biunivocal correspondence with the locally convex topologies on F consistent with $\langle F, G \rangle$ (Chapter III, Exercise 7).

3.2

Theorem. *A (locally convex) topology $\mathfrak{T}$ on F is consistent with a given duality $\langle F, G \rangle$ if and only if $\mathfrak{T}$ is the $\mathfrak{S}$-topology for a saturated class $\mathfrak{S}$, covering G, of $\sigma(G, F)$-relatively compact subsets of G.*

Proof. The necessity of the condition has been noted above; let us prove its sufficiency. Let $\mathfrak{S}$ be a saturated family of subsets of G which covers G and consists of $\sigma(G, F)$-relatively compact sets. It follows from (III, 3.2) that F is a l.c.s. under the $\mathfrak{S}$-topology which, since $\mathfrak{S}$ covers G, is finer than $\sigma(F, G)$. ($\sigma(F, G)$ is the topology of simple convergence.) Hence the dual of F with respect to the $\mathfrak{S}$-topology contains G; we have to show its identity with G. Let f be a linear form on F continuous for the $\mathfrak{S}$-topology; the polar $\{f\}^\circ$ (taken with respect to $\langle F, F^* \rangle$) is, by the continuity of f, a 0-neighborhood for the $\mathfrak{S}$-topology and hence contains a set S°, where S, since $\mathfrak{S}$ is saturated, can be assumed convex, circled, and $\sigma(G, F)$-compact. Thus $f \in S^{\circ\circ}$, where the bipolar is taken with respect to $\langle F, F^* \rangle$, and (1.5) implies that $S = S^{\circ\circ}$, since S, being $\sigma(G, F)$-compact, is compact and hence closed in $(F^*, \sigma(F^*, F))$. It follows that $f \in S \subset G$, completing the proof.

COROLLARY 1. *There exists a finest locally convex topology on F consistent with $\langle F, G \rangle$, namely the topology of uniform convergence on all $\sigma(G, F)$-compact, convex, circled subsets of G.*

This topology on F is called the **Mackey topology** on F with respect to $\langle F, G \rangle$ and denoted by $\tau(F, G)$. The saturated hull $\overline{\mathfrak{C}}$ of the family $\mathfrak{C}$ of all $\sigma(G, F)$-compact, convex, circled subsets of G is obtained by adjoining all subsets of members of $\mathfrak{C}$, but $\overline{\mathfrak{C}}$ must not be confused with the family of all $\sigma(G, F)$-relatively compact subsets of G, this latter family being not saturated unless the convex hull of every relatively compact subset is again relatively compact for $\sigma(G, F)$ (cf. Section 11). On the other hand, if G is quasi-complete for $\sigma(G, F)$ then the words "saturated" and "convex, circled" can be omitted in the statements of (3.2) and Corollary 1, respectively.

COROLLARY 2. *Every $\sigma(F, G)$-bounded subset of F is bounded for $\tau(F, G)$; consequently, the respective families of bounded sets are identical for all locally convex topologies on F consistent with $\langle F, G \rangle$.*

Proof. The family $\mathfrak{C}$ of all convex, circled, $\sigma(G, F)$-compact sets in G is contained in (in fact, identical with) the family $\mathfrak{S}$ of (III, 3.4) if, in (III, 3.4), we take $F = K_0$, for every $C \in \mathfrak{C}$ is $\sigma(G, F)$-complete (cf. (5.5), Corollary 2). Thus the first assertion follows from (III, 3.4); the second is immediate.

In view of the frequent (and often tacit) application of the preceding results, we summarize them once more in the following statement; the reader should keep in mind the particular case where E is a l.c.s., $E = F$, and $E' = G$.

3.3

Let $\langle F, G \rangle$ be a duality. A locally convex topology $\mathfrak{T}$ on F yields the dual G if and only if $\mathfrak{T}$ is finer than $\sigma(F, G)$ and coarser than $\tau(F, G)$; if $\mathfrak{T}$ is such a topology, a convex subset of F is $\mathfrak{T}$-closed if it is $\tau(F, G)$-closed, and any subset of F is $\mathfrak{T}$-bounded if it is $\sigma(F, G)$-bounded.

More generally, if $\mathfrak{T}$ is a locally convex topology on F such that $F(\mathfrak{T})' \subset G$, then $\mathfrak{T}$ is coarser than $\tau(F, G)$; if $\mathfrak{T}$ is such that $F(\mathfrak{T})' \supset G$, then $\mathfrak{T}$ is finer than $\sigma(F, G)$. There exists, on a given vector space F, a finest locally convex topology (Chapter II, Section 6 and Exercise 7), which is consistent with $\langle F, G \rangle$ if and only if $G = F^*$, and is clearly identical with $\tau(F, F^*)$. The coarsest l.c. topology on F is of course the trivial topology $\{\varnothing, F\}$, but there does not necessarily exist a coarsest l.c. topology which is separated; if such a topology $\mathfrak{T}_m$ exists, then $F = (F')^*$ and $\mathfrak{T}_m = \sigma(F, F')$, where $F' = F(\mathfrak{T}_m)'$, and $\mathfrak{T}_m$ is called a **minimal** topology on F (Exercise 6).

A l.c.s. E is called a **Mackey space** if its topology is $\tau(E, E')$. The following result shows that most of the l.c.s. occurring in applications are Mackey spaces.

3.4

If E is a l.c.s. which is either barreled or bornological (hence if E is metrizable), then E is a Mackey space.

Proof. If E is barreled, then by (1.6), Corollary, every bounded, and a fortiori every compact subset of E'_σ, is equicontinuous, which shows the topology $\mathfrak{T}$ of E to be finer than $\tau(E, E')$, hence identical with $\tau(E, E')$ by (3.3). Similar reasoning applies when E is bornological, since every $\tau(E, E')$-neighborhood of 0 absorbs all bounded subsets of E, and hence is a 0-neighborhood for $\mathfrak{T}$.

Let $(E, \mathfrak{T})$ be a l.c.s. and let $(\tilde{E}, \tilde{\mathfrak{T}})$ be its completion. By means of the mapping $f \to \tilde{f}$, where $\tilde{f}$ denotes the continuous extension of $f \in E'$ to $\tilde{E}$, we can identify the dual of $\tilde{E}$ with E'. It is immediate from (I, 1.5) (and a special case of Chapter III, Exercise 5) that under this identification every $\mathfrak{T}$-equicontinuous subset is $\tilde{\mathfrak{T}}$-equicontinuous, and conversely. The following is a simple consequence of this fact.

3.5

If E is a Mackey space, so is its completion $\tilde{E}$.

For if C is a convex circled $\sigma(E', \tilde{E})$-compact subset of E', C is a fortiori $\sigma(E', E)$-compact, hence $\mathfrak{T}$-equicontinuous, since E is a Mackey space; it follows that C is $\tilde{\mathfrak{T}}$-equicontinuous, and hence $\tilde{E}$ is a Mackey space.

4. DUALITY OF PROJECTIVE AND INDUCTIVE TOPOLOGIES

The definition of projective and inductive topologies (Chapter II, Sections 5, 6) suggests that these two types of topologies will occur in pairs on dual systems; the present section is concerned with this sort of duality. We do not approach the subject in the greatest possible generality, but present the duality between induced and quotient topologies and between product and direct sum topologies; this will permit us to make some applications to the duality between projective and inductive limits.

Let $\langle F, G \rangle$ be a dual system, let M be a subspace of F, and let $M°$ be the subspace of G orthogonal to M. Then the restriction of the canonical bilinear form to $M \times G$ is constant on each set $\{(x_0, y)\}$, where $x_0 \in M$ is fixed and y runs through an equivalence class $[y]$ of G mod $M°$. Therefore $(x, [y]) \to f_1(x, [y]) = \langle x, y \rangle$, where $y \in [y]$, is a well-defined bilinear form on $M \times G/M°$; it is easy to see that f_1 places M and $G/M°$ in duality. The dual system $(M, G/M°, f_1)$ will be denoted by $\langle M, G/M° \rangle$.

Let ψ denote the canonical imbedding of M into F, and ϕ the quotient map $G \to G/M°$. It follows from the definition of the dual system $\langle M, G/M° \rangle$ that the identity

$$\langle x, \phi(y) \rangle = \langle \psi(x), y \rangle$$

holds on $M \times G$. This implies that ψ is continuous for $\sigma(M, G/M°)$ and $\sigma(F, G)$, ϕ is continuous for $\sigma(G, F)$ and $\sigma(G/M°, M)$, and that ψ and ϕ are mutually adjoint (Section 2). This observation will be helpful in proving the following theorem.

4.1

Theorem. *Let $\langle F, G \rangle$ be a dual system and let M be a subspace of F. Denote by $\mathfrak{S}_1'$ and $\mathfrak{S}_2'$ saturated families of weakly bounded subsets of G and $G/M°$ for the dualities $\langle F, G \rangle$ and $\langle M, G/M° \rangle$, respectively, and denote by $\mathfrak{T}_1$ and $\mathfrak{T}_2$ the corresponding $\mathfrak{S}$-topologies on F and M. Dually, let $\mathfrak{S}_1$ and $\mathfrak{S}_2$ be saturated families of weakly bounded subsets of F and M, respectively, and denote by $\mathfrak{T}_1'$ and $\mathfrak{T}_2'$ the corresponding $\mathfrak{S}$-topologies on G and $G/M°$. Consider the following assertions:*

(a) $\phi(\mathfrak{S}_1') = \mathfrak{S}_2'$.
(b) $\mathfrak{T}_1$ *induces* $\mathfrak{T}_2$ *on* M.
(c) $\psi^{-1}(\mathfrak{S}_1) = \mathfrak{S}_2$.
(d) $\mathfrak{T}_2'$ *is the quotient topology of* $\mathfrak{T}_1'$.

Then we have the implications (a) ⇒ (b) *and* (d) ⇒ (c); *if* $\mathfrak{T}_1$ *is consistent with* $\langle F, G \rangle$, (b) ⇒ (a); *if* $\mathfrak{T}_1'$ *is consistent with* $\langle F, G \rangle$ *and* M *closed,* (c) ⇒ (d).

Proof. For greater clarity we denote polars with respect to $\langle F, G \rangle$ by $^\circ$, and polars with respect to $\langle M, G/M^\circ \rangle$ by $^\bullet$.

(a) ⇒ (b): If $S_1 \in \mathfrak{S}_1'$ it follows from (2.3)(a) that

$$[\phi(S_1)]^\bullet = \psi^{-1}(S_1^\circ) = S_1^\circ \cap M.$$

As S_1 runs through $\mathfrak{S}_1'$, S_1° runs through a $\mathfrak{T}_1$-neighborhood base of 0 in F; since by assumption $\phi(S_1)$ runs through $\mathfrak{S}_2'$, it is clear that $\mathfrak{T}_1$ induces $\mathfrak{T}_2$ on M.

(d) ⇒ (c): Let $\mathfrak{U}$ be the $\mathfrak{T}_1'$-neighborhood filter of 0 in G. Then $\mathfrak{B} = \phi(\mathfrak{U})$ is the 0-neighborhood filter of the quotient topology on G/M°. Again by (2.3)(a) we have

$$V^\bullet = [\phi(U)]^\bullet = \psi^{-1}(U^\circ) = U^\circ \cap M$$

for all $U \in \mathfrak{U}$. Since U° runs through a fundamental subfamily of $\mathfrak{S}_1$ as U runs through $\mathfrak{U}$, the assumption that $\mathfrak{T}_2'$ be the quotient topology of $\mathfrak{T}_1'$ implies that $\psi^{-1}(\mathfrak{S}_1) = \mathfrak{S}_2$.

(b) ⇒ (a): We assume that $\mathfrak{T}_1$ is consistent with $\langle F, G \rangle$. Denote by $\mathfrak{U}_1$ the family of all closed, convex $\mathfrak{T}_1$-neighborhoods of 0 in F; then $\mathfrak{U}_2 = \mathfrak{U}_1 \cap M$ is a base for the $\mathfrak{T}_2$-neighborhood filter of 0 in M. Notice that since U° ($U \in \mathfrak{U}_1$) is compact, $U^\circ + M^\circ$ is closed for $\sigma(G, F)$ and $\phi(U^\circ)$ is compact (hence closed), ϕ being continuous for $\sigma(G, F)$ and $\sigma(G/M^\circ, M)$; from (1.5), Corollary 2, we obtain

$$\phi(U^\circ) = \phi(U^\circ + M^\circ) = \phi([U \cap \overline{M}]^\circ) = \phi([U \cap M]^\circ) = [U \cap M]^\bullet,$$

where $\overline{M}$ denotes the $\sigma(F, G)$-closure of M. As U runs over $\mathfrak{U}_1$, U° runs over a fundamental subfamily of $\mathfrak{S}_1'$; likewise, $[U \cap M]^\bullet$ runs over a fundamental subfamily of $\mathfrak{S}_2'$. Since both families are saturated, it follows that $\phi(\mathfrak{S}_1') = \mathfrak{S}_2'$.

(c) ⇒ (d): We assume that $\mathfrak{T}_1'$ is consistent with $\langle F, G \rangle$ and that M is closed for $\sigma(F, G)$. Since $\psi(\mathfrak{S}_2) \subset \mathfrak{S}_1$, (2.4) implies that ϕ is continuous for $\mathfrak{T}_1'$ and $\mathfrak{T}_2'$, hence $\mathfrak{T}_2'$ is coarser than the quotient topology of $\mathfrak{T}_1'$. Thus it is sufficient to show that for each closed, convex, circled $S_1 \in \mathfrak{S}_1$, $\phi(S_1^\circ)$ is a $\mathfrak{T}_2'$-neighborhood of 0 in G/M°. If $S_2 = S_1 \cap M$, it follows from (2.3)(a) and (1.5) Corollary 2, that

$$\phi^{-1}(S_2^\bullet) = [\psi(S_2)]^\circ = [S_1 \cap M]^\circ = (S_1^\circ + M^\circ)^-.$$

Here $V = S_1^\circ + M^\circ$ is a $\mathfrak{T}_1'$-neighborhood of 0, and the closure is with respect to $\sigma(G, F)$; since $\mathfrak{T}_1'$ is consistent with $\langle F, G \rangle$ and V is convex, the closure is also with respect to $\mathfrak{T}_1'$. This implies $(S_1^\circ + M^\circ)^- = \overline{V} \subset V + V = 2S_1^\circ + 2M^\circ$. It follows, therefore, from the relation above that $S_2^\bullet \subset 2\phi(S_1^\circ)$, which shows $\phi(S_1^\circ)$ to be a $\mathfrak{T}_2'$-neighborhood of 0 in G/M°.

This completes the proof of (4.1).

REMARK. The consistency with $\langle F, G \rangle$ of the topologies $\mathfrak{T}_1$ and $\mathfrak{T}_1'$ is indispensable for the implications (b) $\Rightarrow$ (a) and (c) $\Rightarrow$ (d) (Exercise 7); also M must be assumed closed for (c) $\Rightarrow$ (d) as we shall see shortly. (4.1) can easily be stated in more general form replacing equality in (a) and (c) by inclusion and changing, accordingly, the statements of (b) and (d) to the corresponding inclusion relations for the $\mathfrak{S}$-topologies.

COROLLARY 1. *If $\langle F, G \rangle$ is a duality and M is a subspace of F, the weak topology $\sigma(M, G/M^\circ)$ is the topology induced on M by $\sigma(F, G)$. On the other hand, $\sigma(G/M^\circ, M)$ is the quotient topology of $\sigma(G, F)$ if and only if M is closed in F.*

Proof. The first assertion follows from (a) $\Rightarrow$ (b) of (4.1) by taking $\mathfrak{S}_1'$ and $\mathfrak{S}_2'$ to be the saturated families generated by all finite subsets of G and G/M°, respectively. The sufficiency part of the second assertion follows similarly from (c) $\Rightarrow$ (d). Conversely, if $\sigma(G/M^\circ, M)$ is the quotient of $\sigma(G, F)$, then we have (since $M^\circ = \overline{M}^\circ$) $\sigma(G/M^\circ, M) = \sigma(G/M^\circ, \overline{M})$ by the preceding, which implies $M = \overline{M}$ (Corollary of (1.2)).

Let E be a l.c.s., let M be a subspace of E, and let $F = E/N$ be a quotient of E; denote by $\psi: M \to E$ and $\phi: E \to E/N$ the canonical maps. $f \to f \circ \psi$ is a linear map of E' onto M' which is onto M' by (II, 4.2) and defines an algebraic isomorphism between M' and E'/M°. Dually, $g \to g \circ \phi$ defines an algebraic isomorphism between F' and $N^\circ \subset E'$. In view of this, the dual of M (respectively, E/N) is frequently identified with E'/M° (respectively, N°). The following is now immediate from Corollary 1.

COROLLARY 2. *Let M be a subspace and let F be a quotient space of the l.c.s. E. The weak topology $\sigma(M, M')$ is the topology induced by $\sigma(E, E')$, and the topology $\sigma(F, F')$ is the quotient topology of $\sigma(E, E')$.*

COROLLARY 3. *If $\langle F, G \rangle$ is a duality and M is a subspace of F, then the Mackey topology $\tau(G/M^\circ, M)$ is the quotient of $\tau(G, F)$ if and only if M is closed. On the other hand, the topology induced on M by $\tau(F, G)$ is coarser than $\tau(M, G/M^\circ)$, but consistent with $\langle M, G/M^\circ \rangle$.*

Proof. The sufficiency part of the first assertion is immediate from the implication (c) $\Rightarrow$ (d) of (4.1). Conversely, if $\tau(G/M^\circ, M)$ is the quotient of $\tau(G, F)$, then $\tau(G/M^\circ, M)$ yields the same continuous linear forms on G/M° as the quotient of $\sigma(G, F)$, which is $\sigma(G/M^\circ, \overline{M})$ by Corollary 1; it follows that $M = \overline{M}$. For the second assertion, note that ϕ is continuous for $\sigma(G, F)$ and $\sigma(G/M^\circ, M)$, which implies $\phi(\mathfrak{S}_1') \subset \mathfrak{S}_2'$ if $\mathfrak{S}_1', \mathfrak{S}_2'$ denote the saturated hulls generated by all convex, circled, weakly compact subsets of G and G/M°, respectively; it follows from (2.4) that ψ is continuous for $\tau(M, G/M^\circ)$ and $\tau(F, G)$, which is equivalent to the last topology being coarser on M than $\tau(M, G/M^\circ)$. The final assertion is clear, since $\tau(F, G)$ is finer than $\sigma(F, G)$.

The final corollary is obtained in analogy to Corollary 2, using (3.4) for the proof of the second assertion.

COROLLARY 4. *Let M be a subspace and let F be a quotient space of the l.c.s. E. The Mackey topology $\tau(F, F')$ is the quotient topology of $\tau(E, E')$; if the restriction of $\tau(E, E')$ to M is metrizable, it is identical with $\tau(M, M')$.*

This last result can be rephrased by saying that every (separated) quotient of a Mackey space is a Mackey space, and that every metrizable subspace of a Mackey space is a Mackey space.

We turn to the duality between products and direct sums. Let $\{\langle F_\alpha, G_\alpha \rangle : \alpha \in A\}$ denote a family of dualities over K and let $F = \prod_\alpha F_\alpha$, $G = \oplus_\alpha G_\alpha$. The bilinear form f on $F \times G$, defined by

$$f(x, y) = \sum_\alpha \langle x_\alpha, y_\alpha \rangle$$

(note that the sum is over an at most finite number of non-zero terms), places F and G in duality; let us denote by $\langle F, G \rangle$ the dual system (F, G, f). As before (Chapter II, Sections 5, 6), we shall identify each F_α with the subspace $F_\alpha \times \{0\}$ of F, and each G_α with the subspace $G_\alpha \oplus \{0\}$ of G; but for greater clarity polars with respect to $\langle F_\alpha, G_\alpha \rangle$ will be denoted by $^\circ$ ($\alpha \in A$) and polars with respect to $\langle F, G \rangle$ by $^\bullet$. We further note that if p_α is the projection $F \to F_\alpha$, q_α the injection $G_\alpha \to G$ ($\alpha \in A$), then

$$\langle p_\alpha x, y_\alpha \rangle = \langle x, q_\alpha y_\alpha \rangle$$

is an identity for $x \in F$, $y_\alpha \in G_\alpha$ and $\alpha \in A$. Hence by (2.1) p_α and q_α are weakly continuous with respect to $\langle F, G \rangle$ and $\langle F_\alpha, G_\alpha \rangle$.

If $\mathfrak{S}_\alpha$ is a family of weakly bounded, circled subsets of F_α ($\alpha \in A$), then it is immediate that each product $S = \prod_\alpha S_\alpha$ is a $\sigma(F, G)$-bounded, circled subset of F; let us denote by $\mathfrak{S} = \prod_\alpha \mathfrak{S}_\alpha$ the family of all such product sets. Clearly, $\mathfrak{S}$ covers F if each $\mathfrak{S}_\alpha$ covers F_α ($\alpha \in A$). Dually, let $\mathfrak{S}'_\alpha$ be a family of weakly bounded, circled subsets of G_α ($\alpha \in A$); then each set $S' = \oplus_{\alpha \in H} S'_\alpha$, where H is any finite subset of A, is circled, and $\sigma(F, G)$-bounded in G; let us denote by $\mathfrak{S}' = \oplus_\alpha \mathfrak{S}'_\alpha$ the family of all such sums. $\mathfrak{S}'$ covers G if each $\mathfrak{S}'_\alpha$ covers G_α ($\alpha \in A$). With this notation we obtain

4.2

The product of the $\mathfrak{S}'_\alpha$-topologies is identical with the $\mathfrak{S}'$-topology on F; dually, the locally convex direct sum of the $\mathfrak{S}_\alpha$-topologies is identical with the $\mathfrak{S}$-topology on G.

Proof. If $S' = \oplus_{\alpha \in H} S'_\alpha$, where H contains $n \geq 1$ elements, a short computation shows that

$$(S')^\bullet \subset \prod_{\alpha \in H} (S'_\alpha)^\circ \times \prod_{\alpha \notin H} F_\alpha \subset n(S')^\bullet,$$

which proves the first assertion.

Dually, let $S = \prod_\alpha S_\alpha$ and assume each S_α ($\alpha \in A$) to be weakly closed, convex, and circled. It is evident that the convex, circled hull $\Gamma_\alpha S_\alpha^\circ$ is contained in $S^\bullet$. Conversely, if $y = (y_\alpha) \in S^\bullet$, then $\sum_\alpha |\langle x_\alpha, y_\alpha \rangle| \leq 1$ for all

$x = (x_\alpha) \in S$; letting $\lambda_\alpha = \sup\{|\langle x_\alpha, y_\alpha\rangle|: x \in S\}$, it follows that $\lambda_\alpha = 0$ except for finitely many $\alpha \in A$, and $\sum_\alpha \lambda_\alpha \leqq 1$. Now $y_\alpha \in \lambda_\alpha S_\alpha^\circ$; hence $y = \sum_\alpha y_\alpha \in \Gamma_\alpha S_\alpha^\circ$, which shows that $S^\bullet = \Gamma_\alpha S_\alpha^\circ$. Since the totality of sets $\{\Gamma_\alpha S_\alpha^\circ\}$ form a 0-neighborhood base for the locally convex direct sum of the $\mathfrak{S}_\alpha$-topologies (cf. (*) preceding (II, 6.2)), this topology is identical with the $\mathfrak{S}$-topology on G.

We apply (4.2) to families of l.c.s.

4.3

Theorem. *Let $\{E_\alpha: \alpha \in A\}$ be a family of l.c.s. and let $E = \prod_\alpha E_\alpha$. The dual E' of E is algebraically isomorphic with $\oplus_\alpha E'_\alpha$, and the following topological identities are valid:*

1. $\sigma(E, E') = \prod_\alpha \sigma(E_\alpha, E'_\alpha)$.
2. $\tau(E, E') = \prod_\alpha \tau(E_\alpha, E'_\alpha)$.
3. $\tau(E', E) = \oplus_\alpha \tau(E'_\alpha, E_\alpha)$.

REMARK. We have $\sigma(E', E) = \oplus_\alpha \sigma(E'_\alpha, E_\alpha)$ if and only if the family $\{E_\alpha\}$ is finite (Exercise 8).

Proof. It is immediate that each $f = (f_\alpha) \in \oplus_\alpha E'_\alpha$ defines a linear form $x \to f(x) = \sum_\alpha f_\alpha(x_\alpha)$ on E which is continuous, since $f = \sum_\alpha f_\alpha \circ p_\alpha$ (the sum having only a finite number of non-zero terms); clearly, this mapping of $\oplus_\alpha E'_\alpha$ into E' is one-to-one. There remains to show that each $g \in E'$ originates in this fashion. There exists a 0-neighborhood U in E on which g is bounded; U can be assumed of the form $\prod_{\alpha \in H} U_\alpha \times \prod_{\alpha \notin H} E_\alpha$ for a suitable finite subset $H \subset A$. Denote by f_α ($\alpha \in A$) the restriction of g to E_α; then, clearly, $f_\alpha \in E'_\alpha$ for all α and $f_\alpha = 0$ if $\alpha \notin H$. Hence for $x \in E$ we obtain

$$g(x) = g\left(\sum_{\alpha \in H} p_\alpha x\right) = \sum_{\alpha \in H} f_\alpha(x_\alpha),$$

which establishes the assertion; E' is hence isomorphic with the algebraic direct sum $\oplus_\alpha E'_\alpha$ by virtue of the duality between products and direct sums introduced above.

It remains to prove the topological propositions.

1. If $\mathfrak{S}'_\alpha$ denotes the family of all finite dimensional, bounded, circled subsets of E'_α ($\alpha \in A$), it is evident that $\mathfrak{S}' = \oplus_\alpha \mathfrak{S}'_\alpha$ is fundamental for the family of all finite dimensional, bounded, circled subsets of $\oplus_\alpha E'_\alpha$; the proposition follows from (4.2).

3. If $\mathfrak{S}_\alpha$ denotes the family of all convex, circled, weakly compact subsets of E_α ($\alpha \in A$), then $\mathfrak{S} = \prod_\alpha \mathfrak{S}_\alpha$ is a fundamental subfamily of the family $\mathfrak{C}$ of all convex, circled, weakly compact subsets of E; in fact, if $C \in \mathfrak{C}$, then $p_\alpha(C) \in \mathfrak{S}_\alpha$, since by virtue of 1, p_α is weakly continuous on E into E_α ($\alpha \in A$), and $\prod_\alpha p_\alpha(C) \in \mathfrak{C}$ again by virtue of 1, above, and the Tychonov theorem which asserts that any product of compact spaces is compact. Thus by (4.2) this $\mathfrak{S}$-topology on E' is $\tau(E', E)$.

2. If $\mathfrak{S}'_\alpha$ denotes the family of all convex, circled, weakly compact subsets of E'_α ($\alpha \in A$), it suffices by (4.2) to show that $\mathfrak{S}' = \oplus_\alpha \mathfrak{S}'_\alpha$ is a fundamental system of convex, circled subsets of E' that are compact for $\sigma(E', E)$. If C is such a set, C is bounded for $\sigma(E', E)$ and hence bounded for $\tau(E', E)$. Thus by 3, above, and (II, 6.3) C is contained in $\oplus_{\alpha \in H} \tilde{p}_\alpha(C)$, where H is a suitable finite subset of A, and where $\tilde{p}_\alpha$ denotes the projection of E' onto E'_α. Since $\tilde{p}_\alpha$ is continuous for $\sigma(E', E)$ ($\tilde{p}_\alpha$ is even continuous for the coarser topology induced on E' by $\prod_\alpha \sigma(E'_\alpha, E_\alpha)$) into $(E'_\alpha, \sigma(E'_\alpha, E_\alpha))$, it follows that $\tilde{p}_\alpha(C) \in \mathfrak{S}'_\alpha$, which is the desired conclusion, since clearly every member of $\mathfrak{S}'$ is convex, circled, and compact for $\sigma(E', E)$.

This completes the proof of (4.3).

COROLLARY 1. *Let $\{E_\alpha : \alpha \in A\}$ be a family of l.c.s. and let E be their locally convex direct sum. E' is algebraically isomorphic with $\prod_\alpha E'_\alpha$, and the following topological identities are valid:*

1. $\tau(E, E') = \oplus_\alpha \tau(E_\alpha, E'_\alpha)$.
2. $\tau(E', E) = \prod_\alpha \tau(E'_\alpha, E_\alpha)$.
3. $\sigma(E', E) = \prod_\alpha \sigma(E'_\alpha, E_\alpha)$.

Proof. It follows readily from (II, 6.1) that the dual E' of E can be identified with $\prod_\alpha E'_\alpha$ by virtue of the canonical duality between products and direct sums; for the remaining assertions it is sufficient to interchange E and E' in (4.3).

COROLLARY 2. *The product, locally convex direct sum, and the inductive limit of a family of Mackey spaces is a Mackey space.*

For products and direct sums the result is immediate from (4.3) and Corollary 1; for inductive limits it follows then from Corollary 4 of (4.1).

(4.3) and Corollary 1 supply an explicit characterization of various families of bounded subsets in the dual of products and l.c. direct sums (Exercise 8; cf. also the last part of the proof of (4.3)). In particular, if $\{E_\alpha\}$ is a family of l.c.s. and S is an equicontinuous subset of the dual $\oplus_\alpha E'_\alpha$ of $\prod_\alpha E_\alpha$, then the projection $\tilde{p}_\alpha(S)$ is equicontinuous in E'_α for each α, and every finite sum of equicontinuous sets is equicontinuous in $\oplus_\alpha E'_\alpha$. Thus from (II, 6.3) and (4.3), 3, it follows that $\mathfrak{S}' = \oplus_\alpha \mathfrak{S}'_\alpha$ is a fundamental family of equicontinuous sets in $\oplus_\alpha E'_\alpha$ if each $\mathfrak{S}'_\alpha$ is such a family in E'_α. A corresponding result holds if "equicontinuous" is replaced by "weakly bounded"; thus, in view of the characterization of equicontinuous sets in the dual of a barreled space (Corollary of (1.6)), the following is proved:

COROLLARY 3. *The product of any family of barreled spaces is barreled.*

Finally we obtain a representation of the dual of a space of continuous linear maps.

COROLLARY 4. *Let E, F be l.c.s. and denote by $\mathscr{L}_s(E, F)$ the space of continuous linear maps of E into F under the topology of simple convergence. The correspondence $\sum x_i \otimes y_i' \to f$, defined by*

$$f(u) = \sum \langle ux_i, y_i' \rangle \qquad (u \in \mathscr{L}(E, F)),$$

is an (algebraic) isomorphism of $E \otimes F'$ onto the dual of $\mathscr{L}_s(E, F)$.

Proof. If $v = \sum x_i \otimes y_i'$, the mapping $v \to f$ is obviously a linear map of $E \otimes F'$ into $\mathscr{L}_s'$, which is also biunivocal, since the bilinear form $(v, u) \to f(u)$ places even the subspace $E' \otimes F$ of $\mathscr{L}(E, F)$ (Chapter III, Section 7) in separated duality with $E \otimes F'$ (Section 1, Example 4, above). There remains to show that this mapping is onto $\mathscr{L}_s'$; since $\mathscr{L}_s(E, F)$ is a subspace of the product space F^E, every $g \in \mathscr{L}_s'$ is the restriction of a continuous linear form on F^E hence, by (4.3), of the form

$$u \to g(u) = \sum \langle ux_i, y_i' \rangle$$

for suitable finite subsets $\{x_i\} \subset E$ and $\{y_i'\} \subset F'$, which completes the proof.

We conclude this section with an application of the preceding results to the duality between projective and inductive limits. Recall that a projective limit $E = \varprojlim g_{\alpha\beta}E_\beta$ (Chapter II, Section 5) is by definition a subspace of $\prod_\beta E_\beta$, namely the subspace $\bigcap_{\alpha \leq \beta} u_{\alpha\beta}^{-1}(0)$, where $u_{\alpha\beta} = p_\alpha - g_{\alpha\beta} \circ p_\beta$ whenever $\alpha \leq \beta$. The projective limit E is called **reduced** if for each α, the projection $p_\alpha(E)$ is dense in E_α. There is no restriction of generality in assuming a projective limit to be reduced: Letting $F_\alpha = p_\alpha(E)^-$ (closure in E_α) and denoting by $\tilde{u}_{\alpha\beta}$ the restriction of $u_{\alpha\beta}$ to $\prod_\beta F_\beta$, E is identical with the subspace $\bigcap_{\alpha \leq \beta} \tilde{u}_{\alpha\beta}^{-1}(0)$ of $\prod_\beta F_\beta$.

Denote by $h_{\beta\alpha}$ the adjoint of $g_{\alpha\beta}$ with respect to the dualities $\langle E_\alpha, E_\alpha' \rangle$ and $\langle E_\beta, E_\beta' \rangle (\alpha \leq \beta)$; it follows (since $g_{\alpha\beta}$ is weakly continuous) from (2.4) that $h_{\beta\alpha}$ is continuous for the weak and Mackey topologies, respectively, on E_β' and E_α'. Moreover, $g_{\alpha\gamma} = g_{\alpha\beta} \circ g_{\beta\gamma} (\alpha \leq \beta \leq \gamma)$ implies $h_{\gamma\alpha} = h_{\gamma\beta} \circ h_{\beta\alpha}$ by (2.2). (Cf. Chapter II, Exercise 9.)

4.4

If $E = \varprojlim g_{\alpha\beta}E_\beta$ is a reduced projective limit of l.c.s., then the dual E', under its Mackey topology $\tau(E', E)$, can be identified with the inductive limit of the family $\{(E_\alpha', \tau(E_\alpha', E_\alpha))\}$ with respect to the adjoint mappings $h_{\beta\alpha}$ of $g_{\alpha\beta}$.

Proof. Let $F = \oplus_\alpha E_\alpha'$, where each E_α' is endowed with $\tau(E_\alpha', E_\alpha)$. By definition $\varinjlim h_{\beta\alpha}E_\alpha'$ is the quotient space F/H_0 (provided H_0 is closed in F), where H_0 is the subspace of F generated by the ranges $v_{\beta\alpha}(E_\alpha')$, where $v_{\beta\alpha} = q_\alpha - q_\beta \circ h_{\beta\alpha}$ $(\alpha \leq \beta)$.

We show that H_0 is the subspace of F orthogonal to E with respect to the duality $\langle \prod_\alpha E_\alpha, F \rangle$. By Corollary 2 of (1.5) E° is the weakly closed, convex hull of $\bigcup_{\alpha \leq \beta} [u_{\alpha\beta}^{-1}(0)]^\circ$, which in view of the corollary of (2.3) is the same as the weakly closed, convex hull of $\bigcup_{\alpha \leq \beta} v_{\beta\alpha}(F)$; this implies $H_0 \subset E^\circ$. Conversely, let $y = (y_\alpha)$ be an element of E°, let H be the finite set of indices such that $\alpha \in H$ if and only if $y_\alpha \neq 0$, and choose an index β such that $\alpha \leq \beta$ for all $\alpha \in H$; finally let x be any element of E. Then we have

$$\langle x, y \rangle = \sum_{\alpha \in H} \langle x_\alpha, y_\alpha \rangle = \sum_{\alpha \in H} \langle g_{\alpha\beta} x_\beta, y_\alpha \rangle = \sum_{\alpha \in H} \langle x_\beta, h_{\beta\alpha} y_\alpha \rangle$$

$$= \langle x_\beta, \sum_{\alpha \in H} h_{\beta\alpha} y_\alpha \rangle = 0;$$

since by assumption x_β runs through a dense subspace of E_β as x runs through E, the preceding relation implies that $\sum_{\alpha \in H} h_{\beta\alpha} y_\alpha = 0$, hence $y = \sum_{\alpha \in H} (q_\alpha - q_\beta \circ h_{\beta\alpha}) y_\alpha \in H_0$.

Thus H_0 is weakly closed in F, hence closed for $\tau(F, \prod_\alpha E_\alpha)$, which by (4.3) is the topology $\bigoplus_\alpha \tau(E_\alpha', E_\alpha)$; thus the inductive limit $\varinjlim h_{\beta\alpha} E_\alpha'$ of the Mackey duals $(E_\alpha', \tau(E_\alpha', E_\alpha))$ exists and by (4.1), Corollary 3, its topology is the topology $\tau(F/H_0, E)$, which proves it to be isomorphic with the Mackey dual $(E', \tau(E', E))$ of E.

With the aid of (4.3), Corollary 1, and (4.1), Corollary 1, we now easily obtain the following dual result for inductive limits:

4.5

Let $E = \varinjlim h_{\beta\alpha} E_\alpha$ be an inductive limit of l.c.s. The weak dual of E is isomorphic with the projective limit of the weak duals $(E_\alpha', \sigma(E_\alpha', E_\alpha))$ with respect to the adjoint maps $g_{\alpha\beta}$ of $h_{\beta\alpha}$ ($\alpha \leq \beta$).

REMARK. If the duals E_α' are endowed with their respective Mackey topologies, then it follows from (4.3), Corollary 1, and (4.1), Corollary 3, that the projective limit of these duals, algebraically identified with E', carries a topology $\mathfrak{T}$ which is consistent with $\langle E, E' \rangle$. Thus if $\mathfrak{T}$ is known to be the Mackey topology (in particular, if $\mathfrak{T}$ is metrizable), then the Mackey dual of E can be identified with the projective limit of the Mackey duals E_α'. See also Exercise 24.

5. STRONG DUAL OF A LOCALLY CONVEX SPACE. BIDUAL. REFLEXIVE SPACES

Let $\langle F, G \rangle$ be a duality. Among the $\mathfrak{S}$-topologies on F, generated by families $\mathfrak{S}$ of $\sigma(G, F)$-bounded subsets of G, we have so far mainly considered those consistent with $\langle F, G \rangle$, in particular, the weak and Mackey topologies $\sigma(F, G)$ and $\tau(F, G)$. If $\mathfrak{S}$ is the family of all weakly bounded subsets of G, the corresponding $\mathfrak{S}$-topology is called the **strong topology** on F (with respect to $\langle F, G \rangle$), and denoted by $\beta(F, G)$. Since a weakly bounded subset of G

is not necessarily relatively compact for $\sigma(G, F)$, $\beta(F, G)$ is, in general, not consistent with $\langle F, G \rangle$ (for an example, cf. Chapter II, Exercise 14; other examples will become obvious from the discussion below).

Let E be a l.c.s. The topologies $\beta(E, E')$ and $\beta(E', E)$ are called the **strong topologies** on E and E', respectively (usually without explicit reference to the duality $\langle E, E' \rangle$); $(E', \beta(E', E))$ is called the **strong dual** of E. The following notation is more convenient: Let E_σ, E_τ, E_β denote the space E under the topologies $\sigma(E, E')$, $\tau(E, E')$, $\beta(E, E')$ respectively; accordingly, let E'_σ, E'_τ, E'_β denote the dual E' of E under $\sigma(E', E)$, $\tau(E', E)$, $\beta(E', E)$, respectively. In using this notation the reader should be cautioned that, in general, $(E_\beta)' \neq E'$ and $(E'_\beta)' \neq E$ (consequently, in general, $E'_\sigma \neq (E'_\beta)_\sigma$). It follows from (4.2) that the strong topology is inherited by products and locally convex direct sums; however, it is not necessarily inherited by subspaces and quotient spaces (Exercise 14). (See also end of Section 7.)

If the strong topology on E' is not consistent with $\langle E, E' \rangle$, it cannot be expected that the respective families of bounded subsets of E'_σ and E'_β are identical; at any rate, the following assertion is true.

5.1

Every convex, circled, compact subset of E'_σ is bounded in E'_β.

In fact, if C is such a set, its polar C° in E is a 0-neighborhood for $\tau(E, E')$, hence absorbs every bounded set $B \subset E$, which implies that B° absorbs $C^{\circ\circ} = C$.

Thus we have, in the dual E' of any l.c.s. E, the inclusions $\mathfrak{E} \subset \mathfrak{C} \subset \mathfrak{B} \subset \mathfrak{B}_\sigma$, where $\mathfrak{E}$ denotes the family of all equicontinuous sets, $\mathfrak{C}$ the family of all sets with weakly compact closed, convex, circled hull, $\mathfrak{B}$ the family of all strongly bounded sets, and $\mathfrak{B}_\sigma$ the family of all weakly bounded sets in E'. All four of these families are saturated in E'_σ; this is obvious for $\mathfrak{B}_\sigma$ and $\mathfrak{B}$ (notice that E'_β possesses a base of $\sigma(E', E)$-closed 0-neighborhoods); for $\mathfrak{E}$ it follows from (1.5) and its corollaries; for $\mathfrak{C}$ it follows quickly from (II, 10.2). In particular, we obtain the following strengthened version of the corollary of (1.6).

5.2

Let E be a l.c.s. If (and only if) E is barreled, then the properties of being equicontinuous, relatively weakly compact, strongly bounded, and weakly bounded are equivalent for any subset of E'. Moreover, if (and only if) E is barreled, then 0-neighborhood bases and fundamental families of bounded sets in E and in E'_β correspond to each other by polarity with respect to $\langle E, E' \rangle$.

The first assertion is clear from the preceding, since by (1.6), Corollary, E is barreled if and only if $\mathfrak{E} = \mathfrak{B}_\sigma$. The second assertion is clear from (1.6), since $(E, \mathfrak{T})$ is barreled if and only if $\mathfrak{T} = \beta(E, E')$, while the family of all barrels in E'_σ is a 0-neighborhood base for $\beta(E', E)$.

There are locally convex spaces for which the families $\mathfrak{E}, \mathfrak{C}, \mathfrak{B}, \mathfrak{B}_\sigma$ are all distinct (Exercise 15); hence the coincidence of some of them indicates certain special properties. We have seen that $\mathfrak{E} = \mathfrak{B}_\sigma$ characterizes barreled spaces, while, obviously, $\mathfrak{E} = \mathfrak{C}$ characterizes Mackey spaces. By (1.6) a subset $B \subset E'$ is weakly bounded if and only if its absolute polar D is a barrel in E; for B to be strongly bounded it is necessary and sufficient, by the bipolar theorem, that D is a barrel absorbing all bounded sets in E (briefly, a bound absorbing barrel). Thus $\mathfrak{E} = \mathfrak{B}$ if and only if every bound absorbing barrel in E is a neighborhood of 0; a l.c.s. with this property is called **infrabarreled**. In particular, every bornological space (and, of course, every barreled space) is infrabarreled. Spaces with $\mathfrak{C} = \mathfrak{B}$ will be discussed below. Let us note the following sufficient condition for $\mathfrak{B} = \mathfrak{B}_\sigma$.

5.3

If E is a quasi-complete l.c.s., then $\mathfrak{B} = \mathfrak{B}_\sigma$; equivalently, every weakly bounded subset of E' is strongly bounded.

Proof. The polar B° of every convex, circled, bounded subset $B \subset E'$ is a barrel in E by (1.6); hence B° absorbs every bounded subset of E which is convex, circled, and complete, by (II, 8.5). But the bounded subsets of E with these properties form a fundamental family of bounded sets, since E is quasi-complete.

COROLLARY. *Every quasi-complete infrabarreled space is barreled.*

Let us consider an example. If $(E, \| \ \|)$ is a normed space and $(E', \| \ \|)$ is the strong dual (as defined in Chapter II, Section 2) whose norm is given by

$$\|x'\| = \sup\{|\langle x, x'\rangle|\colon \|x\| \leqq 1\},$$

the topology on E' defined by this norm is clearly $\beta(E', E)$. Conversely, if (and only if) E is barreled, its topology is $\beta(E, E')$. Since E (whether it it is barreled or not) is infrabarreled, the equicontinuous sets are those bounded in $(E', \| \ \|)$. By the bipolar theorem (or by a direct application of the Hahn-Banach theorem (II, 3.2)), it follows that

$$\|x\| = \sup\{|\langle x, x'\rangle|\colon \|x'\| \leqq 1\};$$

hence the norm of E can be recovered from that of E'. In the latter formula it suffices to take the supremum over any subset whose convex, circled hull is $\sigma(E', E)$-dense in the unit ball of E'. (5.2) above explains why it is not necessary to distinguish between weakly and strongly bounded sets in the dual of a Banach space; if $(E, \| \ \|)$ is not complete, this distinction may well be necessary (Chapter II, Exercise 14).

It is clear from (1.2) that every element of E, by virtue of the duality $\langle E, E'\rangle$, defines a continuous linear form on E'_β. But the existence of a bounded set in E'_σ which is not bounded in E'_β indicates, by (3.2), Corollary 2, that

$\beta(E', E)$ is not consistent with $\langle E, E' \rangle$; hence, in general, the dual of E'_β cannot be identified with E. The dual of E'_β is called the **bidual** of the l.c.s. E and denoted by E''. Thus if, as usual, E and E'' are identified with subspaces of the algebraic dual E'^* of E', we have $E \subset E'' \subset E'^*$, where generally both inclusions are proper. The algebraic isomorphism of E into E'' thus defined is called the **canonical imbedding**, or **evaluation map** of E into E'', and is explicitly given by $x \to f_x$, where f_x is the linear form on E' defined by $f_x(x') = \langle x, x' \rangle$.

The bidual E'' can be usefully topologized in several ways. If E'' is given the $\mathfrak{S}$-topology, where $\mathfrak{S} = \mathfrak{B}$ is the family of strongly bounded subsets of E', E'' is called the **strong bidual** of E. The $\mathfrak{S}$-topology, where $\mathfrak{S} = \mathfrak{E}$ is the family of equicontinuous subsets of E', has the advantage of inducing the given topology on E (for any l.c.s. E), and is often called the **natural topology**. Since $\mathfrak{E} \subset \mathfrak{B}$, the natural topology is always coarser than the strong topology of E''; for the identity of the two topologies (equivalently, for the evaluation map to be a topological isomorphism into the strong bidual), it is evidently necessary and sufficient that E be infrabarreled.

The following characterization of E'' as a subspace of E'^* is useful.

5.4

Consider the canonical inclusions $E \subset E'' \subset E'^*$, where E is a l.c.s. Then E'' is the union of the $\sigma(E'^*, E')$-closures in E'^* of all bounded subsets of E.

Proof. Let us denote polars with respect to $\langle E, E' \rangle$ by $°$ and polars with respect to $\langle E'', E' \rangle$ by $•$; also note that $\sigma(E'^*, E')$ induces $\sigma(E'', E')$ on E''.

If $z \in E''$, then $\{z\}^•$ is a 0-neighborhood for $\beta(E', E)$ by definition of E''; so $\{z\}^• \supset B°$ for some bounded, convex subset B of E containing 0. Now, clearly, $z \in B°^• = B^{••}$, and by (1.5), $B^{••}$ is the $\sigma(E'', E')$-closure of B. Since $B^{••}$ is a $\sigma(E'', E')$-closed, equicontinuous subset of E'', it is $\sigma(E'', E')$-compact, hence identical with the $\sigma(E'^*, E')$-closure of B in E'^*. Conversely, let B be a bounded, convex subset of E containing 0; then $B^{••}$, which is the $\sigma(E'', E')$ closure of B in E'' by (1.5), is equicontinuous, hence $\sigma(E'', E')$-compact, hence $\sigma(E'^*, E')$-closed in E'^*.

COROLLARY. *Every $z \in E''$ is the limit of a $\sigma(E'', E')$-Cauchy filter possessing a base of bounded subsets of E.*

In particular, if E is a normed space and E'' its strong bidual with the standard norm, then the unit ball U of E is $\sigma(E'', E')$-dense in the unit ball $U^{°°}$ of E''. Also if E is any l.c.s. such that E'_β is separable, then it follows from (III, 4.7) and (5.4), Corollary, that every $z \in E''$ is the limit (for $\sigma(E'', E')$) of a weak Cauchy sequence in E; if in addition E is normed, the members of such a sequence can be assumed to be of norm $\leq \|z\|$.

A locally convex space E for which $E = E''$ (more precisely, for which the evaluation map is onto E'') is called **semi-reflexive**; we note that this property depends only on the duality $\langle E, E' \rangle$, and hence is shared by all or by none of the l.c. topologies on E that are consistent with $\langle E, E' \rangle$. It is important to have a number of alternative characterizations of semi-reflexive spaces.

5.5

For any l.c.s. E, the following assertions are equivalent:

(a) *E is semi-reflexive.*
(b) *Every $\beta(E', E)$-continuous linear form on E' is continuous for $\sigma(E', E)$.*
(c) *E'_τ is barreled.*
(d) *Every bounded subset of E is relatively $\sigma(E, E')$-compact.*
(e) *E is quasi-complete under $\sigma(E, E')$.*

Proof. (a) $\Rightarrow$ (b) is immediate from the definition of semi-reflexivity. (b) $\Rightarrow$ (c): (b) implies that the strong topology $\beta(E', E)$ is consistent with $\langle E, E' \rangle$, whence $\beta(E', E) = \tau(E', E)$. Since by (1.6) each barrel in E'_σ (hence by (3.1) each barrel in E'_τ) is a 0-neighborhood for $\beta(E', E)$, it follows that E'_τ is barreled. (c) $\Rightarrow$ (d): If E'_τ is barreled then, again by (1.6), each bounded set in E is equicontinuous as a subset of $E = \mathscr{L}(E'_\tau, K_0)$, hence relatively compact for $\sigma(E, E')$. (d) $\Rightarrow$ (e) is immediate, since each compact subset of E_σ is complete. (e) $\Rightarrow$ (a): The corollary of (5.4) implies that each $z \in E''$ is in E, whence $\psi(E) = E''$ where ψ denotes the evaluation map.

COROLLARY 1. *Every semi-reflexive space is quasi-complete.*

Proof. By (1.6) and (5.5)(c) every bounded subset of E is equicontinuous if E is viewed as the space $\mathscr{L}_\mathfrak{S}(E'_\tau, K_0)$, where $\mathfrak{S}$ is the family of equicontinuous subsets of E'; the assertion is hence a special case of (III, 4.4).

COROLLARY 2. *Every bounded subset of a l.c.s. E is $\sigma(E, E')$-precompact.*

Proof. Let B be a bounded, hence weakly bounded, subset of E. It is evident that the space $E'^*_\sigma = (E'^*, \sigma(E'^*, E'))$ is semi-reflexive. Hence by (5.5)(d) the closure $\bar{B}$ of B in E'^*_σ is compact, hence complete and thus uniformly isomorphic with the completion of the uniform space B, which shows B to be precompact in E_σ.

Examples of semi-reflexive spaces are furnished by all quasi-complete nuclear spaces, for by (III, 7.2), Corollary 2, every closed, bounded subset of such a space is compact hence a fortiori weakly compact. The symmetry between E and E' is not complete for semi-reflexive spaces (in particular, the strong dual of a semi-reflexive space need not be semi-reflexive, since $\beta(E, E')$ need not be consistent with $\langle E, E' \rangle$ (see below)), which is obviously due to the fact that the definition of semi-reflexivity disregards the topological properties of the evaluation map. By contrast, a l.c.s. E is called **reflexive** if the evaluation map is an isomorphism of E onto the strong bidual $(E'_\beta)'_\beta$. Thus E is reflexive if and only if it is semi-reflexive and its topology is $\beta(E, E')$, i.e., if and only if E is semi-reflexive and barreled. In view of (5.5), Corollary 1, and the corollary of (5.3), the requirement that E be barreled can be replaced by the formally weaker requirement that E be infrabarreled, and the semi-reflexivity can be replaced by any one of the equivalent properties listed in (5.5) (e.g., weak quasi-completeness). Of all these possible characterizations of reflexive spaces, we state the following which seems to be the most useful.

5.6

Theorem. *A locally convex space E is reflexive if and only if E is barreled and every bounded subset of E is relatively compact for $\sigma(E, E')$.*

The proof is covered by the preceding remarks.

COROLLARY 1. *The strong dual of a reflexive space is reflexive.*

Proof. If E is reflexive, then $\beta(E', E) = \tau(E', E)$ and $E'_\beta = E'_\tau$ is barreled by (5.5), since E is semi-reflexive. Since, also, E is barreled, each bounded subset of E'_β is relatively compact for $\sigma(E', E)$ by (5.2); hence E'_β is reflexive.

Simple examples show that the converse of Corollary 1 is false, but it can be proved that if E is a quasi-complete Mackey space such that E'_β is semi-reflexive, then E is reflexive (Exercise 18).

COROLLARY 2. *Every semi-reflexive normed space is a reflexive Banach space.*

In fact, if E is a semi-reflexive normed space, E is a Banach space by (5.5), Corollary 1, hence also barreled.

Thus semi-reflexivity and reflexivity agree for normed spaces and, as has been observed earlier, more generally for infrabarreled (in particular, bornological) spaces. One might suspect from this that every semi-reflexive Mackey space is reflexive, but this is false. Indeed let $(E, \mathfrak{T})$ be a barreled space which is not reflexive (e.g., a non-reflexive Banach space); E'_τ is a semi-reflexive Mackey space which is not reflexive, since its strong dual $(E, \mathfrak{T})$ is not reflexive. It is, moreover, clear from (5.5) that this example supplies all semi-reflexive spaces that are not reflexive, in the sense that every such space is isomorphic with the dual E' of a non-reflexive barreled space E, where E' is supplied with a suitable l.c. topology consistent with $\langle E, E' \rangle$. The following result will set in evidence even more clearly the relation between semi-reflexivity and reflexivity, and the complete symmetry between E and E' when E is reflexive.

5.7

Let E be a l.c.s. with dual E'. These assertions are equivalent:

(a) *E_τ is reflexive.*
(b) *E'_τ is reflexive.*
(c) *E_σ and E'_σ are both semi-reflexive.*
(d) *E_τ and E'_τ are both barreled.*

Proof. (a) ⇒ (b): If E_τ is reflexive, then $\beta(E', E)$ is consistent with $\langle E, E' \rangle$, whence $\beta(E', E) = \tau(E', E)$. Thus E'_τ is the strong dual of E; hence E'_τ is reflexive by Corollary 1 of (5.6). (b) ⇒ (c): If E'_τ is reflexive, then clearly E'_σ is semi-reflexive; moreover, by (5.6) E'_τ is barreled, hence every bounded subset of E is relatively $\sigma(E, E')$-compact by (5.2), which shows E_σ to be semi-reflexive by (5.5). (c) ⇒ (d): (c) implies that both $\beta(E', E)$ and $\beta(E, E')$ are consistent with $\langle E, E' \rangle$, hence that $\beta(E', E) = \tau(E', E)$ and $\beta(E, E') = \tau(E, E')$. Since all barrels in E_τ and E'_τ, respectively, are 0-neighborhoods for $\beta(E, E')$ and $\beta(E', E)$, it follows that E_τ and E'_τ are barreled. ∘ (d) ⇒ (a): E_τ is barreled and by (5.5) semi-reflexive, which implies that E_τ is reflexive.

Semi-reflexivity is inherited by closed subspaces (immediate from (5.5)), but, in general, not by quotient spaces (Exercise 20); reflexivity is inherited, in general, neither by closed subspaces (which may fail to be barreled) nor by quotients (which may fail to be semi-reflexive) (Exercise 20). At any rate, if E is a reflexive Banach space, then every closed subspace and every separated quotient of E is a reflexive Banach space (the latter being true, since every bounded subset of the quotient is the canonical image of a bounded set in E, and hence reflexivity is preserved). Moreover, it follows from (4.1), Corollary 3, and the relation $\beta(E', E) = \tau(E', E)$ (which is characteristic of semi-reflexive spaces by (5.5)) that if M is a closed subspace of a semi-reflexive space E, the strong dual of M can be identified with E'_β/M°.

The situation is less complicated in the case of products and locally convex direct sums.

5.8

Let $\{E_\alpha : \alpha \in A\}$ be a family of semi-reflexive (respectively, reflexive) l.c.s.; both the product and the locally convex direct sum of this family is semi-reflexive (respectively, reflexive). Moreover, the projective limit of any family of semi-reflexive l.c.s. is semi-reflexive, and the strict inductive limit of a sequence of reflexive spaces is reflexive.

Proof. Via (5.5)(d) it follows from (4.3), (4.3), Corollary 1, and from the characterizations (I, 5.5) and (II, 6.3) of bounded subsets of products and locally convex direct sums, respectively, that semi-reflexivity is preserved in both cases. Thus by (5.6) reflexivity is also preserved in both cases, since any product of barreled spaces is barreled, (4.3), Corollary 3, and since any locally convex direct sum of barreled spaces is barreled, (II, 7.2), Corollary 1. The assertion concerning projective limits is now clear since a projective limit of a family of l.c.s. is a closed subspace of their product; the assertion concerning inductive limits is clear from the characterization of bounded subsets (II, 6.5) of a strict inductive limit of a sequence of l.c.s., and from (II, 7.2), Corollary 1. It is also clear that mere semi-reflexivity is preserved in this case.

Examples

1. Let $L^p(\mu)$ be the Banach space introduced in Example 2 of Chapter II, Section 2. If $1 < p < +\infty$, it is well known (cf. Day [2]) that the strong dual of $L^p(\mu)$ is norm isomorphic with $L^q(\mu)$, where $p^{-1} + q^{-1} = 1$, the canonical bilinear form of the duality being $([f], [g]) \to \int fg^* \, d\mu$. Thus if $1 < p < +\infty$, $L^p(\mu)$ is reflexive; in particular, the Banach spaces $l^p_d (1 < p < +\infty)$ are reflexive (Exercise 18).

2. Every Hilbert space is reflexive, for every such space H is isomorphic with a space l^2_d, where d is the Hilbert dimension of H (Chapter II, Section 2, Example 5).

3. The one-dimensional l.c.s. K_0 associated with the scalar field K is reflexive. Hence by (5.8) the spaces ω_d and ϕ_d of Section 1, Example 4, are reflexive and the strong duals of one another under the canonical duality of products and direct sums.

4. It has been observed earlier that by virtue of (III, 7.2), Corollary 2, every quasi-complete nuclear space E is semi-reflexive; if, in addition, E is barreled (in particular, if E is a nuclear (F)-space), then E is reflexive. Since all spaces enumerated in Chapter III, Section 8, are barreled, each of these spaces furnishes an example of a non-normable reflexive space.

The spaces of Chapter III, Section 8, are locally convex spaces which are not only reflexive but such that every closed, bounded subset is compact for the strong topology. A reflexive l.c.s. in which every closed, bounded subset is compact, is called a **Montel space**, or briefly (M)-space. The permanence properties of (M)-spaces are virtually the same as for reflexive spaces; in particular, it is evident that any product, any l.c. direct sum, and every strict inductive limit of a sequence, of (M)-spaces is again an (M)-space. The same is true for the strong dual:

5.9

The strong dual of a Montel space is a Montel space.

Proof. If E is a Montel space, then E'_β is reflexive by (5.6), Corollary 1. Since E is barreled, every strongly bounded subset of E' is equicontinuous; thus if B is a strongly bounded and closed, convex subset of E', B is $\sigma(E', E)$-compact, and hence by (III, 4.5), compact for the topology $\mathfrak{T}_c$ of compact convergence. But $\mathfrak{T}_c = \beta(E', E)$, since E being a Montel space, every closed, bounded subset of E is compact.

6. DUAL CHARACTERIZATION OF COMPLETENESS. METRIZABLE SPACES. THEOREMS OF GROTHENDIECK, BANACH-DIEUDONNE, AND KREIN-ŠMULIAN

Let E be a locally convex space. What can be said about the completeness of E' for a given $\mathfrak{S}$-topology? Dually, what can be said in terms of E' about completeness properties of E? Let us examine the weak dual E'_σ. The algebraic dual E^* is complete for $\sigma(E^*, E)$, since it is a closed subspace of the complete space K_0^E. On the other hand, E' is dense in E^*_σ by (1.3), Corollary, and E^*_σ induces $\sigma(E', E)$ on E'; hence E^*_σ is (isomorphic with) the completion of E'_σ. Thus E'_σ is complete if and only if $E' = E^*$, and E_σ is complete if and only if $E = E'^*$; it is not difficult to infer that if E is an infinite dimensional metrizable l.c.s., E'_σ is never complete (cf. Exercises 6, 21). Before proving Grothendieck's dual characterization of completeness, we record the following completeness properties of the duals of barreled and bornological spaces.

6.1

Let E be a l.c.s. If E is barreled, E' is quasi-complete for every $\mathfrak{S}$-topology, where $\mathfrak{S}$ is a family of bounded sets covering E. If E is bornological, the strong dual E'_β is complete.

Proof. If E is barreled, by (5.2) the bounded subsets of E' are the same for all $\mathfrak{S}$-topologies in question and each bounded set is equicontinuous; the first assertion is hence a special case of (III, 4.4). Now every strong Cauchy filter in E' converges pointwise to an element $f \in E^*$, and f is bounded on bounded subsets $B \subset E$, since on each B the convergence is uniform; hence if E is bornological, we have $f \in E'$ by (II, 8.3).

REMARK. Komura [1] asserts that there exist reflexive spaces that are not complete; this shows that, in general, the strong dual of a barreled space is not complete. On the other hand, the preceding result on bornological spaces can be considerably strengthened (Chapter III, Exercise 8): If $\mathfrak{S}$ is a family of bounded subsets of a bornological space E such that the range of each null sequence in E is contained in a suitable $S \in \mathfrak{S}$, then E' is complete for the $\mathfrak{S}$-topology.

The following basic theorem is due to Grothendieck [1].

6.2

Theorem. *Let $(E, \mathfrak{T})$ be a l.c.s. and let $\mathfrak{S}$ be a saturated family of bounded sets covering E. For E' to be complete under the $\mathfrak{S}$-topology, it is necessary and sufficient that every linear form f on E which is $\mathfrak{T}$-continuous on each $S \in \mathfrak{S}$, be continuous on $(E, \mathfrak{T})$.*

Proof. The condition is sufficient. For let $\mathfrak{F}$ be a Cauchy filter in E' with respect to the $\mathfrak{S}$-topology; since $\mathfrak{S}$ covers E, $\mathfrak{F}$ converges pointwise to a linear form $f \in E^*$, the convergence being uniform on each $S \in \mathfrak{S}$. Hence for each S the restriction f_S of f to S, being the uniform limit of $\mathfrak{T}$-continuous scalar functions on S, is continuous for $\mathfrak{T}$; by hypothesis it follows that $f \in E'$.

The condition is necessary. For this it suffices to show that each $f \in E^*$ such that the restriction f_S $(S \in \mathfrak{S})$ is $\mathfrak{T}$-continuous can be approximated, uniformly on S, by elements $g \in E'$. Let $S \in \mathfrak{S}$ and $\varepsilon > 0$ be given; since $\mathfrak{S}$ is saturated, we can assume S is convex, circled, and $\sigma(E, E')$-closed. Let $f \in E^*$ and let f_S be $\mathfrak{T}$-continuous at $0 \in S$; there exists a convex, circled, closed (hence $\sigma(E, E')$-closed) $\mathfrak{T}$-neighborhood U of 0 such that $|f(x)| \leq \varepsilon$ whenever $x \in S \cap U$. This is equivalent to $f \in \varepsilon(S \cap U)^\circ$, where the polar is taken with respect to $\langle E, E^* \rangle$. Since S and U are a fortiori $\sigma(E, E^*)$-closed, it follows from (1.5), Corollary 2, that $(S \cap U)^\circ$ is contained in the $\sigma(E^*, E)$-closure of $U^\circ + S^\circ$. But U° is compact and S° closed for $\sigma(E^*, E)$, hence by (I, 1.1) $U^\circ + S^\circ$ is closed, and hence $f \in \varepsilon(U^\circ + S^\circ)$. It follows that for some $g \in \varepsilon U^\circ \subset E'$, $f - g \in \varepsilon S^\circ$, which means that $|f(x) - g(x)| \leq \varepsilon$ whenever $x \in S$, completing the proof.

COROLLARY 1. *Let $\langle F, G \rangle$ be a duality and let $\mathfrak{S}$ be a saturated family, covering F, of weakly bounded subsets of F. Denote by G_1 the vector space of all $f \in F^*$ whose restrictions to each $S \in \mathfrak{S}$ are weakly continuous, and endow G_1 with the $\mathfrak{S}$-topology. Then G_1 is a complete l.c.s. in which G is dense.*

Proof. Supposing, for the moment, that for each $f \in G_1$ and $S \in \mathfrak{S}$, $f(S)$ is bounded, it is clear that G_1 is a l.c.s. under the $\mathfrak{S}$-topology (Chapter III, Section 3), and obviously complete. Now if $\bar{G}$ denotes the closure of G in G_1, $\bar{G}$ is complete under the $\mathfrak{S}$-topology; hence (6.2) shows that $\bar{G} = G_1$. To complete the proof it suffices to show that $f(S)$ is bounded if $f \in G_1$ and S is a convex, circled member of $\mathfrak{S}$. Since f is weakly continuous at $0 \in S$, and since $x, y \in S$ imply $\frac{1}{2}(x - y) \in S$, the identity $|f(x) - f(y)| = 2|f(x - y)/2|$ shows that f is uniformly weakly continuous on S. Hence $f(S)$ is bounded, since S is weakly precompact by (5.5), Corollary 2.

Let us note also that if S is a closed, convex, circled (not necessarily bounded) subset of a l.c.s. $(E, \mathfrak{T})$ and f is a linear form on E whose restriction to S is $\mathfrak{T}$-continuous at 0, then f is uniformly weakly continuous on S; in fact, by the second part of the proof of (6.2), f_S is the uniform limit of uniformly weakly continuous functions on S.

COROLLARY 2. *The following propositions on a l.c.s. E are equivalent:*

(a) *E is complete.*
(b) *Every linear form on E' which is $\sigma(E', E)$-continuous on every equicontinuous subset of E' is $\sigma(E', E)$-continuous on all of E'.*
(c) *Every hyperplane H in E' such that $H \cap A$ is weakly ($\sigma(E', E)$-)closed in A for each equicontinuous subset A of E', is closed in E'_σ.*

Proof. (a) $\Leftrightarrow$ (b) is immediate from (6.2), since the topology of E is the $\mathfrak{S}$-topology, where $\mathfrak{S}$ is the (saturated) family of equicontinuous subsets of E'_σ, (1.5), Corollary 3, and E is the dual of E'_σ. (c) $\Rightarrow$ (b): Let $H = \{x' \in E': f(x') = \alpha\}$; then $H \cap A = \{x' \in A: f(x') = \alpha\}$ is weakly closed in A if the restriction f_A is $\sigma(E', E)$-continuous. (b) $\Rightarrow$ (c): Assume that $H = \{x' \in E': f(x') = \alpha\}$ is a hyperplane in E' such that $H \cap A$ is closed in A for $\sigma(E', E)$ whenever A is equicontinuous. To show that f is $\sigma(E', E)$-continuous on A, it suffices to show that f_A is continuous at $0 \in A$ whenever A is convex and circled. Notice first that since $H \cap A + x_0' = (H + x_0') \cap (A + x_0')$ for all $x_0' \in E'$, and since $A + x_0'$ is equicontinuous if A is, $\{x' \in A: f(x') = \beta\}$ is closed in A for each $\beta \in K$, and equicontinuous set A. If f_A were not continuous at $0 \in A$, there would exist an infinite subset $B \subset A$ such that $0 \in \bar{B}$ and such that $f(x) = \beta_0$, $\beta_0 \neq 0$, for all $x \in B$, which conflicts with the fact that $\{x' \in A: f(x') = \beta_0\}$ is closed in A.

COROLLARY 3. *Let E be a separable, complete l.c.s. and let f be a linear form on the dual E'. For f to be $\sigma(E', E)$-continuous it suffices that $\lim_n f(x_n') = 0$ whenever $\{x_n'\}$ is a null sequence for $\sigma(E', E)$.*

Proof. This is immediate from Corollary 2, since by (1.7), the equicontinuous subsets of E' are metrizable for the topology induced by $\sigma(E', E)$. (It is even sufficient that f converge to 0 on every weak null sequence whose range is equicontinuous.)

The preceding results lead one to ask whether there exists, on the dual E' of a l.c.s. E, a finest topology $\mathfrak{T}_f$ which agrees with $\sigma(E', E)$ on each equicontinuous set. If so, then by Grothendieck's theorem (6.2), Corollary 1, the completion of E consists exactly of the linear forms on E' that are continuous for $\mathfrak{T}_f$. It is indeed easy to see that such a topology exists: If $\mathfrak{G}$ is the family of all subsets of E' such that for every $G \in \mathfrak{G}$ and equicontinuous set A, $G \cap A$ is open in A for $\sigma(E', E)$ then $\mathfrak{G}$ is evidently invariant under the formation of finite intersections and arbitrary unions; hence $\mathfrak{G}$ is the family of open sets for the topology in question. It is clear, moreover, that this topology is translation-invariant (since $\sigma(E', E)$ and the family $\mathfrak{E}$ of equicontinuous subsets of E' are translation-invariant), and has a 0-neighborhood base consisting of radial and circled sets (Exercise 22). However, in general, property (a) of (I, 1.2) fails (Komura [1]); it would hold automatically if $\mathfrak{T}_f$ were necessarily locally convex, but this is false as an earlier example due to Collins [1] shows (Exercise 22). It is hence of interest to determine cases in which $(E', \mathfrak{T}_f)$ is a t.v.s. and, in particular, in which it is a l.c.s. (Note that $\mathfrak{T}_f$ is Hausdorff, since it is finer than $\sigma(E', E)$ and, in fact, finer than the topology of precompact convergence, (III, 4.5).) It was known to Banach that $\mathfrak{T}_f$ is the topology of compact convergence if E is a Banach space; Dieudonné [1] proved the result by a new method which is applicable to metrizable l.c.s. The following lemma is the critical step in the proof.

LEMMA. *Let E be a metrizable l.c.s., $\{U_n: n \in N\}$ a 0-neighborhood base in E consisting of a decreasing sequence of closed, convex sets and let G be an open $\mathfrak{T}_f$-neighborhood of 0 in E'. There exists a sequence $\{F_n: n = 0, 1, 2, \ldots\}$ of non-empty finite subsets of E having these properties:*

(i) $F_n \subset U_n (n \in N)$.

(ii) $H_n^\circ \cap U_n^\circ \subset G (n \in N)$ *where* $H_n = \bigcup_{k < n} F_k$.

Proof. The proof uses induction with respect to n. Supplementing the base $\{U_n: n \in N\}$ by setting $U_0 = E$ and assuming that the sets $F_k \subset U_k$ ($k = 0, 1, \dots, m-1$) have been selected to satisfy (ii) for $n = 1, \dots, m$, we shall show that there exists a non-empty finite set $F_m \subset U_m$ such that $(F_m \cup H_m)^\circ \cap U_{m+1}^\circ \subset G$. To include the existence proof for F_0 in the general induction step, we set $H_0 = \varnothing$; then clearly (ii) is satisfied for $n = 0$.

Now suppose that $(F \cup H_m)^\circ \cap U_{m+1}^\circ$ is not contained in G for any non-empty finite subset $F \subset U_m$. Since $G^\sim = E \sim G$ is $\mathfrak{T}_f$-closed and U_{m+1}° is equicontinuous in E', the set $G_1 = G^\sim \cap U_{m+1}^\circ$ is closed in U_{m+1}° for the topology induced by $\mathfrak{T}_f$, hence for the topology induced by $\sigma(E', E)$; this implies that G_1 is $\sigma(E', E)$-compact. Now the non-empty finite subsets $F \subset U_m$ are directed upward by inclusion; hence the sets $(F \cup H_m)^\circ \cap G_1 = F^\circ \cap H_m^\circ \cap G_1$ form a filter base of closed subsets of G_1. From the compactness of G_1 it follows that the intersection of all these sets contains an element x', which is consequently an element of $U_m^\circ \cap H_m^\circ \cap G_1$; for by Remark 3 preceding (1.4), $U_m^\circ = \bigcap F^\circ$, where F runs through all non-empty finite subsets $F \subset U_m$. This is contradictory, since $U_m^\circ \cap H_m^\circ \subset G$ by assumption and since $G_1 \subset G^\sim$; the lemma is proved.

The following is the theorem of Banach-Dieudonné.

6.3

Theorem. *Let E be a metrizable l.c.s. and let $\mathfrak{S}$ be the family of sets formed by the ranges of all null sequences in E. The $\mathfrak{S}$-topology is the topology of precompact convergence and the finest topology on E' that agrees with $\sigma(E', E)$ on each equicontinuous subset of E'.*

Proof. Since the range S of any null sequence is relatively compact in E, hence the closed, convex, circled hull of S precompact, (II, 4.3), the $\mathfrak{S}$-topology is coarser on E' than is the topology $\mathfrak{T}_{pc}$ of precompact convergence. On the other hand, the topology $\mathfrak{T}_f$ is finer than $\mathfrak{T}_{pc}$ by (III, 4.5); hence it suffices to show that the $\mathfrak{S}$-topology is finer than $\mathfrak{T}_f$. In view of the translation invariance of $\mathfrak{T}_f$ this is immediate from the lemma above: If G is an open $\mathfrak{T}_f$-neighborhood of 0 in E', let $S = \bigcup_0^\infty F_n$, where F_n are as in the lemma, then $S^\circ \cap U_n^\circ \subset G$ for all $n \in N$, and hence $S^\circ \subset G$, since $\bigcup_1^\infty U_n^\circ = E'$. Since, clearly, $S \in \mathfrak{S}$, the proof is complete.

COROLLARY 1. *If E is a metrizable l.c.s., each precompact subset of E is contained in the closed, convex, circled hull of a suitable null sequence.*

COROLLARY 2. *If E is an (F)-space, the topology $\mathfrak{T}_f$ on E' is the topology of compact convergence (hence consistent with $\langle E, E' \rangle$).*

It is now easy to prove the theorem of Krein-Šmulian (Krein-Šmulian [1]).

6.4

Theorem. *A metrizable l.c.s. E is complete if and only if a convex set $M \subset E'$ is $\sigma(E', E)$-closed whenever $M \cap U°$ is $\sigma(E', E)$-closed for every 0-neighborhood U in E.*

Proof. Since every hyperplane H in E' is convex, the sufficiency of the condition is clear from (6.2), Corollary 2. To prove its necessity, let E be complete and let M be a convex subset of E' such that $M \cap U°$ is $\sigma(E', E)$-closed for every 0-neighborhood U in E. Since $\mathfrak{T}_f$ is consistent with $\langle E, E' \rangle$ by Corollary 2 above, it suffices to show that M is $\mathfrak{T}_f$-closed. Denote by $M^\sim$ the complement of M in E'; the assumption implies that $M^\sim \cap U°$ is open in $U°$ for $\sigma(E', E)$ and hence for $\mathfrak{T}_f$. Since every equicontinuous set $A \subset E'$ is contained in a suitable $U°$, it follows that $M^\sim \cap A$ is open in A for $\mathfrak{T}_f$, and hence $M^\sim$ is open for $\mathfrak{T}_f$.

COROLLARY. *Let E be a Banach space, let M be a subspace of E', and let B be the dual unit ball $\{x': \|x'\| \leq 1\}$. If $M \cap B$ is closed in E'_σ, then M is closed in E'_σ.*

Proof. The assumption implies that $\rho(M \cap B) = M \cap \rho B$ is closed in E' for all $\rho > 0$ so that (6.4) applies. (It is evidently sufficient that the condition of (6.4) be satisfied for the members U of an arbitrary 0-neighborhood base in E.)

Thus the weak dual of an (F)-space possesses a number of striking properties; in contrast with this, and in contrast with the strong dual of a Banach space, the strong dual of an (F)-space E has a structure, in general, much more complicated than that of E. For instance, if E is a metrizable l.c.s., then E'_β is not metrizable unless E is normable; in fact, E'_β possesses a fundamental sequence $\{B_n\}$ of bounded sets which we can assume are convex, circled, and closed; if E'_β (which is complete, (6.1)) were metrizable, one of the B_n would be a 0-neighborhood and hence E'_β would be normable. Thus the strong bidual would be normable and hence E would be normable, since E is infrabarreled (Section 5). Also, E'_β need not be barreled (hence not infrabarreled or bornological), even if E is an (F)-space; for an example we refer the reader to Exercise 20 (see also Grothendieck [10] and Köthe [1]). Nevertheless, the fact that the strong dual of a metrizable l.c.s. possesses a fundamental family of bounded sets which is countable, has some important consequences which will be derived now. In the proofs of (6.5) through (6.7) we follow Kelley-Namioka [1].

6.5

If E is a metrizable l.c.s. and $\{V_n\}$ is a sequence of convex 0-neighborhoods in E'_β such that $V = \bigcap_1^\infty V_n$ absorbs strongly bounded sets, then V is a 0-neighborhood in E'_β.

Proof. Let $\{U_n: n \in N\}$ be a 0-neighborhood base in E; then the sequence of polars $B_n = U_n°$ constitutes a fundamental family of strongly bounded (equivalently, equicontinuous) subsets of E'. It is sufficient to show that V contains a set W which is radial, convex, and closed in E'_σ; for then W is the polar of a bounded set in E, hence a 0-neighborhood in E'_β. By hypothesis there exists, for each $n \in N$, $\rho_n > 0$ such that $2\rho_n B_n \subset V$ and a barrel D_n in E'_σ such that $2D_n \subset V_n$. Now the convex hull C_n of $\bigcup_1^n \rho_k B_k$ is $\sigma(E', E)$-compact by (II, 10.2), and hence $W_n = C_n + D_n$ is convex and $\sigma(E', E)$-closed by (I, 1.1); obviously $W_n \subset V_n$. It is easily verified that $W = \bigcap_1^\infty W_n$ absorbs each B_n and hence is radial; clearly, W is convex, $\sigma(E', E)$-closed, and contained in V.

The property of E'_β expressed by (6.5) is evidently equivalent to the following: If $\{M_n\}$ is a sequence of equicontinuous subsets of E'' such that $M = \bigcup_n M_n$ is strongly bounded (i.e., $\beta(E'', E')$-bounded), then M is equicontinuous. Since every finite subset of E'' is equicontinuous, we obtain the following corollary.

COROLLARY 1. *Every countable, bounded subset of the strong bidual E'' of a metrizable l.c.s. E is equicontinuous.*

In view of (III, 4.3) and the obvious fact that the strong bidual of a metrizable l.c.s. E is again metrizable, this implies:

COROLLARY 2. *The strong bidual E'' of a metrizable l.c.s. E is an (F)-space, and semi-complete for $\sigma(E'', E')$.*

(6.5) can also be viewed as a weakened form of the property defining bornological spaces. As has been observed above, the strong dual of a metrizable (hence bornological) l.c.s. is not necessarily bornological, but the following proposition is valid.

6.6

For the strong dual of a metrizable l.c.s. E, the following properties are equivalent:

(a) E'_β *is bornological.*
(b) E'_β *is infrabarreled.*
(c) E'_β *is barreled.*

Proof. (a) $\Rightarrow$ (b) is immediate (and true for any l.c.s., Section 5), (b) $\Rightarrow$ (c) follows from the corollary of (5.3), since E'_β is complete by (6.1). (c) $\Rightarrow$ (a): It suffices to show that each convex circled subset $C \subset E'$ that absorbs strongly bounded sets contains a barrel in E'_β. Let $\{B_n\}$ be a fundamental sequence of $\sigma(E', E)$-compact, convex, circled equicontinuous sets; since each B_n is strongly bounded (in fact, $\{B_n\}$ is a fundamental family of strongly bounded sets), there exists $\rho_n > 0$ such that $2\rho_n B_n \subset C$ $(n \in N)$. Let C_n be the convex hull of $\bigcup_1^n \rho_k B_k$; each C_n is convex, circled, and $\sigma(E', E)$-compact by (II, 10.2), and $C_0 = \bigcup_1^\infty C_n$ is a convex, circled, radial set such that $2C_0 \subset C$. Thus the proof will be complete if we show that $\bar{C}_0 \subset 2C_0$, the closure being taken for $\beta(E', E)$. Let $x' \notin 2C_0$. Then for each $n \in N$, there exists a convex circled 0-neighborhood V_n in E' such that $(x' + V_n) \cap C_n = \varnothing$, for C_n is strongly closed. Let $W_n = V_n + C_n$. Then by (6.5) $W = \bigcap_1^\infty W_n$ is a strong 0-neighborhood, since W absorbs each B_n ($n \in N$). On the other hand, $x' \notin 2C_0$ implies $(x' + W_n) \cap C_n = \varnothing$ for all $n \in N$, whence $(x' + W) \cap C_0 = \varnothing$; it follows that $x' \notin \bar{C}_0$.

COROLLARY 1. *The strong dual of every reflexive Fréchet space is bornological.*

This is immediate from (5.6) and (5.6), Corollary 1.

COROLLARY 2. *If the strong dual of a metrizable l.c.s. is separable, then it is bornological.*

Proof. It is sufficient to show that E'_β is infrabarreled. Hence let D be a barrel in E'_β that absorbs strongly bounded sets, and let $\{x'_n\}$ be a countable dense subset of $E' \sim D$. By (II, 9.2) there exists a closed (real) semi-space H_n such that $x'_n \notin H_n$ and D is contained in the interior of H_n. Hence $D \subset \bigcap_n H_n$ and $U = \bigcap_n H_n$ is a 0-neighborhood in E'_β in view of (6.5). Now the complement of the interior $\mathring{U}$ of U contains $\{x'_n\}$ and is closed; hence $D \supset \mathring{U}$, which proves the assertion.

The property (6.5) of strong duals of metrizable spaces led Grothendieck [10] to introduce a special class of locally convex spaces: A l.c.s. E is called a (DF)-**space** if E possesses a fundamental sequence of bounded sets, and if every strongly bounded countable union of equicontinuous subsets of E' is equicontinuous. Every strong dual of a metrizable l.c.s. is a (DF)-space, but not conversely; the class comprises all normable spaces and, more generally, all infrabarreled spaces possessing a fundamental sequence of bounded sets (Exercise 24). For a detailed study of these spaces the reader is referred to Grothendieck [10]; these spaces also play a considerable part in the theory of topological tensor products (Grothendieck [13]). As an important example of the properties of (DF)-spaces, we prove that the topology of a (DF)-space can be "localized" in a fashion analogous to the localization of the topology of compact convergence in the dual of an (F)-space.

6.7

Let E be a (DF)-space. A convex, circled subset V of E is a neighborhood of 0 if (and only if) for every convex, circled bounded subset $B \subset E$, $B \cap V$ is a 0-neighborhood in B.

Proof. The necessity of the condition being trivial, suppose that $\{B_n\}$ is an increasing sequence of bounded, convex, circled sets which is fundamental, and that for each n, $B_n \cap V$ is a 0-neighborhood in B_n. There exists, for each $n \in N$, a convex, circled 0-neighborhood U_n in E satisfying $B_n \cap U_n \subset B_n \cap V$.

Let W_n be the closure of $B_n \cap V + \frac{1}{2}U_n$; then $W_n \subset B_n \cap V + U_n$, hence

$$B_n \cap W_n \subset B_n \cap V + (2B_n) \cap U_n \subset B_n \cap V + 2(B_n \cap U_n) \subset 3(B_n \cap V);$$

letting $W = \bigcap_1^\infty W_n$, it follows that $B_n \cap W \subset 3B_n \cap 3V$ which, in view of $\bigcup_1^\infty B_n = \bigcup_1^\infty 3B_n = E$, implies that $W \subset 3V$. On the other hand, W is closed, convex, circled, and absorbs each B_n, since for suitable $\rho_n > 0$, $B_n \subset \rho_n(B_n \cap U_n)$

$$\subset \rho_n(B_n \cap V) \subset \rho_n(B_{n+p} \cap V) \subset \rho_n W_{n+p} \quad \text{for all} \quad p \in N. \quad \text{Hence} \quad \bigcup_1^\infty W_n^\circ \text{ is}$$

strongly bounded and therefore equicontinuous in E'. It follows that W and hence $V \supset \frac{1}{3}W$ is a 0-neighborhood in E.

COROLLARY. *A linear map of a* (DF)-*space E into a l.c.s. F is continuous if its restriction to each bounded subset of E is continuous.*

7. ADJOINTS OF CLOSED LINEAR MAPPINGS

Let E, F be l.c.s., let E_0 be a dense subspace of E, and let u be a linear map with domain $D_u = E_0$ and with values in F. Consider the set F_0' of elements $y' \in F'$ for which the linear form $x \to \langle ux, y' \rangle$ is continuous on E_0; F_0' is non-empty, since it contains 0, and is, clearly, a subspace of F'. Since E_0 is dense in E, the form $x \to \langle ux, y' \rangle$ has for each $y' \in F_0'$ a unique continuous extension to E, which is thus an element $x' \in E'$. Let us denote this mapping $y' \to x'$ by v; obviously, v is a linear map with domain $D_v = F_0'$ and it follows that the relation

$$\langle ux, y' \rangle = \langle x, vy' \rangle$$

is an identity on $D_u \times D_v$. If u is continuous on E_0 into F, then, clearly, $D_v = F'$ and (since the dual of E_0 can be identified with E') v is the adjoint of u as defined in Section 2; it is thus consistent to call the mapping v just defined the **adjoint map** (briefly, adjoint) of u. Hence every linear map u, defined on a dense subspace of E (for convenience, we shall say that u is densely defined *in E*), with values in F possesses a well-defined adjoint with domain $D_v \subset F'$; notice that $D_v = (u^*)^{-1}(E') \cap F'$, where u^* is the algebraic adjoint of u and E' is identified with a subspace of E_0^*. We are asking for conditions under which D_v is dense in F_σ'. Recall that the map u is called closed if its graph $G = \{(x, ux): x \in D_u\}$ is a closed subspace of $E \times F$; in view of (4.3) and the convexity of G, this is equivalent to G being closed in $E_\sigma \times F_\sigma$.

7.1

Let E, F be l.c.s. and let u be a linear map densely defined in E with values in F. The graph of the adjoint v of u is closed in $F_\sigma' \times E_\sigma'$; for the domain D_v to be dense in F_σ', it is necessary and sufficient that u have a closed extension. If this is the case, then the adjoint $\bar{u}$ of v is the smallest closed extension of u.

Proof. Note first that by (4.3), the dual of $E \times F$ can be identified with $E' \times F'$, where the canonical bilinear form on $(E \times F) \times (E' \times F')$ is given by $\langle (x, y), (x', y') \rangle = \langle x, x' \rangle + \langle y, y' \rangle$; further, the mapping $\chi: (y', x') \to (-x', y')$ is an isomorphism of $F'_\sigma \times E'_\sigma$ onto $E'_\sigma \times F'_\sigma$. Let G be the graph of u and let H be the graph of v; the identity $\langle ux, y' \rangle - \langle x, vy' \rangle = 0$, valid on $D_u \times D_v$, shows that $G^\circ = \chi(H)$; hence $\chi(H)$ is closed in $E'_\sigma \times F'_\sigma$ and, therefore, H is closed in $F'_\sigma \times E'_\sigma$.

Assume now that D_v is dense in F'_σ. Then the graph G_1 of the adjoint $\bar{u}$ of v is closed in $E_\sigma \times F_\sigma$ (hence in $E \times F$) by the preceding, and, clearly, $D_{\bar{u}}$ contains D_u; thus $\bar{u}$ is a closed extension of u. Also $\chi'(G_1) = H^\circ$, where χ' denotes the adjoint of χ (Section 2); in view of the fact that H and G_1 are closed, it follows from (2.3)(a),(c) that $\chi'(G_1) = H^\circ$ is equivalent with $G_1 = \chi(H)^\circ$. Thus $G_1 = G^{\circ\circ}$ which, by the bipolar theorem (1.5), shows G_1 to be the closure of G in $E \times F$; hence $\bar{u}$ is the smallest (in an obvious sense) closed extension of u.

Since the domain of the adjoint of an extension of u is contained in D_v, the proof will be complete if we show that D_v is dense in F'_σ whenever u is closed. Let $y \in D_v^\circ$; then $(y, 0) \in H^\circ$ and $(0, y) \in \chi(H)^\circ$. If G is closed, then $\chi(H)^\circ = G^{\circ\circ} = G$ by the bipolar theorem, and $(0, y) \in G$ implies $y = 0$; hence D_v is dense in F'_σ.

We shall extend the notation introduced in Section 2 and denote by u' the adjoint of u whenever u is a densely defined closed linear map.

COROLLARY. *Let u be a closed linear map, densely defined in E, with values in F. Then u' has the same properties with respect to F'_σ and E'_σ, and $u'' = (u')' = u$.*

It is often convenient to reduce the study of a linear map u to the case where u is biunivocal; if u is a continuous linear map on a t.v.s. E into a t.v.s. F with null space N, the biunivocal map u_0 of E/N into F associated with u (Chapter III, Section 1) is continuous (and conversely). Likewise, a linear map u is open if and only if u_0 is open. It is a useful fact that closedness and density of domain is reflected in a similar manner by the associated biunivocal maps. Only in the following statement, E, F are not assumed to be l.c.s.

7.2

Let E, F be t.v.s., let u be a linear map with domain $D_u \subset E$, null space $N \subset D_u$, and range in F. Denote by ϕ the canonical map $E \to E/N$, and by u_0 the biunivocal map of $\phi(D_u)$ into F associated with u. Then the graph of u_0 is closed in $(E/N) \times F$ if and only if the graph of u is closed in $E \times F$. Moreover, if u is densely defined, then so is u_0; and if u is closed and F is separated, then N is closed in E.

Proof. We identify $(E/N) \times F$ canonically with $(E \times F)/(N \times \{0\})$ and denote by ϕ_1 the quotient map $E \times F \to (E \times F)/(N \times \{0\})$. If G and G_0

are the respective graphs of u and u_0, it is evident that $\phi_1(G) = G_0$; since $N \times \{0\} \subset G$, it follows that $G = \phi_1^{-1}(G_0)$ and $G_0^{\sim} = \phi_1(G^{\sim})$, where $\sim$ denotes complementation. Since ϕ_1 is continuous and open, the two latter relations show that G is closed exactly when G_0 is. If D_u is dense in E, then $\phi(D_u)$ is dense in E/N, since ϕ is continuous and onto. If F is separated (equivalently, if $\{0\}$ is closed in F), then $N \times \{0\}$ is clearly closed in G, hence in $E \times F$ if G is closed. Thus N is closed in E in this case. This completes the proof.

Thus if u is a closed linear map with domain E_0 dense in E and values in F, the mapping u_0 on $\phi(E_0)$ (which can be identified with E_0/N) into F is again closed and densely defined in E/N. If u is canonically decomposed, $u = \psi \circ u_0 \circ \phi_0$, where $\phi_0: E_0 \to E_0/N$ and $\psi: u(E_0) \to F$ are the canonical maps, it follows, as before, that u is continuous (respectively, open) if and only if u_0 is continuous (respectively, open). Weakly open linear maps with closed graph are characterized by the following dual property; E, F are again supposed to be l.c.s.

7.3

Let u be a linear map, densely defined in E, into F with closed graph. Then u is weakly open if and only if the range of its adjoint u' is closed in E'_σ.

Proof. By (7.2) and Corollary 2 of (4.1), it can be supposed that u is one–to–one and onto F; denote by E_0 the domain of u. If u is weakly open, then u^{-1} is continuous for $\sigma(F, F')$ and $\sigma(E_0, E')$; hence u^{-1} has an adjoint v which maps E' into F'. Since it is clear that u' is one–to–one and that $v = (u')^{-1}$, it follows that the range of u' is E' and hence is closed. Conversely, if the range H of u' is closed in E'_σ, then $H = E'$. For $x \in H^\circ$ (polar with respect to $\langle E, E' \rangle$) implies $x \in D_u = E_0$ in view of $u'' = u$; hence $x = 0$ because $D_{u'}$ is dense in F'_σ and u is one–to–one; it follows that $H^\circ = \{0\}$ and $H = H^{\circ\circ} = E'$. Now if $U = \{x \in E_0: |\langle x, x_i' \rangle| \leq 1, \ i = 1, ..., n\}$ is a weak 0-neighborhood in E_0, there exist elements $y_i' \in D_{u'}$ such that $u'(y_i') = x_i'$ for all i; it follows that $u(U) = V$, where $V = \{y \in F: |\langle y, y_i' \rangle| \leq 1, \ i = 1, ..., n\}$, and hence that u is weakly open.

COROLLARY. *Let u be a weakly continuous linear map of E into F. For u to be a weak homomorphism, it is necessary and sufficient that its adjoint u' have a closed range in E'_σ.*

In view of (2.1), the foregoing results can be given a more symmetrical form: Let u be a closed linear map with range in F and domain dense in E. Then u is weakly continuous if and only if the domain of u' is closed in F'_σ, and u is weakly open if and only if the range of u' is closed in E'_σ.

The following proposition relates the continuity properties of a linear map u with properties of its adjoint v; u is not assumed to be closed, and v is the adjoint of u, with domain $D_v = F'_0$, as defined at the beginning of this section.

7.4

Let E, F be l.c.s., let u be a linear map of E into F, and let v be the adjoint of u. Consider these statements:

(a)   *u is continuous.*
(b$_1$) *u is continuous for the Mackey topologies on E, F.*
(b$_2$) *u is weakly continuous.*
(b$_3$) *v is defined on all of F'.*
(b$_4$) *v is continuous on F'_σ into E'_σ.*
(c)   *v has a continuous extension $F'_\beta \to E'_\beta$, and D_v is dense in F'_σ.*

One has the implications: $(b_1) \Leftrightarrow (b_2) \Leftrightarrow (b_3) \Leftrightarrow (b_4)$, $(a) \Rightarrow (b_1)$ *and* $(b_4) \Rightarrow (c)$. *If F is semi-reflexive, then* $(c) \Rightarrow (b_4)$, *and if E is a Mackey space, then* $(b_1) \Rightarrow (a)$.

Proof. $(a) \Rightarrow (b_2)$ and $(b_1) \Rightarrow (b_2)$: For each $y' \in F'$, $x \to \langle ux, y' \rangle$ is a continuous, hence weakly continuous linear form on E, which is equivalent to the weak continuity of u. $(b_2) \Rightarrow (b_3)$: Immediate from (2.1). $(b_3) \Rightarrow (b_4)$: Since $u(E) \subset F$ and u is the adjoint of v with respect to the dualities $\langle E, E' \rangle$ and $\langle F, F' \rangle$, v is continuous for $\sigma(F', F)$ and $\sigma(E', E)$ by (2.1). $(b_4) \Rightarrow (b_1)$: v maps convex, circled, compact subsets of F'_σ onto convex, circled, compact subsets of E'_σ, and u is the adjoint of v; the assertion follows from (2.4). $(b_4) \Rightarrow (c)$: It suffices to prove $(b_2) \Rightarrow (c)$, and this is immediate from (2.4), since u maps bounded sets onto bounded sets. If F is semi-reflexive, then $F'_\beta = F'_\tau$ by (5.5) and hence v is continuous on F'_τ into E'_β, and a fortiori for the Mackey topologies on F' and E'; by the proven equivalence of (b_1) and (b_2), it follows that $(c) \Rightarrow (b_4)$ when F is semi-reflexive. If E is a Mackey space, then, clearly, $(b_1) \Rightarrow (a)$, since the topology of F is coarser than $\tau(F, F')$.

Among the consequences of the preceding result, let us note the following: If u is a linear map of a Mackey space E into a l.c.s. F with an adjoint defined on all of F', u is necessarily continuous. (This contains a classical theorem due to Hellinger and Toeplitz: Every self-adjoint transformation of a Hilbert space which is defined everywhere, is continuous.) The following proposition gives a dual characterization of topological homomorphisms.

7.5

Let E, F be l.c.s. with respective duals E', F', and let u be a closed linear map of E into F. Then u is a topological homomorphism if and only if the domain and range of u' are closed in F'_σ and E'_σ, respectively, and u' maps the equicontinuous subsets of F' onto the equicontinuous subsets of its range.

Proof. Denote by N and M, respectively, the null space and range of u. We decompose $u = \psi \circ u_0 \circ \phi$ where, as usual, ϕ is the quotient map $E \to E/N$ (note that N is closed), ψ the canonical imbedding $M \to F$. u is a topological homomorphism if and only if u_0 is an isomorphism for the quotient and induced topologies $\mathfrak{T}_1$ on E/N and $\mathfrak{T}_2$ on M, respectively.

From (4.1), (a)$\Rightarrow$(b) it follows that $\mathfrak{T}_2$ is the $\mathfrak{S}_2$-topology, where $\mathfrak{S}_2$ is the family of all canonical images in F'/M° of equicontinuous subsets of F'; similarly by (4.1), (c)$\Rightarrow$(d), $\mathfrak{T}_1$ is the $\mathfrak{S}_1$-topology, where $\mathfrak{S}_1$ is the family of all equicontinuous subsets of E' that are contained in N°.

The condition is necessary. For if u_0 is an isomorphism of $(E/N, \mathfrak{T}_1)$ onto $(M, \mathfrak{T}_2)$, then by (7.4) u_0 is an isomorphism for $\sigma(E/N, N^\circ)$ and $\sigma(M, F'/M^\circ)$, and u_0' is a weak isomorphism of F'/M° onto N°. Clearly, $D_{u'}$ is closed in F_σ' since $D_{u'} = F'$ by the continuity of u, and $u'(F')$ is closed in E_σ', since $u'(F') = u_0'(F'/M^\circ) = N^\circ$. Moreover, $u_0'(\mathfrak{S}_2) = \mathfrak{S}_1$ by (2.4) as asserted.

The condition is sufficient. By the observation following the corollary of (7.3), the assumption implies that u is a weak homomorphism; hence, since $\sigma(E/N, N^\circ)$ is the quotient of $\sigma(E, E')$, u_0 is an isomorphism of E/N onto M for the topologies $\sigma(E/N, N^\circ)$ and $\sigma(M, F'/M^\circ)$. By (7.4) u_0' is an isomorphism of F'/M° onto N° for the topologies $\sigma(F'/M^\circ, M)$ and $\sigma(N^\circ, E/N)$, and we have $u_0'(\mathfrak{S}_2) = \mathfrak{S}_1$ by hypothesis; it follows from (2.4) that both u_0 and u_0^{-1} are continuous for $\mathfrak{T}_1$ and $\mathfrak{T}_2$, and hence u is a topological homomorphism.

COROLLARY. *Every topological homomorphism of E into F is a weak homomorphism.*

This is immediate from the corollary of (7.3). The converse of the last corollary is false as can be seen considering the identity map $E_\tau \to E_\sigma$, where E is a l.c.s. for which $\tau(E, E') \neq \sigma(E, E')$. But even for the Mackey topologies on E and F, a weak homomorphism of E into F is not necessarily open (Exercise 26). The remainder of this section will provide conditions under which it can be concluded that a continuous linear map is open, assuming this property for the weak topologies, or of the adjoint.

7.6

Let E be a Mackey space and let u be a weak homomorphism of E into F such that the subspace $u(E)$ of F is a Mackey space. Then u is a topological homomorphism of E into F.

Proof. Let $N = u^{-1}(0)$ and $M = u(E)$; by (7.5) it suffices to show that whenever B is a convex, circled, compact subset of E_σ' contained in N°, then $B = u'(A)$, where A is equicontinuous in F'. Since M is a Mackey space under the topology induced by F, (4.1) (b)$\Rightarrow$(a) implies that every convex, circled, $\sigma(F'/M^\circ, M)$-compact subset B_1 of F'/M° is the canonical image of an equicontinuous subset A of F'. The assertion follows now from the fact that the map u_0' is an isomorphism of F'/M° onto N° for the topologies $\sigma(F'/M^\circ, M)$ and $\sigma(N^\circ, E/N)$, since u_0 is a weak isomorphism of E/N onto M by hypothesis.

The preceding result applies, in particular, whenever F is a metrizable l.c.s., (3.4). When both E and F are (F)-spaces, we obtain the following set of equivalences with the aid of Banach's homomorphism theorem.

7.7

If E, F are Fréchet spaces and u is a continuous (equivalently, closed) linear map of E into F, the following properties of u are equivalent:

(a) *u is a topological homomorphism.*
(b) *u is a weak homomorphism.*
(c) *u has a closed range.*
(d) *u' is a homomorphism for $\sigma(F', F)$ and $\sigma(E', E)$.*
(e) *u' has a range closed for $\sigma(E', E)$.*

Proof. (a) $\Leftrightarrow$ (b) by (7.5), Corollary, and by (7.6). Corollary 1 of (III, 2.1) shows that (a) $\Leftrightarrow$ (c). Finally, (b) $\Leftrightarrow$ (e) and (c) $\Leftrightarrow$ (d) are immediate from the corollary of (7.3).

However, the equivalent conditions of (7.7) do not imply that u' is a homomorphism for the strong topologies on F' and E', even if $u(E) = F$; the reason for this is the fact that, letting $N = u^{-1}(0)$, $\beta(N^\circ, E/N)$ is in general strictly finer than the topology on N° induced by $\beta(E', E)$ (Exercise 14). In the converse direction the following assertion can be made (Dieudonné-Schwartz [1]).

7.8

If E, F are Fréchet spaces, and u is a continuous linear map of E into F whose adjoint u' is an isomorphism of F'_β into E'_β, then $u(E) = F$ (hence u is a homomorphism).

Proof. Since u' is one-to-one, $u(E)$ is dense in F; hence in view of (7.7), (e) $\Rightarrow$ (c) it suffices to show that $u'(F')$ is closed in E'_σ. Let U be any 0-neighborhood in E; U° is closed and bounded in E'_σ and in E'_β. Since u' is a strong isomorphism, $B = (u')^{-1}(U^\circ)$ is bounded in F'_β (hence in F'_σ) and is closed in F'_σ by the weak continuity of u'; thus B is $\sigma(F', F)$-compact. It follows that $u'(B) = U^\circ \cap u'(F')$ is compact, hence closed in E'_σ. Since U was arbitrary, $u'(F')$ is closed in E'_σ by the Krein-Šmulian theorem (6.4).

Let us add a few remarks on normed spaces. If E and F are normed spaces with respective unit balls B and C, the norm $u \to \|u\| = \sup\{\|ux\|: x \in B\}$ is the natural norm on $\mathscr{L}(E, F)$; it generates the topology of bounded convergence (Chapter III, Section 3). As has been pointed out earlier, the spaces E'_β, F'_β, and consequently $\mathscr{L}(F'_\beta, E'_\beta)$, are Banach spaces under their natural norms. If $u \in \mathscr{L}(E, F)$, then $u' \in \mathscr{L}(F'_\beta, E'_\beta)$ by (7.4) and, by the bipolar theorem,

$$\|u'\| = \sup\{|\langle x, u'y'\rangle|: x \in B, y' \in C^\circ\}$$
$$= \sup\{|\langle ux, y'\rangle|: x \in B, y' \in C^\circ\} = \|u\|;$$

hence $u \to u'$ is a norm isomorphism of $\mathscr{L}(E, F)$ into $\mathscr{L}(F'_\beta, E'_\beta)$, a fact that has been used repeatedly in Section 9 of Chapter III. Let us note the following facts on the strong duals of subspaces and quotients of a normed space E:

(i) *If M is a subspace of E, the Banach space M'_β is norm isomorphic with the normed quotient $E'_\beta/M°$.*

(ii) *If N is a closed subspace of E, then the Banach space $(E/N)'_\beta$, strong dual of the normed space E/N, is norm isomorphic with the subspace $N°$ of the Banach space E'_β.*

If ψ, ϕ denote the canonical maps $M \to E$ and $E \to E/N$, respectively, the norm isomorphisms in question are provided by the biunivocal map ψ'_0 associated with ψ' (Chapter III, Section 1) and by ϕ', respectively; the detailed verification is omitted. In the present circumstances, $\beta(N°, E/N)$ is induced by $\beta(E', E)$ and $\beta(E'/M°, M)$ is the quotient of $\beta(E', E)$. For normed spaces E, F (and, of course, for normable spaces E, F), we obtain the following supplement of (7.7) (cf. Exercise 27):

7.9

Let E, F be normed spaces and $u \in \mathscr{L}(E, F)$. If u is a homomorphism, then its adjoint u' is a strong homomorphism; the converse is true whenever E is complete.

Proof. If u is a homomorphism and $N = u^{-1}(0)$, then the biunivocal map u_0 associated with u is an isomorphism of E/N onto $M \subset F$, where $M = u(E)$. It follows that u'_0 is an isomorphism of $F'/M°$ onto $N°$ for $\beta(F'/M°, M)$ and $\beta(N°, E/N)$, which shows, in view of (i) and (ii) above, that u' is a strong homomorphism.

Conversely, if u' is a strong homomorphism, then $G = u'(F')$ is closed in E'_β. To show that G is closed in E'_σ (equivalently, that $G = N°$) it suffices, by the corollary of (6.4), to prove that $G \cap B°$ (B the unit ball of E) is $\sigma(E', E)$-closed. Since u'_0 is a strong isomorphism of $F'/M°$ onto G and each strongly bounded subset of $F'/M°$ is the canonical image of a strongly bounded subset of F', we have $G \cap B° = u'(H)$, where H is a strongly bounded, hence equicontinuous subset of F'. But the weak closure $\bar{H}$ is compact in F'_σ; since u' is weakly continuous, $u'(\bar{H})$ is compact in E'_σ and obviously equals $G \cap B°$, which completes the proof.

8. THE GENERAL OPEN MAPPING AND CLOSED GRAPH THEOREMS

A continuous linear map of E onto F is necessarily open whenever E and F are Fréchet spaces: This is the essential content, for locally convex spaces, of Banach's homomorphism theorem (Chapter III, Section 2). Even though (III, 2.2) shows that this important result continues to hold when E, F are (LF)-spaces (hence neither metrizable nor Baire spaces, cf. also Chapter III, Exercise 4), the homomorphism theorem, and with it the closed graph theorem, for a long time appeared to be intrinsically of category type and on the whole unrelated to the recent theory of locally convex spaces—in particular, unrelated to duality. Ptak [1], [7] was the first to recognize an intimate relationship with the theorem of Krein-Šmulian and, more generally, with the dual characterization of completeness. We present in this section the essential results of the theory and some related results due to Collins [1] and Mahowald [1].

We define a l.c.s. E to be B-**complete** (or a **Ptak space**) if a subspace $Q \subset E'$ is closed for $\sigma(E', E)$ whenever $Q \cap A$ is $\sigma(E', E)$-closed in A for each equicontinuous set $A \subset E'$; E is said to be B$_r$-**complete** if every *dense* subspace Q of E'_σ such that $Q \cap A$ is $\sigma(E', E)$-closed in A for each equicontinuous set $A \subset E'$, is closed in E'_σ (hence identical with E'). It is immediate that in both definitions the family of all equicontinuous subsets of E' can be replaced by any fundamental subfamily, in particular, by the polars U° of the members U of a 0-neighborhood base in E.

Examples

1. Every (F)-space E is B-complete by the theorem of Krein-Šmulian, (6.4).
2. The Mackey dual E'_τ of an (F)-space E is B-complete (direct verification); in particular, the strong dual of a reflexive (F)-space is B-complete. (For concrete examples, see Chapter III, Section 8.)
3. Every weakly complete l.c.s. is B-complete. In fact, one has $E = E'^*$ (see beginning of Section 6) which shows that $\tau(E', E)$ is the finest locally convex topology on E' (Chapter II, Section 6 and Exercise 7); hence every subspace Q of E' is closed for $\tau(E', E)$ and, by (3.1), for $\sigma(E', E)$. Note that E is weakly complete if and only if it is isomorphic with a product K_0^A of one-dimensional spaces K_0 (cf. Exercise 6).
4. It will be seen shortly that every closed subspace and every separated quotient of a B-complete space is B-complete.

8.1

Every B-*complete space is* B$_r$-*complete, and every* B$_r$-*complete space is complete.*

Proof. The first assertion being trivial, suppose that E is B$_r$-complete and that H is a hyperplane in E' such that $H \cap A$ is $\sigma(E', E)$-closed in A for each equicontinuous set $A \subset E'$. To show that H is closed, it suffices to show that some translate of H is closed in E'_σ; hence we can assume that $0 \in H$. If H were not closed, it would be dense by (I, 4.2) and hence $H = E'$, since E is B$_r$-complete, which is impossible; thus H is closed and the assertion follows from (6.2), Corollary 3.

The following result is due to Collins [1].

8.2

Every closed subspace of a Ptak space (respectively, B$_r$-*complete space) is a Ptak space (respectively,* B$_r$-*complete).*

Proof. Let M be a closed subspace of E. We identify the dual of M with $E'/M°$ and conclude from (4.1), Corollary 1, that the weak topology $\sigma(E'/M°, M)$ is the quotient of $\sigma(E', E)$; denote by ψ' the quotient map $E' \to E'/M°$. We give the proof for both assertions simultaneously, understanding by Q any subspace of $E'/M°$ or a weakly dense subspace of $E'/M°$, accordingly as E is assumed to be a Ptak space or only B_r-complete.

Let Q be such that $Q \cap V°$ is weakly closed in $E'/M°$ for any $V \in \mathfrak{B}$ and let $\mathfrak{B}$ be a 0-neighborhood base in M; we have to show that Q is closed for $\sigma(E'/M°, M)$. We can assume that $\mathfrak{B} = \{U \cap M : U \in \mathfrak{U}\}$, where $\mathfrak{U}$ is a base of closed, convex 0-neighborhoods in E; it follows that $\psi'(U°) = V°$ for $V = U \cap M$, $U \in \mathfrak{U}$, by (1.5), Corollary 2. $P = (\psi')^{-1}(Q)$ is a subspace of E'; since ψ' is continuous, $(\psi')^{-1}(V° \cap Q) = (U° + M°) \cap P$ is closed in E'_σ. Since $U°$ is compact for $\sigma(E', E)$, $U°$ is closed in $U° + M°$, and hence $U° \cap P$ is closed in $(U° + M°) \cap P$. Since the latter set is closed in E'_σ, so is $U° \cap P$. Hence P, which is dense if Q is, is closed in E'_σ by hypothesis and since $M° \subset P$, it follows that $\psi'(P) = Q$ is closed for the quotient topology $\sigma(E'/M°, M)$, which completes the proof.

Next we prove Ptak's central result. To formulate it conveniently, the following definition is useful: Let E, F be t.v.s.; a linear map u of E into F is called **nearly open** if for each 0-neighborhood $U \subset E$, $u(U)$ is dense in some 0-neighborhood in $u(E)$. Evidently u is nearly open if and only if it maps each open subset $G \subset E$ into the interior (taken in $u(E)$) of $\overline{u(G)}$ (cf. Exercise 28). If E, F are l.c.s., then for u to be nearly open it suffices that for each convex 0-neighborhood $U \subset E$, $u(U)$ be weakly dense in some 0-neighborhood in $u(E)$. Note also that every linear map u of a l.c.s. E onto a barreled space F is nearly open.

8.3

Theorem. *Consider the following properties of a locally convex space E:*

(a) *E is a Ptak space.*

(b) *Every continuous, nearly open linear map of E into any l.c.s. F is a topological homomorphism.*

(c) *E is B_r-complete.*

(d) *Every biunivocal, continuous, and nearly open linear map of E into any l.c.s. F is an isomorphism.*

Then (a) $\Leftrightarrow$ (b) *and* (c) $\Leftrightarrow$ (d).

Proof. It will be sufficient to prove the equivalence of (a) and (b); a proof of (c) $\Leftrightarrow$ (d) will then be obtained, in view of the corollary of (2.3), by restricting u to biunivocal maps and Q to dense subspaces of E'.

(a) $\Rightarrow$ (b): Let u be a continuous, nearly open linear map of E into F; we can assume that $u(E) = F$. If $N = u^{-1}(0)$ and u_0 is the continuous linear map of E/N onto F associated with u, then u_0 is nearly open. Since $E \to E/N$ is open, it suffices to show that u_0 is open. Let U be any closed, convex 0-neighborhood in E/N; if $V = u_0(U)$, then $\overline{V}$ is a 0-neighborhood in F. Denote by u_0' the adjoint of u_0; by the corollary of (2.3) $Q = u_0'(F')$ is a dense subspace of $N^\circ \subset E_\sigma'$. By (2.3) (a) we have

$$(u_0')^{-1}(U^\circ) = [u_0(U)]^\circ = V^\circ = \overline{V}^\circ.$$

Thus V° is closed and equicontinuous, and therefore compact in F_σ'; u_0' being continuous for $\sigma(F', F)$ and $\sigma(E', E)$, it follows that $u_0'(V^\circ) = U^\circ \cap Q$ is compact and hence closed in E_σ'. U being arbitrary, the assumed B-completeness of E implies that Q is closed in E_σ' (hence $Q = N^\circ$); thus u_0 is a weak isomorphism by the corollary of (7.3). Since U is closed in E/N, it is weakly closed (by convexity), which implies that $V = u_0(U)$ is weakly closed in F, whence $V = \overline{V}$; it follows that u_0 is open.

  (b) $\Rightarrow$ (a): Let $\mathfrak{U}$ be the family of all convex, circled 0-neighborhoods in E, and let Q be a subspace of E' such that $Q \cap U^\circ$ is $\sigma(E', E)$-closed for each $U \in \mathfrak{U}$; we have to show that Q is closed in E_σ'. Denoting polars with respect to $\langle E, E' \rangle$ by $^\circ$, let F be the quotient space E/Q° without topology; the canonical bilinear form on $F \times Q$ places F and Q in duality (Section 4). Consider the family $\mathfrak{W} = \{(U^\circ \cap Q)^\circ : U \in \mathfrak{U}\}$ of subsets of E and the locally convex topology $\mathfrak{T}$ on F for which $\phi(\mathfrak{W})$ is a neighborhood base of 0, where ϕ is the quotient map $E \to E/Q^\circ$. With respect to the duality $\langle F, Q \rangle$, $\mathfrak{T}$ is the $\mathfrak{S}$-topology on F, where $\mathfrak{S} = \{U^\circ \cap Q : U \in \mathfrak{U}\}$; by assumption, each $S \in \mathfrak{S}$ is closed, hence compact in E_σ' and, therefore, compact for $\sigma(Q, F)$. Clearly $\mathfrak{S}$ covers Q and the family of all subsets of sets $S \in \mathfrak{S}$ is saturated under $\sigma(Q, F)$; hence (3.2) implies that $\mathfrak{T}$ is consistent with $\langle F, Q \rangle$. Now if $U \in \mathfrak{U}$, the polar of $\phi(U)$ with respect to $\langle F, Q \rangle$ is $Q \cap U^\circ$, whence $\phi[(U^\circ \cap Q)^\circ]$ is the $\mathfrak{T}$-closure of $\phi(U)$ by the bipolar theorem, which shows that ϕ is a nearly open linear map of E onto $(F, \mathfrak{T})$. Since $\mathfrak{T}$ is coarser on F than the quotient topology of E/Q°, ϕ is also continuous and hence open by hypothesis; this proves $\mathfrak{T}$ to be the quotient topology of E/Q°. From (4.1), Corollary 1 and Corollary 2, it follows now that Q is necessarily closed in E_σ'.

  The theorem is proved.

  Among the corollaries of the foregoing result, the most striking is the following which extends Banach's classical theorem substantially with respect to both the domain and the range spaces.

COROLLARY 1. (Homomorphism Theorem). *Every continuous linear map of a Ptak space E onto a barreled space F is a topological homomorphism.*

Proof. Since $u(E) = F$, for each convex, circled 0-neighborhood $U \subset E$ the closure of $u(U)$ is a barrel in F, whence u is nearly open. Thus u is open since E is a Ptak space.

COROLLARY 2. *Let E be a Ptak space and let F be a l.c.s. such that $u(E) = F$ for some continuous, nearly open linear map u; F is a Ptak space.*

Proof. Let G be any l.c.s. and let v be a continuous, nearly open linear map of F into G. Since u is open, $v \circ u$ is nearly open and hence also open, E being a Ptak space; G and v being arbitrary, it follows that F is a Ptak space.

Applying Corollary 2 to the canonical map $E \to E/N$, where N is a closed subspace of E, we obtain

COROLLARY 3. *Every separated quotient of a Ptak space is a Ptak space.*

In other words, B-completeness is preserved by quotients over closed subspaces, a property not shared by the weaker concept of completeness (Exercise 11). For an enlightening discussion of these questions the reader is referred to Ptak [7].

To obtain a correspondingly generalized closed graph theorem, we need an open mapping theorem for not necessarily continuous linear maps with closed graph. The following general open mapping theorem, which contains (8.3), Corollary 1, as a special case, is also due to Ptak.

8.4

Theorem. *Let E be a Ptak space and let u be a nearly open linear map with domain dense in E and range in any l.c.s. F, and such that the graph of u is closed in $E \times F$. Then u is open. If, in addition, u is one–to–one it suffices for the conclusion that E be B_r-complete.*

Proof. The general case can be reduced to the case in which u is one–to–one, for by (7.2) the null space N of u is closed in E, E/N is a Ptak space by (8.3), Corollary 3, and the biunivocal map associated with u is by (7.2) densely defined, has a closed graph, and is obviously nearly open if u is. We assume, hence, that u is biunivocal with domain E_0 dense in E and graph closed in $E \times F$, that E is B_r-complete, and finally that $u(E_0)$ is dense in F, which is clearly no restriction of generality.

Let u' be the adjoint of u with domain F_0' dense in F_σ', (7.1), and let $Q = u'(F_0')$; Q is dense in E_σ' (cf. proof of (7.3)). Denote by U any closed, convex 0-neighborhood in E; then $U^\circ = (U \cap E_0)^\circ$ by (1.5), since $U \cap E_0$ is dense, hence weakly dense in U. By (2.1) u is continuous for $\sigma(E_0, Q)$ and $\sigma(F, F_0')$, and u' is continuous for $\sigma(F_0', F)$ and $\sigma(Q, E_0)$. Letting $V = u(U \cap E_0)$, the closure $\overline{V}$ is a 0-neighborhood in $u(E_0)$, since u is nearly open by hypothesis; if V° is the polar of V with respect to $\langle F, F' \rangle$, then $V^\circ \subset F_0'$, for $y' \in V^\circ$ implies that $x \to \operatorname{Re} \langle ux, y' \rangle$ is $\leqq 1$ on $U \cap E_0$ and hence continuous on E_0. It follows that V° is compact for $\sigma(F_0', F)$; hence $u'(V^\circ) = U^\circ \cap Q$ is compact for $\sigma(Q, E_0)$ by the continuity of u' for these topologies. Since E_0 is dense in E and $U^\circ \subset E'$ is equicontinuous, (III, 4.5) implies that $U^\circ \cap Q$ is compact for $\sigma(E', E)$. Since E is B_r-complete by hypothesis, Q is closed in E_σ' (hence $Q = E'$), which by (7.3) implies that u is weakly open, or equivalently that u^{-1} is weakly continuous. Since $\overline{V}$ (by convexity) is the weak closure of V in $u(E_0)$, we have

$$U \cap E_0 = u^{-1}(V) \subset u^{-1}(\bar{V}) \subset U \cap E_0,$$

for $U \cap E_0$ is weakly closed in E_0. Hence $u(U \cap E_0) = V = \bar{V}$, which proves u to be an open map, thus completing the proof.

The following general closed graph theorem (cf. Robertson-Robertson [1]) is now an easy consequence of the preceding results.

8.5

Theorem. *Let E be barreled and let F be B_r-complete. If u is a linear map of E into F with closed graph, then u is continuous.*

Proof. As in the foregoing proof, the general case can be reduced to the case in which u is one–to–one. For $N = u^{-1}(0)$ is closed in E and the bi-univocal map of E/N into F associated with u has a closed graph by (7.2), and E/N is barreled; it can moreover be assumed that $u(E)$ is dense in F, since by (8.2), each closed subspace of a B_r-complete space is B_r-complete.

Suppose, hence, that u is one–to–one onto a dense subspace of F. Then since u is closed, u^{-1} is a closed linear map, densely defined in F, onto E, which is nearly open, since E is barreled; by (8.4) u^{-1} is open, hence u continuous.

COROLLARY. *Let E be a l.c.s. that is barreled and B_r-complete. If E is the algebraic direct sum of two closed subspaces M, N, then the sum is topological: $E = M \oplus N$.*

Proof. Let p be the projection of E onto M vanishing on N. Since M is B_r-complete by (8.2), it suffices to show that the graph of p is closed in $E \times M$. But this is quickly seen to be equivalent with the closedness of N.

We conclude this brief account with a result, due to Mahowald [1], suggesting that the open mapping and closed graph theorems, in their general forms (8.4) and (8.5), have been extended to the natural limit of their validity.

8.6

Let E be a l.c.s. such that for every Banach space F, a closed linear map of E into F is necessarily continuous; then E is barreled.

Proof. Let D be any barrel in E, and denote by ϕ the canonical map of E into the Banach space $\tilde{E}_D$ (for notation, see Chapter III, Section 7); one has $D = \{x \in E: \|\phi(x)\| \le 1\}$, since D is closed. We have to show that D is a neighborhood of 0; in view of the hypothesis on E, it suffices to show that the graph G of ϕ is closed in $E \times \tilde{E}_D$.

Now if $(x_0, y_0) \notin G$, then $\|\phi(x_0) - y_0\| > 2\varepsilon$ for a suitable $\varepsilon > 0$; since $\phi(E)$ is dense in $\tilde{E}_D$, there exists $y_1 \in \phi(E)$ such that $\|y_0 - y_1\| < \varepsilon$, and it follows that $\|\phi(x_0) - y_1\| > \varepsilon$. The set $A = \{x \in E: \|\phi(x) - y_1\| \le \varepsilon\}$ is closed in E, since it is a translate of εD; hence $W = E \sim A$ is open. Denoting by B_ε the open ball $\{y \in \tilde{E}_D: \|y - y_1\| < \varepsilon\}$, it follows that $(W \times B_\varepsilon) \cap G = \varnothing$. Since $(x_0, y_0) \in W \times B_\varepsilon$, we conclude that G is closed in $E \times \tilde{E}_D$.

9. TENSOR PRODUCTS AND NUCLEAR SPACES

We consider briefly the relations between spaces of linear mappings (Chapter III, Section 3) and spaces of bilinear forms (Chapter III, Section 5). Let E, F be l.c.s. with respective duals E', F'; as before, we shall use subscripts when referring to the most frequent topologies (as in E'_σ, E'_τ, E'_β etc. (cf. Section 5)).

If v is a separately continuous bilinear form on $E \times F$, it is quickly seen that the formula

$$v(x, y) = \langle ux, y \rangle = \langle x, u'y \rangle$$

defines a continuous linear map $u \in \mathcal{L}(E, F'_\sigma)$ with adjoint $u' \in \mathcal{L}(F, E'_\sigma)$ (cf. (7.4)); in fact, the mappings $v \to u$ and $v \to u'$ are algebraic isomorphisms of $\mathfrak{B}(E, F)$ onto $\mathcal{L}(E, F'_\sigma)$ and of $\mathfrak{B}(E, F)$ onto $\mathcal{L}(F, E'_\sigma)$, respectively. By virtue of these isomorphisms, which are called **canonical**, we shall frequently identify spaces of linear maps with spaces of bilinear forms. We note that under $v \to u$ the space $\mathfrak{B}(E, F)$ of continuous bilinear forms on $E \times F$ is carried onto the subspace of $\mathcal{L}(E, F'_\sigma)$, each of whose elements maps a suitable 0-neighborhood in E onto an equicontinuous set in F'.

Furthermore, to a given $\mathfrak{S} \times \mathfrak{T}$-topology on $\mathfrak{B}(E, F)$ (Chapter III, Section 5) there corresponds on $\mathcal{L}(E, F'_\sigma)$ the topology of uniform convergence on $\mathfrak{S}$ with respect to the $\mathfrak{T}$-topology on F'. Conversely, to a given $\mathfrak{S}$-topology on $\mathcal{L}(E, F)$ (which is, in general, a proper subspace of $\mathcal{L}(E, F_\sigma)$) there corresponds the $\mathfrak{S} \times \mathfrak{T}$-topology on the canonical image of $\mathcal{L}(E, F)$ in $\mathfrak{B}(E, F'_\sigma)$, $\mathfrak{T}$ denoting the family of all equicontinuous subsets of F'.

However, from (7.4) it follows that $\mathcal{L}(E, F_\sigma) = \mathcal{L}(E_\tau, F)$; hence $\mathcal{L}(E, F)$ can be identified with $\mathfrak{B}(E, F'_\sigma)$ if E is a Mackey space. Dually, $\mathcal{L}(E', F) = \mathcal{L}(E'_\sigma, F_\sigma)$, hence $\mathcal{L}(E'_\tau, F)$ can be identified with $\mathfrak{B}(E'_\sigma, F'_\sigma)$, and if $\mathfrak{S}$ is the family of all equicontinuous subsets of E', then the $\mathfrak{S}$-topology on $\mathcal{L}(E', F)$ corresponds to the topology of bi-equicontinuous convergence on $\mathfrak{B}(E'_\sigma, F'_\sigma)$; under this topology, $\mathcal{L}(E'_\tau, F)$ will be denoted by $\mathcal{L}_e(E'_\tau, F)$.

9.1

Let E, F be l.c.s. not equal to $\{0\}$. The space $\mathfrak{B}_e(E'_\sigma, F'_\sigma)$ (equivalently, $\mathcal{L}_e(E'_\tau, F)$) is complete if and only if both E and F are complete. In this case $E \widetilde{\otimes} F$ can be identified with the closure of $E \otimes F$ in $\mathcal{L}_e(E'_\tau, F)$.

Proof. Let f be a bilinear form on $E' \times F'$ which is the limit of a filter in $\mathfrak{B}(E'_\sigma, F'_\sigma)$, uniformly on each product $S \times T$ where S, T are arbitrary equicontinuous subsets of E', F' respectively. It follows that for each $y' \in F'$, the partial map $f_{y'}$ is $\sigma(E', E)$-continuous on S, whence $f_{y'}$ is a continuous linear form on E'_σ by (6.2), Corollary 2, if E is complete; likewise $f_{x'}$ ($x' \in E'$) is continuous on F'_σ if F is complete, which proves the condition to be sufficient. Conversely, if $\mathfrak{B}_e(E'_\sigma, F'_\sigma)$ is complete and $x \in E$, $y \in F$ are non-zero elements, then the closed subspaces $x \otimes F$ (note that $x \otimes F$ is closed in $\mathcal{L}_e(E'_\tau, F)$) and

$E \otimes y$ of $\mathfrak{B}_e(E'_\sigma, F'_\sigma)$ are complete, and isomorphic with F and E respectively; hence E and F are complete if $\mathfrak{B}_e(E'_\sigma, F'_\sigma)$ is complete. The final assertion is also clear, since $E \tilde{\otimes} F$ is by definition the completion of the subspace $E \otimes F$ of $\mathfrak{B}_e(E'_\sigma, F'_\sigma)$ (Chapter III, Section 6).

Recall that the dual of $E \otimes F$ for the projective (respectively, inductive) tensor product topology is $\mathscr{B}(E, F)$ (respectively, $\mathfrak{B}(E, F)$); our next objective is to determine the dual of $E \otimes F$ for the topology of bi-equicontinuous convergence (equivalently, the dual of $E \tilde{\otimes} F$). To begin with, a base of 0-neighborhoods for this topology is formed by the polars, with respect to the duality between $E \otimes F$ and $E' \otimes F' \subset B(E, F)$ (cf. Chapter III, end of Section 6), of the sets $S \otimes T$, where S, T are arbitrary equicontinuous subsets of E', F' respectively. Hence the bipolars $(\Gamma S \otimes T)^{\circ\circ}$ in $B(E, F)$ (equivalently, the closures $(\Gamma S \otimes T)^-$ for $\sigma(B(E, F), E \otimes F)$) form a fundamental family of equicontinuous sets in the algebraic dual $B(E, F)$ of $E \otimes F$. Since, clearly, each set $S \otimes T$ is equicontinuous for the projective topology, the dual $\mathscr{J}(E, F)$ of $E \tilde{\otimes} F$ is a subspace of $\mathscr{B}(E, F)$.

Recall also (Chapter II, Section 2, Example 3) that a Radon measure on a compact space X is a continuous linear form $\mu \in \mathscr{M}(X) = \mathscr{C}(X)'$, where $\mathscr{C}(X)$ is the Banach space of scalar-valued continuous functions on X. It is customary to write $\mu(f) = \langle f, \mu \rangle = \int_X f d\mu$; $\|\mu\|$ denotes the norm of μ in the strong dual of $\mathscr{C}(X)$. In the following proposition, $S \times T$ is the compact product of S and T under the induced weak topologies.

9.2

The dual $\mathscr{J}(E, F)$ of $E \tilde{\otimes} F$ consists exactly of those elements $v \in \mathscr{B}(E, F)$ that can be represented in the form

$$u \to v(u) = \langle u, v \rangle = \int_{S \times T} u_0(x', y') d\mu(x', y'),$$

where S, T are suitable closed, equicontinuous subsets of E'_σ, F'_σ respectively, and where u_0 is the restriction of the bilinear form u on $E' \times F'$ to $S \times T$. If A is an equicontinuous subset of $\mathscr{J}(E, F)$, the elements $v \in A$ can be represented with $S \times T$ fixed and μ running through a norm bounded subset of $\mathscr{M}(S \times T)$.

Proof. Since each $u \in E \otimes F$, viewed as a bilinear form on $E' \times F'$, has a restriction to $S \times T$ which is continuous for the topology induced by $E'_\sigma \times F'_\sigma$, the same is true for any $u \in E \tilde{\otimes} F$, since the latter are limits, uniformly on each product $S \times T$ of equicontinuous sets, of elements of $E \otimes F$. Hence $u_0 \in \mathscr{C}(S \times T)$, and the integral defines a linear form v on $E \tilde{\otimes} F$. Moreover, if W denotes the 0-neighborhood in $E \tilde{\otimes} F$ which is the polar (with respect to the duality $\langle E \tilde{\otimes} F, E' \otimes F' \rangle$) of $\Gamma S \otimes T$, it follows that $|v(u)| \leq \|\mu\|$ whenever $u \in W$; hence $v \in \mathscr{J}(E, F)$.

For the remainder it suffices to prove that any equicontinuous set $A \subset \mathscr{J}(E, F)$ can be represented as asserted. If A is equicontinuous, there exist compact equicontinuous subsets S, T of E'_σ, F'_σ, respectively, such that $A \subset (\Gamma S \otimes T)^{\circ\circ} = W^\circ$, where W is the polar of $\Gamma S \otimes T$, hence a 0-neighborhood in $E \tilde{\otimes} F$. Consider the map $u \to u_0$ of $E \tilde{\otimes} F$ into $\mathscr{C}(S \times T)$; the associated map ψ of $(E \tilde{\otimes} F)_W$ (Chapter III, Section 7) into $\mathscr{C}(S \times T)$ is a norm isomorphism. Hence the adjoint ψ' maps $\mathscr{M}(S \times T)$ homomorphically onto $[\mathscr{J}(E, F)]_{W^\circ}$ by (7.9), and by (7.5) the set A, being equicontinuous in $[\mathscr{J}(E, F)]_{W^\circ}$, is the image under ψ' of an equicontinuous subset of $\mathscr{M}(S \times T)$, which completes the proof.

The elements of $\mathscr{J}(E, F)$ are called **integral bilinear forms** on $E \times F$, and the linear maps $u \in \mathscr{L}(E, F'_\sigma)$ and $u' \in \mathscr{L}(F, E'_\sigma)$ originating from a bilinear form $v \in \mathscr{J}(E, F)$ are called **integral linear maps**. It follows that an integral map $u \in \mathscr{L}(E, F'_\sigma)$ is of the form

$$x \to u(x) = \int_{S \times T} \langle x, x' \rangle y' \, d\mu(x', y')$$

for suitable weakly closed, equicontinuous sets S, T. The integral $\int f d\mu$ (called weak integral) is defined as the linear form $y \to \int \langle y, f \rangle d\mu$ in F, which is certainly continuous whenever f is a continuous function on $S \times T$ into F'_σ such that $f(S \times T)$ is contained in a compact, convex, circled subset of F'_σ.

It is evident from the preceding that a nuclear map $x \to \sum_{i=1}^{\infty} \lambda_i \langle x, x'_i \rangle y'_i$ of E into F'_σ (cf. (III, 7.1)) is integral if the sequence $\{y'_i\}$ is equicontinuous in F'. We shall see shortly that whenever E is a nuclear space, and F is any l.c.s., then $\mathscr{J}(E, F) = \mathscr{B}(E, F)$ and every integral map in $\mathscr{L}(E, F'_\sigma)$ is nuclear. We first prove a dual characterization of nuclear spaces for which, in turn, the following lemma is needed.

LEMMA 1. *Let E, F be l.c.s. and let u be a linear map of E into F which maps a suitable 0-neighborhood in E into a weakly compact subset of F. Then the second adjoint u'' maps the bidual E'' into $F \subset F''$.*

Proof. Let U be a convex 0-neighborhood in E such that $u(U) \subset C$, where C is weakly compact in F. Since u'' is continuous for $\sigma(E'', E')$ and $\sigma(F'', F')$, and since E is $\sigma(E'', E')$-dense in E'' by the bipolar theorem, it follows that u'' is the continuous extension of u to E'' into F'' with respect to the topologies $\sigma(E'', E')$ and $\sigma(F'', F')$. Since U is $\sigma(E'', E')$-dense in $U^{\circ\circ}$ and C is weakly compact, it follows that $u''(U^{\circ\circ}) \subset C$. But $U^{\circ\circ}$, being a 0-neighborhood for the natural topology of E'', is radial, whence $u''(E'') \subset F$.

Let us point out that in the remainder of this section, frequent use will be made of the notational devices introduced at the beginning of Section 7 of Chapter III (cf. also Chapter III, Exercise 3). In particular, whenever E is a l.c.s. and B a convex, circled, and bounded subset of E'_σ, we shall write E'_B in place of $[E']_B$; we set $E'_B = \{0\}$ if $B = \varnothing$.

9.3

A l.c.s. E is nuclear if and only if for each closed, convex, circled subset $A \subset E'_\sigma$ which is equicontinuous, there exists another set B with these properties, and such that $A \subset B$ and the canonical imbedding $E'_A \to E'_B$ is nuclear.

Proof. The condition is necessary. In fact, $U = A^\circ$ is a convex, circled 0-neighborhood in E; if E is nuclear, there exists by (III, 7.2) a convex, circled 0-neighborhood $V \subset U$ such that the canonical map $\phi_{U,V} \colon \tilde{E}_V \to \tilde{E}_U$ is nuclear. Let $B = V^\circ$; it is immediate that $A \subset B$ and that $\psi_{B,A} \colon E'_A \to E'_B$, being the adjoint of $\phi_{U,V}$, is nuclear.

The condition is sufficient. Let U be a given convex, circled 0-neighborhood in E; by (III, 7.2) it suffices to show that there exists another such 0-neighborhood V for which $V \subset U$ and $\phi_{U,V}$ is nuclear. Put $A = U^\circ$; by hypothesis, there exist closed, convex, circled, and equicontinuous subsets B, C of E'_σ such that $A \subset B \subset C$ and such that the canonical maps $\psi_{C,B}$ and $\psi_{B,A}$ are both nuclear. Let $W = B^\circ$, $V = C^\circ$ (polars with respect to $\langle E, E' \rangle$), and denote by F, G, H the strong duals of E'_C, E'_B, E'_A, respectively; F, G, H are the respective strong biduals of $\tilde{E}_V, \tilde{E}_W, \tilde{E}_U$ and the second adjoints $\phi''_{W,V}$ and $\phi''_{U,W}$ are evidently nuclear. By Lemma 1 $\phi''_{U,W}$ maps G into $\tilde{E}_U$, since $\phi_{U,W}$ is compact, being the restriction to $\tilde{E}_W$ of a nuclear (hence compact) map. Now by (III, 7.1) $\phi''_{W,V}$ is of the form $\sum_1^\infty \lambda_n f_n \otimes y_n$, where $(\lambda_n) \in l^1$, $\{y_n\}$ is a bounded sequence in G, and $\{f_n\}$ is an equicontinuous sequence of linear forms on F. Let $z_n = \phi''_{U,W}(y_n)$ and let g_n be the restriction of f_n to $\tilde{E}_V$ $(n \in N)$; then $\{z_n\}$ is bounded in $\tilde{E}_U$ and $\{g_n\}$ equicontinuous on $\tilde{E}_V$. Since $\phi_{U,V}$ is the restriction of $\phi''_{U,W} \circ \phi''_{W,V}$ to $\tilde{E}_V$, it is of the form $\sum_1^\infty \lambda_n g_n \otimes z_n$ and hence nuclear, which completes the proof.

9.4

Theorem. *Let E be a nuclear space, let F be any locally convex space, and endow $E \otimes F$ with its projective tensor product topology. The canonical imbedding of $E \otimes F$ into $\mathfrak{B}(E'_\sigma, F'_\sigma)$ is a topological isomorphism onto a dense subspace of $\mathfrak{B}_e(E'_\sigma, F'_\sigma)$.*

REMARK. The property expressed by the theorem is actually characteristic of nuclear spaces and is used by Grothendieck ([13], chap. II, def. 4 and theor. 6) to define nuclear spaces. We shall not show here that the validity of the assertion, for a given locally convex space E and any l.c.s. F, implies that E is nuclear. For it will be shown below (Section 10) that for E to be nuclear, it is sufficient that the assertion of (9.4) hold for $F = l^1$.

Proof of (9.4). It is clear that the canonical imbedding of $E \otimes F$ into $\mathfrak{B}(E'_\sigma, F'_\sigma)$ is an algebraic isomorphism (cf. Chapter III, discussion preceding (6.3)). The proof consists now of two steps: first we show that $E \otimes F$ is dense in $\mathfrak{B}_e(E'_\sigma, F'_\sigma)$; second that $\mathfrak{B}_e(E'_\sigma, F'_\sigma)$ induces the projective topology on $E \otimes F$.

1. Let us identify $\mathfrak{B}_e(E'_\sigma, F'_\sigma)$ with $\mathscr{L}_e(E'_t, F)$ as before. It is then sufficient to show that, given $u \in \mathscr{L}(E'_t, F)$, a convex, circled, closed equicontinuous set $A \subset E'_t$ and a 0-neighborhood V in F, there exists $u_0 \in E \otimes F$ such that $u(x') - u_0(x') \in V$ for all $x' \in A$. Let $U = A^\circ$ and let $W \subset U$ be a 0-neighborhood in E which is convex, circled, and such that $\tilde{E}_W \to \tilde{E}_U$ is nuclear, say,

$$\phi_{U,W} = \sum_1^\infty \lambda_i y'_i \otimes z_i,$$ where $(\lambda_i) \in l^1$, and $\{y'_i\}$, $\{z_i\}$ are bounded sequences in E'_B ($B = W^\circ$) and $\tilde{E}_U$, respectively. It follows that the canonical imbedding of E'_A into E'_B is of the form $\sum_1^\infty \lambda_i z_i \otimes y'_i$. Since E_U is dense in $\tilde{E}_U$, each z_i (viewed as a linear form on E'_A) can uniformly on A be approximated by (the restrictions to E'_A of) suitable elements $x \in E$; hence if $\varepsilon > 0$ is preassigned, there exists an integer n and elements $x_i \in E$ ($i = 1, \ldots, n$) such that

$$\sum_{i=1}^n \lambda_i \langle x_i, x' \rangle y'_i - x' \in \varepsilon B$$

whenever $x' \in A$. Now if ε is so chosen that $\varepsilon u(B) \subset V$, it follows that $\sum_1^n \lambda_i \langle x_i, x' \rangle u(y'_i) - u(x') \in V$ for $x' \in A$; hence $u_0 = \sum_1^n \lambda_i x_i \otimes u(y'_i)$ satisfies the requirement.

2. Since the dual of $E \otimes F$ for the projective topology is $\mathscr{B}(E, F)$, it is sufficient to show that each equicontinuous set Q in $\mathscr{B}(E, F)$ is contained and equicontinuous in $\mathscr{J}(E, F) = [\mathfrak{B}_e(E'_\sigma, F'_\sigma)]'$. Viewing $\mathscr{B}(E, F)$ as a subspace of $\mathscr{L}(E, F'_\sigma)$, the equicontinuity of Q is tantamount to the existence of a 0-neighborhood $U \subset E$ and an equicontinuous set $B \subset F'$ such that $u(U) \subset B$ whenever $u \in Q$. Of course we can assume that U and B are convex and circled and that B is compact in F'_σ. The map $\tilde{u}$ of $\tilde{E}_U$ into F'_B associated with $u \in Q$, is of norm ≤ 1 in $\mathscr{L}(\tilde{E}_U, F'_B)$. Since E is nuclear, the canonical map $E \to \tilde{E}_U$ is nuclear by (III, 7.2), Corollary 1, say, of the form $\phi_U = \sum_{i=1}^\infty \lambda_i x'_i \otimes y_i$, where it can be assumed that $x'_i \in V^\circ$ and $y_i \in \phi_U(U)^-$ for all $i \in N$, and $\sum_1^\infty |\lambda_i| = c < +\infty$. Since $u = \psi_B \circ \tilde{u} \circ \phi_U$, it follows that

$$u = \sum_{i=1}^\infty \lambda_i x'_i \otimes \tilde{u}(y_i),$$

which shows that u is integral, hence (in view of the identification of $\mathscr{B}(E, F)$ with a subspace of $\mathscr{L}(E, F'_\sigma)$) that $u \in \mathscr{J}(E, F)$. Moreover, $\tilde{u}(y_i) \in B$ for all $u \in Q$ and all $i \in N$; hence the preceding formula shows that $Q \subset c\,(\lceil V^\circ \otimes B)^-$, where the closure is with respect to $\sigma(\mathscr{J}(E, F), E \otimes F)$. Since $\lceil(V^\circ \otimes B)$ is equicontinuous in the dual $\mathscr{J}(E, F)$ of $\mathfrak{B}_e(E'_\sigma, F'_\sigma)$, the same holds for $(\lceil V^\circ \otimes B)^-$ by (III, 4.3) and (III, 4.5), and hence for Q.

The proof is complete.

COROLLARY 1. *If E is a complete nuclear space and if F is any complete l.c.s., then $E \overset{\sim}{\otimes} F$ can be canonically identified with $\mathfrak{B}_e(E'_\sigma, F'_\sigma)$ and with $\mathscr{L}_e(E'_\tau, F)$.*

This is immediate in view of (9.1); let us remark also that if E is nuclear and precompact subsets of E'_τ are equicontinuous, then E'_τ possesses the approximation property by (III, 9.1). More importantly, the identity map of $E \otimes F$ (which is continuous for the projective topology on the domain, and the topology of bi-equicontinuous convergence on the range) is an isomorphism if E is nuclear. Hence:

COROLLARY 2. *If E is nuclear and if F is any l.c.s., then the canonical mapping of $E \overset{\sim}{\otimes} F$ into $E \overset{\sim}{\otimes} F$ is a topological isomorphism of the first space onto the second.*

From Part 2 of the proof of (9.4), we obtain the **kernel theorem**:

9.5

If E is nuclear and if F is locally convex, then every $v \in \mathscr{B}(E, F)$ originates from a space $E'_A \overset{\sim}{\otimes} F'_B$, where A, B are suitable equicontinuous subsets of E', F', respectively. Equivalently, every continuous bilinear form v on $E \times F$ is of the form

$$(x, y) \to v(x, y) = \sum_{i=1}^{\infty} \lambda_i \langle x, x'_i \rangle \langle y, y'_i \rangle,$$

where $(\lambda_i) \in l^1$ and $\{x'_i\}$, $\{y'_i\}$ are equicontinuous sequences.

Note that by (III, 6.5), these sequences can even be supposed to be null sequences in E'_A, F'_B, respectively. We are now in a position to establish two results that will furnish a large number of additional examples of nuclear spaces.

9.6

Theorem. *The strong dual of every nuclear (F)-space is nuclear.*

Proof. If E is a nuclear (F)-space, then E is reflexive (in fact, an (M)-space) by (III, 7.2), Corollary 2; hence E'_β is reflexive, (5.6), Corollary 1, which implies that $E'_\beta = E'_\tau$. By (III, 7.2) it suffices to show that every $u \in \mathscr{L}(E'_\tau, F)$ is nuclear, F being an arbitrary Banach space. Now $\mathscr{L}(E'_\tau, F)$ can be identified with $\mathfrak{B}(E'_\sigma, F'_\sigma)$ and, by (9.4), Corollaries 1 and 2, with $E \overset{\sim}{\otimes} F$, since E, F are complete. The assertion follows now from (III, 6.4) and (III, 7.1).

REMARK. The converse of (9.6) is also true: If E is an (F)-space whose strong dual is nuclear, then E is nuclear; equivalently, the strong dual of a complete nuclear (DF)-space is nuclear (Exercise 33). In general, however, the strong dual of a nuclear space fails to be nuclear: the product K_0^d (d any cardinal) is nuclear by (III, 7.4), but its strong dual fails to be nuclear if d is uncountable (Exercise 31).

It follows from (III, 7.4) that if E is any l.c.s. and F is a nuclear space, then $\mathscr{L}_s(E, F)$ (topology of simple convergence) is nuclear; in fact, $\mathscr{L}_s(E, F)$ is isomorphic with a subspace of the nuclear space F^E. Under additional assumptions on E, a corresponding result holds for $\mathscr{L}_b(E, F)$ (topology of bounded convergence).

9.7

Let E be a semi-reflexive space whose strong dual is nuclear and let F be any nuclear space. Then $\mathscr{L}_b(E, F)$ is a nuclear space.

Proof. Since E is semi-reflexive (Section 5), each bounded subset of E is an equicontinuous subset in the dual of E'_β, and hence $\mathscr{L}_b(E, F)$ can be identified with a subspace of $\mathfrak{B}_e(E_\sigma, F'_\sigma)$. Since E'_β is nuclear, (9.4) implies that the completion of $\mathfrak{B}_e(E_\sigma, F'_\sigma)$ can be identified with $E'_\beta \,\tilde{\otimes}\, F$. Now $E'_\beta \,\tilde{\otimes}\, F$ is nuclear by (III, 7.5) and hence $\mathscr{L}_b(E, F)$ is nuclear by (III, 7.4). Compare Exercise 34.

Examples

1. From (9.6) it follows that the strong duals of the nuclear (F)-spaces enumerated in Chapter III, Section 8, are nuclear; in particular, the spaces $\mathscr{D}'_C$ and $\mathscr{S}'$ are nuclear, and also $\mathscr{H}'$ is nuclear.
2. The space $\mathscr{D}'$ of distributions, strong dual of $\mathscr{D}$, is nuclear. Since $\mathscr{D}$ (Chapter II, Section 6, Example 2) is the strict inductive limit of a sequence of spaces $\mathscr{D}_{G_m}$, each bounded subset of $\mathscr{D}$ is contained in a suitable space $\mathscr{D}_{G_m}$ by (II, 6.5), so (4.1) (a) $\Rightarrow$ (b) implies that the strong dual $\mathscr{D}'$ is isomorphic with a subspace of $\prod_{m=1}^{\infty} \mathscr{D}'_{G_m}$; since the latter is nuclear, $\mathscr{D}'$ is nuclear. This is an example of the situation indicated in the remarks following (4.5).
3. (9.7) implies now that the spaces $\mathscr{L}_b(\mathscr{D})$, $\mathscr{L}_b(\mathscr{D}')$, $\mathscr{L}_b(\mathscr{S})$, etc. are nuclear. This implies that the (F)-space $\mathscr{E}$ (Chapter III, Section 8, Example 3), as well as its strong dual $\mathscr{E}'$, is nuclear. For each $f \in \mathscr{E}$ defines, by virtue of the multiplication operator $g \to fg$, a continuous endomorphism of $\mathscr{D}$; the corresponding imbedding $\mathscr{E} \to \mathscr{L}_b(\mathscr{D})$ is an isomorphism. Since $\mathscr{E}$ and $\mathscr{E}'$ are reflexive, it also follows that $\mathscr{L}_b(\mathscr{E})$ and $\mathscr{L}_b(\mathscr{E}')$ are nuclear spaces.

If E, F are l.c.s., the dual of $E \,\tilde{\otimes}\, F$ is $\mathscr{B}(E, F)$, and it is clear that the $\mathfrak{B}_1 \times \mathfrak{B}_2$-topology ($\mathfrak{B}_1, \mathfrak{B}_2$ the respective families of all bounded subsets of E, F) on $\mathscr{B}(E, F)$, also called the topology of **bi-bounded convergence**, is coarser than the strong topology $\beta(\mathscr{B}(E, F), E \,\tilde{\otimes}\, F)$; it is natural to ask for conditions under which the two topologies agree. This is manifestly true if E, F are normed spaces; but even if E, F are non-normable (F)-spaces, the answer seems to be unknown (Grothendieck [13], chap. I, §1, "problème des topologies"). Alternatively the problem is this: Given a bounded subset $B \subset E \,\tilde{\otimes}\, F$, do there exist bounded subsets $B_1 \subset E$ and $B_2 \subset F$ such that $B \subset (\Gamma B_1 \otimes B_2)^-$? Grothendieck gave an affirmative answer if E, F are both (DF)-spaces, or if E, F are (F)-spaces, one of which is nuclear. To establish the result, we need the following simple lemma.

LEMMA 2. *If E is a metrizable l.c.s. and $\{B_n\}$ is a sequence of bounded subsets of E, there exists a sequence $\{\mu_n\}$ of positive numbers such that $B = \bigcup_1^{\infty} \mu_n B_n$ is bounded in E.*

Proof. We can assume the topology of E is generated by an increasing sequence $\{p_n\}$ of semi-norms, and that each B_n contains a point x_n such that $p_n(x_n) > 0$. Define μ_n by $\mu_n^{-1} = \sup\{p_n(x): x \in B_n\}$. For any semi-norm p_k of the sequence, we obtain $p_k(x) \leqq 1$ whenever $x \in \bigcup_{n \geqq k} \mu_n B_n$; hence $\sup\{p_k(x): x \in B\} < +\infty$. Since this holds for all k, it follows that B is bounded.

9.8

Let E, F be l.c.s. such that either E and F are (DF)-spaces, or such that E and F are (F)-spaces and E is nuclear. Then the topology of the strong dual $\mathscr{B}_\beta(E, F)$ of $E \,\tilde{\otimes}\, F$ agrees with the topology of bi-bounded convergence.

Proof. We first prove the assertion, assuming that both E and F are (DF)-spaces. Then clearly $E \times F$ is a (DF)-space and the topology of bi-bounded convergence on $\mathscr{B}(E, F)$ is the topology of bounded convergence in $E \times F$, hence metrizable. Since this topology is coarser than $\beta(\mathscr{B}(E, F), E \,\tilde{\otimes}\, F)$, it suffices to show that the identity map of $\mathscr{B}_b(E, F)$ onto $\mathscr{B}_\beta(E, F)$ (obvious notation) is continuous; for this, in turn, it suffices that every null sequence $\{f_n\}$ in $\mathscr{B}_b(E, F)$ be bounded in $\mathscr{B}_\beta(E, F)$ (cf. Chapter II, Exercise 17). Let $Z = \{\alpha: |\alpha| \leqq 1\}$ be the unit disk in the scalar field K, then $\bigcap_1^{\infty} f_n^{-1}(Z)$ is a countable intersection of convex 0-neighborhoods in $E \times F$ which absorbs bounded sets, since $\{f_n\}$ is bounded in $\mathscr{B}_b(E, F)$, and hence a 0-neighborhood by the defining property of (DF)-spaces. It follows that $\{f_n\}$ is equicontinuous on $E \times F$, and hence equicontinuous in the dual $\mathscr{B}(E, F)$ of $E \,\tilde{\otimes}\, F$ by the corollary of (III, 6.2) and, therefore, bounded in $\mathscr{B}_\beta(E, F)$.

Turning to the second part of the proof, we assume that E, F are (F)-spaces and that E is nuclear. We shall show that for every bounded subset B of $E \,\tilde{\otimes}\, F$, there exist bounded sets $B_1 \subset E$, $B_2 \subset F$ such that $B \subset (\Gamma B_1 \otimes B_2)^-$. The assumptions imply that E is reflexive (hence $E_\tau' = E_\beta'$) and, by (9.4), that $E \,\tilde{\otimes}\, F$ can be identified with $\mathscr{B}_e(E_\sigma', F_\sigma')$, hence with $\mathscr{L}_e(E_\tau', F)$, and hence with $\mathscr{L}_b(E_\beta', F)$. Viewing B as a bounded set of linear maps in $\mathscr{L}_b(E_\beta', F)$, we see that $B(G)$ is bounded in F for each bounded subset G of E_β'; now if $\{G_n\}$ is a fundamental sequence of bounded subsets of E_β' (E_β' is a (DF)-space) and $B(G_n) = H_n$ ($n \in N$), by Lemma 2 there exists a sequence $\{\mu_n\}$ of positive numbers such that $\bigcup_1^{\infty} \mu_n H_n \subset H$, where H is a bounded, closed, convex, circled subset of F.

It is immediate that each $u \in B$ maps E_β' into F_H; in fact, since for each $u \in B$, $u^{-1}(H)$ is a barrel in E_β' and since E_β' (being reflexive) is barreled, it follows that $u \in \mathscr{L}(E'_\beta, F_H)$. Moreover, B is clearly simply bounded in $\mathscr{L}(E'_\beta, F_H)$ and hence equicontinuous; consequently, there exists a convex, circled 0-neighborhood V in E'_β such that $B(V) \subset H$. Now denote by E_1 the Banach space which is the completion of $[E']_V$; since E'_β is nuclear by (9.6), the canonical map ϕ_V of E'_β into E_1 is nuclear, say of the form $\sum_1^\infty \lambda_i x_i \otimes y_i$, where $\sum_1^\infty |\lambda_i| \leq 1$, and $\{x_i\}$, $\{y_i\}$ are bounded sequences in E, E_1, respectively. If $\tilde{u}$ denotes the linear map of E_1 into F_H associated with $u \in B$, then the family $B_2 = \{\tilde{u}(y_i): u \in B, i \in N\}$ is bounded in F_H (hence in F), and we obtain

$$u = \sum_1^\infty \lambda_i x_i \otimes \tilde{u}(y_i)$$

for all $u \in B$. Letting $B_1 = \{x_i: i \in N\}$, it follows that $B \subset (\Gamma B_1 \otimes B_2)^-$, since the series for u converges in $E \tilde{\otimes} F$; the proof is complete.

We conclude this section with an explicit characterization of the strong dual and bidual of $E \tilde{\otimes} F$ if E, F are (F)-spaces of which at least one is nuclear. As all the results in this section, the theorem is due to Grothendieck ([13], chap. II, theor. 12) and holds also when E is nuclear and E, F are (DF)-spaces that are not strong duals of (F)-spaces (Exercise 32). We need another lemma.

LEMMA 3. *Let E, F be l.c.s., let F''_e be the bidual of F under its natural topology (of uniform convergence on the equicontinuous subsets of F'), and let F''_σ be the space F'' under $\sigma(F'', F')$. Then every continuous bilinear form v on $E \times F$ possesses a unique extension $\bar{v}$ to $E \times F''$ which is a bilinear form continuous on $E \times F''_e$ and separately continuous on $E_\sigma \times F''_\sigma$; $v \to \bar{v}$ is an isomorphism of $\mathscr{B}(E, F)$ onto $\mathscr{B}(E, F''_e) \cap \mathfrak{B}(E_\sigma, F''_\sigma)$.*

Proof. Let us identify $\mathscr{B}(E, F)$ with the subspace of $\mathscr{L}(F, E'_\sigma)$ whose elements map a suitable 0-neighborhood in F onto an equicontinuous subset of E' (both sets depending on the map in question), and $\mathscr{B}(E, F''_e)$ with the corresponding subspace of $\mathscr{L}(F''_e, E'_\sigma)$. Then the argument used in the proof of Lemma 1 shows that $u \to u''$ is the (algebraic) isomorphism that we are seeking.

9.9

Theorem. *Let E, F be (F)-spaces, E being nuclear, and denote by F'' the strong bidual of F. Then the strong dual (respectively, the strong bidual) of $E \tilde{\otimes} F$ can be identified with $E'_\beta \tilde{\otimes} F'_\beta$ (respectively, with $E \tilde{\otimes} F''$).*

Proof. By (9.8) the strong dual of $E \tilde{\otimes} F$ is $\mathscr{B}_b(E, F)$ (topology of bi-bounded convergence). Now $\mathscr{B}(E, F'') = \mathfrak{B}(E, F'')$ by (III, 5.1); hence Lemma 3 shows that $\mathscr{B}(E, F)$ can be identified with $\mathfrak{B}(E_\sigma, F''_\sigma)$, and under this identification the topology of bi-bounded convergence is identical with the topology of bi-equicontinuous convergence; moreover, $\mathfrak{B}_e(E_\sigma, F''_\sigma)$ can be identified with $E'_\beta \tilde{\otimes} F'_\beta$ by (9.4), since E'_β is nuclear by (9.6). This proves the first assertion.

For the proof of the second proposition, put $H = E'_\beta$ and $G = F'_\beta$. We use (9.8) again to see that the strong dual of $H \mathbin{\tilde\otimes} G$ is $\mathscr{B}_b(H, G)$; then, in view of Lemma 3, $\mathscr{B}_b(H, G)$ can be identified with a subspace of $\mathfrak{B}_e(H_\sigma, G''_\sigma)$. Now since $H'_\beta = E$ (note that E is reflexive) is nuclear, it follows from (9.4) that $\mathscr{B}_b(H, G)$ can be identified with a subspace of $H'_\beta \mathbin{\tilde\otimes} G'_\beta = E \mathbin{\tilde\otimes} F''$. On the other hand, (III, 6.4) shows that every element of $E \mathbin{\tilde\otimes} F''$ defines a bilinear form on $H \times G$, which is continuous (cf. (6.5), Corollary 1), whence the proposition follows.

In the following corollaries of (9.9), we do not repeat the assumptions that E be a nuclear (F)-space and F an arbitrary (F)-space; the assertions hold equally when E, F are (DF)-spaces (E being assumed nuclear) (Exercise 32).

COROLLARY 1. *If, also, F is reflexive, then $E \mathbin{\tilde\otimes} F$ (and hence $E'_\beta \mathbin{\tilde\otimes} F'_\beta$) is a reflexive space.*

COROLLARY 2. *The strong dual of $\mathscr{L}_b(E'_\beta, F)$ can be identified with $E'_\beta \mathbin{\tilde\otimes} F'_\beta$; in particular, if F is reflexive, then $\mathscr{L}_b(E'_\beta, F)$ is reflexive.*

COROLLARY 3. *Every separately continuous bilinear form on $E'_\beta \times F'_\beta$ is continuous.*

10. NUCLEAR SPACES AND ABSOLUTE SUMMABILITY

The present section gives a characterization of nuclear spaces which is based on the concept of Radon measure on a compact space. This characterization shows that under suitable assumptions, nuclearity of a l.c.s. E is equivalent to the absolute summability of arbitrary summable families in E (cf. Chapter III, Exercise 23), and will give us access to several important results; among them are the converse of Theorem (9.4) and the theorem of Dvoretzky-Rogers [1] as proved by Grothendieck [13]. The results of this section (except where other references are given) are due to A. Pietsch [3]–[5].

Recall that a Radon measure (Chapter II, Example 3) on a compact space X is a continuous linear form on the (real or complex) Banach space $\mathscr{C}(X)$; a positive Radon measure is a $\mu \in \mathscr{C}(X)'$ such that $\mu(f) = \int f d\mu \geqq 0$ whenever $f(t) \geqq 0$ for all $t \in X$. We shall need the following well-known result, which is also an easy consequence of (V, 7.4), Corollary 2.

LEMMA 1. *If v is a Radon measure on the compact space X, there exists a positive Radon measure μ on X such that $|v(f)| \leqq \mu(|f|)$ for all $f \in \mathscr{C}(X)$.*

Further, if μ is a positive Radon measure on X then $(f, g) \to \int fg^* d\mu$ is a semi-definite Hermitian form on $\mathscr{C}(X) \times \mathscr{C}(X)$; hence the semi-norm $f \to p(f) = (\int |f|^2 d\mu)^{\frac{1}{2}}$ generates a locally convex topology on $\mathscr{C}(X)$ such that the associated Hausdorff t.v.s. is a pre-Hilbert space E_p; the completion $\tilde{E}_p$ is a Hilbert space isomorphic with $L^2(\mu)$ (Chapter II, Section 2, Examples 2, 3, and 5).

We also need the following result due to K. Maurin [1]. Let H_1 and H_2 be Hilbert spaces over K; a linear map $u \in \mathscr{L}(H_1, H_2)$ is called a *Hilbert-Schmidt transformation* if there exists an orthonormal basis $\{x_\alpha: \alpha \in A\}$ of H_1 such that $\sum_\alpha \|u(x_\alpha)\|^2 < +\infty$. It is clear that such a map is compact, and that $u(x_\alpha) = 0$, save for a countable number of indices $\alpha \in A$. It will follow from the subsequent proof that the value of the sum is independent of the choice of the orthonormal basis $\{x_\alpha: \alpha \in A\}$ and, in fact, equal to $\sum_\beta \|u^*(y_\beta)\|^2$ for any orthonormal basis $\{y_\beta: \beta \in B\}$ of H_2.

LEMMA 2. *Let H_i ($i = 1, 2, 3$) be Hilbert spaces and u, v be Hilbert-Schmidt transformations of H_1 into H_2 and of H_2 into H_3, respectively. Then the composite map $w = v \circ u$ of H_1 into H_3 is nuclear.*

Proof. Denote by $\{x_\alpha: \alpha \in A\}$, $\{y_\beta: \beta \in B\}$ orthonormal bases of H_1, H_2, respectively, and by $[\ , \ ]$ the inner product of H_2. If for the moment u is any continuous linear map of H_1 into H_2 with conjugate u^*, we obtain, in view of the identity $u(x_\alpha) = \sum_\beta [u(x_\alpha), y_\beta] y_\beta$ ($\alpha \in A$), the equality

$$\sum_\alpha \|u(x_\alpha)\|^2 = \sum_{\alpha,\beta} |[u(x_\alpha), y_\beta]|^2 = \sum_{\alpha,\beta} |[x_\alpha, u^*(y_\beta)]|^2$$
$$= \sum_\beta \|u^*(y_\beta)\|^2.$$

It follows that the value of the first and last term is independent of the basis $\{x_\alpha\}$ and $\{y_\beta\}$, respectively, and hence that u is a Hilbert-Schmidt transformation if and only if u^* is.

Now let u, v be the maps mentioned in the statement of the lemma. Denote by $\{y_n: n \in N\}$ an orthonormal basis of the range $u(H_1)$, which is clearly separable. For each $x \in H_1$ we obtain

$$w(x) = \sum_{n=1}^{\infty} [u(x), y_n] v(y_n) = \sum_{n=1}^{\infty} [x, u^*(y_n)] v(y_n).$$

Since by Schwarz' inequality

$$\sum_{n=1}^{\infty} \|u^*(y_n)\| \, \|v(y_n)\| \leq \left(\sum_{n=1}^{\infty} \|u^*(y_n)\|^2 \cdot \sum_{n=1}^{\infty} \|v(y_n)\|^2 \right)^{\frac{1}{2}},$$

the left-hand term is finite, it follows from (III, 7.1) that w is nuclear.

Turning to the subject of this section, let us agree on the following definitions. Let E be a locally convex space. A semi-norm p on E is called **prenuclear** if there exists a closed, equicontinuous subset A of E'_σ and a positive Radon measure μ on the $(\sigma(E', E)$-) compact space A such that

$$p(x) \leq \int_A |\langle x, x' \rangle| \, d\mu(x') \qquad (x \in E).$$

A subset $B \subset E'$ is called **prenuclear** if there exists a closed equicontinuous subset A of E'_σ and a positive Radon measure μ on A such that

$$|\langle x, b \rangle| \leq \int_A |\langle x, x' \rangle| \, d\mu(x') \qquad (x \in E, b \in B).$$

(Equivalently, a subset $B \subset E'$ is prenuclear if and only if $x \to \sup\{|\langle x, b \rangle|:$ $b \in B\}$ is a prenuclear semi-norm on E.) It is easy to see that a prenuclear semi-norm is necessarily continuous, and that a prenuclear subset of E' is necessarily equicontinuous. A family $\{x'_\alpha: \alpha \in A\}$ will be called **prenuclear** if its range is a prenuclear subset of E'. The following result is still of an auxiliary character. (Cf. (III, 7.3), Corollary 2.)

10.1

Let E be a l.c.s. on which every continuous semi-norm is prenuclear. There exists a neighborhood base $\mathfrak{U}$ of 0 in E such that for each $U \in \mathfrak{U}$, $\tilde{E}_U$ is norm isomorphic with a Hilbert subspace of $L^2(\mu)$, where μ is a positive Radon measure on a suitable closed equicontinuous subset of E'_σ.

Proof. It suffices to show that each closed, convex, circled 0-neighborhood W in E contains a 0-neighborhood U with the desired property. If p_W denotes the gauge of W, p_W is prenuclear by hypothesis; hence we have

$$p_W(x) \leqq \int_A |\langle x, x' \rangle| d\mu(x') \qquad (x \in E),$$

where A and μ can be so chosen that $\|\mu\| = 1$. Define the 0-neighborhood U by

$$U = \{x \in E: \int_A |\langle x, x' \rangle|^2 \, d\mu(x') \leqq 1\};$$

U is convex and circled, and by Schwarz' inequality $x \in U$ implies $p_W(x) \leqq 1$. Hence we have $U \subset W$, and evidently the gauge of U is given by

$$p_U(x) = \left(\int_A |\langle x, x' \rangle|^2 \, d\mu(x') \right)^{\frac{1}{2}} \qquad (x \in E),$$

which proves the assertion.

This leads to the announced characterization of nuclear spaces.

10.2

A locally convex space E is nuclear if and only if every continuous semi-norm on E is prenuclear, or equivalently, if and only if every equicontinuous subset of E' is prenuclear.

Proof. The condition is necessary. Let E be nuclear and let p be a continuous semi-norm on E. If $U = \{x: p(x) \leqq 1\}$, then by (III, 7.2) the canonical map $E \to \tilde{E}_U$ is nuclear, say $\phi_U = \sum_{i=1}^{\infty} \lambda_i x'_i \otimes \hat{x}_i$, where $\|\hat{x}_i\| \leqq 1$ in $\tilde{E}_U$ $(i \in N)$, $\{x'_i\} \subset V^\circ$ for a suitable 0-neighborhood V in E, and $(\lambda_i) \in l^1$. Now $\sum_1^{\infty} |\lambda_i| |\langle x, x'_i \rangle| \leqq 1$ implies $\|\phi_U(x)\| \leqq 1$; hence (since U is closed) $x \in U$, or equivalently, $p(x) \leqq 1$. It follows that $p(x) \leqq \sum_1^{\infty} |\lambda_i| |\langle x, x'_i \rangle|$ for all $x \in E$.

Since $f \to \sum_1^\infty |\lambda_i| f(x_i')$ is clearly a positive Radon measure on V°, we conclude that p is prenuclear.

The condition is sufficient. Let $\mathfrak{U}$ be a neighborhood base of 0 in E having the property described in (10.1); in view of Lemma 2, it suffices to show that for each $U \in \mathfrak{U}$, there exists $V \in \mathfrak{U}$, $V \subset U$, such that $\tilde{E}_V \to \tilde{E}_U$ is a Hilbert-Schmidt transformation. For, if this is correct, we can select $W \in \mathfrak{U}$, $W \subset V$ such that the canonical map $\tilde{E}_W \to \tilde{E}_U$ is the composite of two Hilbert-Schmidt transformations, and hence nuclear; the nuclearity of E is then a consequence of (III, 7.2). Thus let $U \in \mathfrak{U}$ be given so that

$$\|\phi_U(x)\|^2 = \int_A |\langle x, x'\rangle|^2 \, d\mu(x') \qquad (x \in E);$$

here A denotes, as before, a weakly closed, equicontinuous subset of the dual E' of E. Now choose $V \in \mathfrak{U}$ to satisfy $V \subset U \cap A^\circ$; it follows that $A \subset A^{\circ\circ} \subset V^\circ$. If f_A denotes the restriction of $f \in \mathscr{C}(V^\circ)$ to A, $f \to \int f_A d\mu$ is a positive Radon measure ν on V°. Further, let $\{\hat{x}_\alpha : \alpha \in A\}$ be an orthonormal basis of $\tilde{E}_V$; we have to show that $\sum_{\alpha \in A} \|\phi_{U,V}(\hat{x}_\alpha)\|^2$ is finite. Since $\tilde{E}_V$ is a Hilbert space, there exists a conjugate-linear isomorphism $x' \to z$ of $\tilde{E}_V^\circ$ onto $\tilde{E}_V$ such that $\langle \hat{x}, x'\rangle = [\hat{x}, z]$ for all $\hat{x} \in \tilde{E}_V$; $x' \to z$ induces a homeomorphism of V° onto the unit ball B of $\tilde{E}_V$ with respect to the weak and norm topologies. Denote by ν' the Radon measure on B obtained from ν under $x' \to z$. Since E_V is dense in $\tilde{E}_V$, we conclude that

$$\|\phi_{U,V}(\hat{x}_\alpha)\|^2 = \int_A |\langle \hat{x}_\alpha, x'\rangle|^2 \, d\mu(x') = \int_{V^\circ} |\langle \hat{x}_\alpha, x'\rangle|^2 \, d\nu(x')$$

$$= \int_B |[\hat{x}_\alpha, z]|^2 \, d\nu'(z).$$

Now if H is any finite subset of A, we have $\sum_{\alpha \in H} |[\hat{x}_\alpha, z]|^2 \leq [z, z]^2$ by Bessel's inequality, and the sum is ≤ 1 whenever $z \in B$; it follows that

$$\sum_{\alpha \in H} \|\phi_{U,V}(\hat{x}_\alpha)\|^2 = \int_B \sum_{\alpha \in H} |[\hat{x}_\alpha, z]|^2 \, d\nu'(z) \leq \nu'(B)$$

for any finite $H \subset A$. Clearly, this shows that $\phi_{U,V}$ is a Hilbert-Schmidt transformation, thus completing the proof.

Let E be a l.c.s.; a family $\{x_\alpha : \alpha \in A\}$ in E is called **summable** (Chapter III, Exercise 23) if $\lim_H x_H$ exists in E, where $x_H = \sum_{\alpha \in H} x_\alpha$ and H runs through the family of all finite subsets of A directed by inclusion $\subset$; the limit $x \in E$ is then denoted by $\sum_{\alpha \in A} x_\alpha$, or briefly by $\sum_\alpha x_\alpha$. (If A is infinite, the limit can equivalently be taken along the filter of subsets of A with finite complement.)

A summable family $\{x_\alpha: \alpha \in A\}$ in E is **absolutely summable** if for every continuous semi-norm p on E, the family $\{p(x_\alpha): \alpha \in A\}$ is summable in $\mathbf{R}$. It is well known that the two notions coincide when E is finite-dimensional (Chapter III, Exercise 23); we shall prove below that within the class of (F)-spaces, the identity of the respective sets of summable and absolutely summable sequences characterizes nuclear spaces.

Now let A be a fixed non-empty index set and let E be a given l.c.s.; the set of all absolutely summable families $\mathbf{x} = \{x_\alpha: \alpha \in A\}$ in E can evidently be identified with a subspace S_a of the algebraic product E^A. Denote by $\mathfrak{U}$ any fixed base of convex circled 0-neighborhoods in E, and by r_U the gauge of $U \in \mathfrak{U}$; clearly, the mapping

$$\mathbf{x} \to p_U(\mathbf{x}) = \sum_\alpha r_U(x_\alpha)$$

is a semi-norm on S_a, and the family of semi-norms $\{p_U: U \in \mathfrak{U}\}$ generates a topology under which S_a is a l.c.s. that will be denoted by $l^1[A, E]$. Similarly, the set of all summable families $\mathbf{x} = \{x_\alpha: \alpha \in A\}$ can be identified with a subspace S of E^A, and for each $U \in \mathfrak{U}$ the mapping

$$\mathbf{x} \to q_U(\mathbf{x}) = \sup\{\sum_\alpha|\langle x_\alpha, x'\rangle|: x' \in U^\circ\}$$

is a semi-norm on S; the family of semi-norms $\{q_U: U \in \mathfrak{U}\}$ generates a topology under which S is a l.c.s. that will be denoted by $l^1(A, E)$. The obvious inclusion $S_a \subset S$ defines a canonical imbedding of $l^1[A, E]$ into $l^1(A, E)$ which is continuous, since $q_U(\mathbf{x}) \leq p_U(\mathbf{x})$ for each $\mathbf{x} \in S_a$ and $U \in \mathfrak{U}$. Our next objective is to find a representation of the duals $l^1[A, E]'$ and $l^1(A, E)'$. (See also Exercise 35.)

10.3

The dual of $l^1[A, E]$ can be identified with the subspace of $(E')^A$, each of whose elements constitutes an equicontinuous family $\mathbf{x}' = \{x'_\alpha: \alpha \in A\}$ in E', the canonical bilinear form being given by $\langle \mathbf{x}, \mathbf{x}' \rangle = \sum_\alpha \langle x_\alpha, x'_\alpha \rangle$.

Proof. If $\mathbf{x}' = \{x'_\alpha: \alpha \in A\}$ is an equicontinuous family, then $x'_\alpha \in U^\circ$ $(\alpha \in A)$ for a suitable $U \in \mathfrak{U}$ and we have $(\mathbf{x} \in S_a)$

$$|\sum_\alpha\langle x_\alpha, x'_\alpha\rangle| \leq \sum_\alpha|\langle x_\alpha, x'_\alpha\rangle| \leq \sum_\alpha r_U(x_\alpha) = p_U(\mathbf{x}),$$

which shows that $\mathbf{x} \to \langle \mathbf{x}, \mathbf{x}' \rangle$ is well defined on S_a and continuous on $l^1[A, E]$; it is, moreover, evident that $\langle \mathbf{x}, \mathbf{x}' \rangle$ defines a duality between S_a and the space of equicontinuous families. There remains to show that every $f \in l^1[A, E]'$ can be so represented. Choose $U \in \mathfrak{U}$ such that $|f(\mathbf{x})| \leq p_U(\mathbf{x})$ $(\mathbf{x} \in S_a)$ and let, for any $z \in E$, $\mathbf{z}^{(\alpha)}$ be the family $\{\delta_{\alpha\beta}z: \beta \in A\}$; it is clear that for each $\alpha \in A$, $z \to f(\mathbf{z}^{(\alpha)})$ is an element $x'_\alpha \in E'$, and that $\{x'_\alpha: \alpha \in A\}$ is a family in E' with range in U°, hence equicontinuous. Moreover, if $\mathbf{x} = \{x_\alpha: \alpha \in A\} \in S_a$, then $\{\mathbf{x}^{(\alpha)}: \alpha \in A\}$, where $\mathbf{x}^{(\alpha)} = \{\delta_{\alpha\beta}x_\alpha: \beta \in A\}$, is a summable family in $l^1[A, E]$ such that $\mathbf{x} = \sum_\alpha \mathbf{x}^{(\alpha)}$. It follows from the continuity of f that $f(\mathbf{x}) = \sum_\alpha f(\mathbf{x}^{(\alpha)}) = \sum_\alpha \langle x_\alpha, x'_\alpha \rangle$, and the proof is complete.

Denoting by $l^1(A)$ the Banach space of summable scalar families $\xi = \{\xi_\alpha: \alpha \in A\}$ under its natural norm $\xi \to \|\xi\| = \sum_\alpha |\xi_\alpha|$, we obtain this corollary.

COROLLARY. *The dual of $l^1(A)$ can be identified with the space of all bounded scalar families $\eta = \{\eta_\alpha: \alpha \in A\}$, the canonical bilinear form being given by* $\langle \xi, \eta \rangle = \sum_\alpha \xi_\alpha \eta_\alpha$.

10.4

The dual of $l^1(A, E)$ can be identified with the subspace of $(E')^A$, each of whose elements constitutes a prenuclear family $\mathbf{x}' = \{x'_\alpha: \alpha \in A\}$ *in E', the canonical bilinear form being given by* $\langle \mathbf{x}, \mathbf{x}' \rangle = \sum_\alpha \langle x_\alpha, x'_\alpha \rangle$.

Proof. If $\mathbf{x}' = \{x'_\alpha: \alpha \in A\}$ is a prenuclear family in E', there exists a $U \in \mathfrak{U}$ and a positive Radon measure μ on $U°$ such that

$$|\langle x, x'_\alpha \rangle| \leq \int_{U°} |\langle x, x' \rangle| \, d\mu(x') \qquad (x \in E, \alpha \in A).$$

Now if $\mathbf{x} = \{x_\alpha: \alpha \in A\} \in S$, the definition of summability implies that $\sum_\alpha |\langle x_\alpha, x' \rangle|$ converges uniformly for $x' \in U°$ (hence to a continuous function on $U°$), whence

$$\sum_\alpha |\langle x_\alpha, x'_\alpha \rangle| \leq \sum_\alpha \int_{U°} |\langle x_\alpha, x' \rangle| \, d\mu(x') = \int_{U°} \sum_\alpha |\langle x_\alpha, x' \rangle| \, d\mu(x')$$

$$\leq \|\mu\| q_U(\mathbf{x}).$$

Hence $\mathbf{x} \to \langle \mathbf{x}, \mathbf{x}' \rangle$ is well defined on S and is a continuous linear form on $l^1(A, E)$; evidently, $(\mathbf{x}, \mathbf{x}') \to \langle \mathbf{x}, \mathbf{x}' \rangle$ places S in duality with the space of all prenuclear families.

Conversely, if $f \in l^1(A, E)'$, we define $x'_\alpha \in E'$ $(\alpha \in A)$ as in the preceding proof; there remains to show that $\{x'_\alpha: \alpha \in A\}$ is a prenuclear family. To this end, denote by Z the unit disk of the scalar field K and consider the compact space $Z^A \times U°$, where $U \in \mathfrak{U}$ is chosen such that $|f(\mathbf{x})| \leq q_U(\mathbf{x})$ for all $\mathbf{x} \in S$. Since the convergence of $\sum_\alpha |\langle x_\alpha, x' \rangle|$ is uniform with respect to $x' \in U°$, the formula

$$h_\mathbf{x}[(\lambda_\alpha), x'] = \sum_\alpha \lambda_\alpha \langle x_\alpha, x' \rangle$$

defines, for each $\mathbf{x} \in S$, a function $h_\mathbf{x} \in \mathscr{C}(Z^A \times U°)$. The mapping $\mathbf{x} \to h_\mathbf{x}$ of $l^1(A, E)$ into $\mathscr{C}(Z^A \times U°)$ is obviously linear and such that $\|h_\mathbf{x}\| = q_U(\mathbf{x})$; thus $h_\mathbf{x} \to f(\mathbf{x})$ is a continuous linear form defined on the range of $\mathbf{x} \to h_\mathbf{x}$ and can be extended, by the Hahn-Banach theorem, to a continuous linear form v on $\mathscr{C}(Z^A \times U°)$. v is a Radon measure on $Z^A \times U°$; hence by Lemma 1 there exists a positive Radon measure μ on $Z^A \times U°$ such that $|v(h)| \leq \mu(|h|)$ for all $h \in \mathscr{C}(Z^A \times U°)$. Now $x'_\alpha \in E'$ is the linear form $z \to f(\mathbf{z}^{(\alpha)})$, where $z \in E$ and $\mathbf{z}^{(\alpha)} = \{\delta_{\alpha\beta} z: \beta \in A\} \in l^1(A, E)$. We obtain, for all $\alpha \in A$,

$$|\langle z, x'_\alpha \rangle| = |f(\mathbf{z}^{(\alpha)})| = |v(h_{\mathbf{z}^{(\alpha)}})| \leq \mu(g),$$

where g is the function on $Z^A \times U^\circ$ given by $g((\lambda_\alpha), x') = |\langle z, x' \rangle|$. Now $\mathscr{C}(U^\circ)$ can clearly be identified with the subspace of $\mathscr{C}(Z^A \times U^\circ)$ containing all functions independent of $(\lambda_\alpha) \in Z^A$; under this identification, μ induces a positive Radon measure μ' on U°, and the function g becomes an element of $\mathscr{C}(U^\circ)$. Thus the preceding shows that

$$|\langle z, x'_\alpha \rangle| \leqq \mu(g) = \mu'(g) = \int_{U^\circ} |\langle z, x' \rangle| \, d\mu'(x')$$

for all $\alpha \in A$ and all $z \in E$. Hence $\{x'_\alpha : \alpha \in A\}$ is a prenuclear family, and the proof is complete.

It is remarkable that the algebraic and topological relations between $l^1[A, E]$ and $l^1(A, E)$ are determined, for any infinite index set A, by the corresponding relations for $A = N$.

10.5

Let E be any l.c.s. If the canonical imbedding of $l^1[A, E]$ into $l^1(A, E)$ is an algebraic (respectively, topological) isomorphism of the first space onto the second for $A = N$, the same is true for any (non-empty) index set A.

Proof. Suppose that each summable sequence in E is absolutely summable, and assume that $\{x_\alpha : \alpha \in A\}$ is a summable family in E which is not absolutely summable; then A is necessarily infinite, and there exists $U \in \mathfrak{U}$ such that $\sum_\alpha r_U(x_\alpha) = +\infty$. It follows that $\sum_k r_U(x_{\alpha_k}) = +\infty$ for some countably infinite subset $\{\alpha_1, \alpha_2, \ldots\}$ of A; this is contradictory since, clearly, $\{x_{\alpha_1}, -x_{\alpha_1}, \ldots\}$ is a summable sequence. Assume now that $l^1[N, E] = l^1(N, E)$ algebraically and topologically; then for given $U \in \mathfrak{U}$, there exists $V \in \mathfrak{U}$ such that $p_U(\mathbf{x}) \leqq q_V(\mathbf{x})$ whenever $\mathbf{x} \in l^1(N, E)$. Now if $\mathbf{x} = \{x_\alpha : \alpha \in A\}$ is any summable family in E and H is any finite subset of A, we obtain

$$\sum_{\alpha \in H} r_U(x_\alpha) = p_U(\mathbf{y}) \leqq q_V(\mathbf{y}) \leqq q_V(\mathbf{x}),$$

where $\mathbf{y} = \{y_\alpha : \alpha \in A\}$ is such that $y_\alpha = x_\alpha$ for $\alpha \in H$ and $y_\alpha = 0$ for $\alpha \notin H$; for $\mathbf{y}$ can be viewed as an element of $l^1(N, E)$. This shows that $p_U(\mathbf{x}) \leqq q_V(\mathbf{x})$ for all $x \in l^1(A, E)$, where $A \neq \varnothing$ is arbitrary, and the proof is complete.

We are now prepared to prove the principal theorem of this section characterizing nuclear spaces in terms of summability. However, let us first establish the connection of the preceding material with the theory of topological tensor products.

The mapping $(\xi, x) \to \{\xi_\alpha x : \alpha \in A\}$ of $l^1(A) \times E$ into $l^1[A, E]$ is evidently bilinear, and hence defines a linear mapping of $l^1(A) \otimes E$ into $l^1[A, E]$ (and into $l^1(A, E)$). With the aid of (10.3) it is easy to see that this linear mapping is an algebraic isomorphism; it will be called the **canonical imbedding** of $l^1(A) \otimes E$ in $l^1[A, E]$ (respectively, in $l^1(A, E)$).

10.6

The canonical imbedding of $l^1(A) \otimes E$ in $l^1[A, E]$ is an isomorphism for the projective topology on $l^1(A) \otimes E$, and the canonical imbedding of $l^1(A) \otimes E$ in $l^1(A, E)$ is an isomorphism for the topology of bi-equicontinuous convergence on $l^1(A) \otimes E$. Moreover, the canonical image of $l^1(A) \otimes E$ is dense in both $l^1[A, E]$ and $l^1(A, E)$.

Proof. The last assertion is almost immediate from the definition of the semi-norms p_U and q_U, and the first assertion is a special case of (III, 6.5). Hence let us identify $l^1(A) \otimes E$ algebraically with its canonical image in $l^1(A, E)$ and show that the induced topology is the topology of bi-equicontinuous convergence. By the corollary of (10.3), the dual of the Banach space $l^1(A)$ is the space of bounded scalar families. If B is the unit ball of $l^1(A)$, its polar B° (under the weak topology) can be identified with the compact space Z^A, where $Z = \{\lambda: |\lambda| \leq 1\}$ is the unit disk in K. Let $\sum_i \xi_i \otimes x_i$ be any element of $l^1(A) \otimes E$ and let $U \in \mathfrak{U}$. By definition of q_U we have

$$q_U\left(\sum \xi_i \otimes x_i\right) = \sup_{x' \in U^\circ} \sum_\alpha \left|\sum_i \xi_\alpha^{(i)} \langle x_i, x' \rangle\right|.$$

For each $\alpha \in A$ and each $x' \in U^\circ$, there exists $\eta_\alpha \in Z$ such that

$$\sum_i \xi_\alpha^{(i)} \eta_\alpha \langle x_i, x' \rangle = \left|\sum_i \xi_\alpha^{(i)} \langle x_i, x' \rangle\right|;$$

this implies

$$\sup_{x' \in U^\circ} \sum_\alpha \left|\sum_i \xi_\alpha^{(i)} \langle x_i, x' \rangle\right| = \sup_{x' \in U^\circ, \eta \in Z^A} \left|\sum_\alpha \sum_i \xi_\alpha^{(i)} \eta_\alpha \langle x_i, x' \rangle\right|$$

and, therefore, the relation

$$q_U\left(\sum \xi_i \otimes x_i\right) = \sup_{x' \in U^\circ, \eta \in B^\circ} \left|\left\langle \sum \xi_i \otimes x_i, \eta \otimes x' \right\rangle\right|.$$

This shows that the semi-norms q_U ($U \in \mathfrak{U}$) generate on $l^1(A) \otimes E$ the topology of uniform convergence on the sets $B^\circ \otimes U^\circ$ ($U \in \mathfrak{U}$), thus completing the proof.

Let us note that another proof of the last assertion can be obtained by (9.2); in fact, if $\{x_\alpha': \alpha \in A\}$ is an equicontinuous family in E', we can verify that this family is prenuclear if and only if $(\xi, x) \rightarrow \sum_\alpha \xi_\alpha \langle x, x_\alpha' \rangle$ is an integral bilinear form on $l^1(A) \times E$, and that in this fashion equicontinuous sets of prenuclear families in E' correspond to equicontinuous subsets of $\mathcal{I}(l^1(A), E)$.

The reader will have no difficulty verifying that for arbitrary $A \neq \emptyset$, $l^1[A, E]$ and $l^1(A, E)$ are complete if and only if E is; for $l^1(A, E)$, use (6.2), Corollary 2(b). We obtain these corollaries.

COROLLARY 1. *If E is complete, then $l^1(A) \tilde{\otimes} E$ is isomorphic with $l^1[A, E]$ under the continuous extension of the canonical imbedding of $l^1(A) \otimes E$ in $l^1[A, E]$; in the same fashion, $l^1(A) \tilde{\otimes} E$ is isomorphic with $l^1(A, E)$.*

The isomorphisms thus established are again called **canonical**.

COROLLARY 2. *Let E be a l.c.s. and let $\tilde{E}$ be its completion. Under the canonical isomorphism of $l^1(A) \otimes E$ with $l^1[A, \tilde{E}]$ and of $l^1(A) \tilde{\otimes} E$ with $l^1(A, \tilde{E})$, the canonical map of $l^1(A) \otimes E$ into $l^1(A) \tilde{\otimes} E$ corresponds to the canonical imbedding of $l^1[A, \tilde{E}]$ in $l^1(A, \tilde{E})$ and is, therefore, biunivocal.*

The following is the central result of Pietsch [5].

10.7

Theorem. *A locally convex space E is nuclear if and only if the canonical imbedding of $l^1[N, E]$ in $l^1(N, E)$ is a topological isomorphism of the first space onto the second.*

Proof. In view of (10.2) through (10.5), we have the following chain of implications: E is nuclear $\Rightarrow$ every equicontinuous subset of E' is prenuclear $\Rightarrow$ every equicontinuous subset of $l^1[N, E]'$ is contained in and equicontinuous in $l^1(N, E)' \Rightarrow l^1[N, E]$ is a dense topological vector subspace of $l^1(N, E) \Rightarrow l^1[N, E] = l^1(N, E)$ (cf. proof of (10.5)) $\Rightarrow l^1[A, E] = l^1(A, E)$ for any $A \neq \varnothing \Rightarrow$ every equicontinuous family in E' is prenuclear $\Rightarrow$ every equicontinuous subset of E' is prenuclear $\Rightarrow E$ is nuclear. The proof is complete.

This theorem has several important corollaries; the first of these is obtained from (10.6), Corollary 2, in view of the fact that a l.c.s. is nuclear if and only if its completion is nuclear.

COROLLARY 1. (Grothendieck [13].) *A locally convex space E is nuclear if and only if the canonical map of $l^1 \otimes E$ into $l^1 \tilde{\otimes} E$ is a topological isomorphism of the first space onto the second.*

This corollary implies, in particular, that the property established in Theorem (9.4) characterizes nuclear spaces. If E is an (F)-space, then clearly $l^1[N, E]$ and $l^1(N, E)$ are (F)-spaces; hence if the two spaces are algebraically identical, the canonical imbedding (which is continuous) is a topological isomorphism by Banach's theorem, (III, 2.1), Corollary 1. This implies:

COROLLARY 2. *An (F)-space E is nuclear if and only if every summable sequence in E is absolutely summable.*

We have observed earlier (Chapter III, Section 7) that a nuclear Banach space is finite dimensional, since it is locally compact; hence from Corollary 2 we obtain the following theorem, known as the theorem of Dvoretzky-Rogers [1].

COROLLARY 3. *A Banach space in which every summable sequence is absolutely summable is finite dimensional.*

We remark in conclusion that the algebraic identity of the spaces $l^1[N, E]$ and $l^1(N, E)$ implies the identity of their respective topologies whenever $l^1[N, E]$ is infrabarreled (Exercise 36); hence the completeness of E is dispensable in Corollaries 2 and 3.

11. WEAK COMPACTNESS. THEOREMS OF EBERLEIN AND KREIN

If S is a metrizable topological space, compactness of a subset A can be characterized by the existence, for each sequence in A, of a subsequence converging to a point in A. In more general circumstances this description fails, and several variations of the notion of compactness prove useful. It is the purpose of this section to prove some important characterizations of compact subsets (especially for the weak topology) of a l.c.s. by seemingly weaker properties, and to derive a deep criterion for the compactness of the convex closure of a compact set. For further information, we refer the interested reader to the literature cited below; detailed accounts can be found in Köthe [5] and (with emphasis on normed spaces) in Day [2]. See also Kelley-Namioka [1] and the very interesting paper of James [3].

Let us recall the following definitions, S denoting a Hausdorff topological space: A subset A of S is called **countably compact** if every sequence in A has a cluster point in A (equivalently, if each countable open cover of A has a finite subcover); A is called **sequentially compact** if each sequence in A possesses a subsequence converging to a point in A. It is immediate that compactness of A and sequential compactness of A both imply countable compactness of A; in general, no other implications are valid among these notions (cf. Exercise 37).

We begin with the following result, which is preparatory but of considerable interest in itself (cf. Eberlein [1], Grothendieck [6]). Denote by (Y, d) a metric space which is locally compact and countable at infinity, by X a compact space, and by $\mathscr{C}_Y(X)$ the subset of Y^X whose elements are continuous on X into Y. Y^X is endowed with the topology of simple convergence.

11.1

Theorem. *Let H be a subset of $\mathscr{C}_Y(X)$ such that each sequence in H has a cluster point in $\mathscr{C}_Y(X)$ (for the topology of simple convergence). Then the closure $\overline{H}$ in Y^X is compact and contained in $\mathscr{C}_Y(X)$, and each element of $\overline{H}$ is the limit of a sequence in H.*

Before proving the theorem let us note these corollaries.

COROLLARY 1. *If $H \subset \mathscr{C}_Y(X)$ is countably compact for the topology of simple convergence, then H is compact and sequentially compact.*

COROLLARY 2 (Eberlein [1]). *Each weakly countably compact subset of a Banach space E is weakly compact and weakly sequentially compact.*

Proof. Take (Y, d) to be the scalar field K of E under its usual absolute value, and X to be the dual unit ball under $\sigma(E', E)$. The map $z \to h(z) = f$, which orders to each $z \in E'^*$ its restriction f to X, is a homeomorphism of $(E'^*, \sigma(E'^*, E'))$ onto a closed subspace Q of K^X. By (6.2), $h(E) = Q \cap \mathscr{C}_K(X)$; hence the assertion follows from Corollary 1.

Proof of (11.1). It is clear that for each $\because X$, the set $\{f(t): f \in H\}$ is relatively compact in Y; otherwise, there would exist $t \in X$ and a sequence $\{f_n: n \in N\}$ in H such that $\lim_n f_n(t) = \infty$, and this sequence would have no cluster point in Y^X. It follows now from Tychonov's theorem that H is relatively compact in Y^X.

We show next that $\bar{H} \subset \mathscr{C}_Y(X)$. Assume to the contrary, that there exists a $g \in \bar{H}$ not continuous on X. There exists a non-empty subset $M \subset X$, a $t_0 \in \bar{M}$, and an $\varepsilon > 0$ such that $d(g(t), g(t_0)) > 4\varepsilon$ whenever $t \in M$. We define the sequences $\{t_0, t_1, t_2, ...\}$ in X and $\{f_1, f_2, ...\}$ in H inductively, as follows: Since $g \in \bar{H}$, it is possible to choose an $f_1 \in H$ so that $d(f_1(t_0), g(t_0)) < \varepsilon$. After $\{t_0, ..., t_{n-1}\}$ and $\{f_1, ..., f_n\}$ have been selected, choose $t_n \in M \cap M_1 \cap \cdots \cap M_n$, where $M_v = \{t \in X: d(f_v(t), f_v(t_0)) < \varepsilon\}$ $(v = 1, ..., n)$. Then choose $f_{n+1} \in H$ so that $d(f_{n+1}(t_v), g(t_v)) < \varepsilon$ $(v = 0, 1, ..., n)$, which is possible, since $g \in \bar{H}$. From this construction we obtain the following inequalities:

$$d(g(t_n), g(t_0)) > 4\varepsilon \qquad (n \in N). \tag{1}$$

$$d(f_n(t_v), f_n(t_0)) < \varepsilon \qquad \text{for all } v \geq n, n \in N. \tag{2}$$

$$d(f_n(t_v), g(t_v)) < \varepsilon \qquad \text{whenever } 0 \leq v \leq n - 1, n \in N. \tag{3}$$

By hypothesis, the sequence $\{f_n: n \in N\}$ has a cluster point $h \in \mathscr{C}_Y(X)$; then from (3) it follows that $d(h(t_v), g(t_v)) \leq \varepsilon$ for all $v \geq 0$, and from this and (1) we conclude that $d(h(t_n), h(t_0)) > 2\varepsilon$ for all $n \in N$. Since X is compact, the sequence $\{t_n: n \in N\}$ has a cluster point $s \in X$, and the continuity of h implies that

$$d(h(s), h(t_0)) \geq 2\varepsilon. \tag{4}$$

Since h is a cluster point of $\{f_n\}$, there exists an $m \in N$ such that $d(f_m(t_0), h(t_0)) + d(f_m(s), h(s)) < \varepsilon$. By (2) and the continuity of f_m, we have $d(f_m(s), f_m(t_0)) \leq \varepsilon$. The last two inequalities yield $d(h(s), h(t_0)) < 2\varepsilon$, contradicting (4). Hence the assumption $g \notin \mathscr{C}_Y(X)$ is absurd, and it follows that $\bar{H} \subset \mathscr{C}_Y(X)$.

There remains to show that each $g \in \bar{H}$ is the limit of a sequence in H. As an intermediate step, we observe that g is a cluster point of a sequence $\{g_n: n \in N\}$ in H. In fact, if $n \in N$ is fixed and $(t_1, ..., t_n)$ is a given n-tuple of points in X, there exists (since $g \in \bar{H}$) a function $h \in H$ such that (*) $d(g(t_v), h(t_v)) < n^{-1}$ $(v = 1, ..., n)$; since the topological product X^n is compact and g is continuous, we can find a finite subset $H_n \subset H$ such that whatever $(t_1, ..., t_n) \in X^n$, the relation (*) is satisfied for at least one $h \in H_n$. Now if such a set H_n is selected for each $n \in N$ and if $\{g_n: n \in N\}$ is a sequence with range $\bigcup_n H_n$, then clearly g is a cluster point of $\{g_n: n \in N\}$.

The proof will now be completed by showing that a given cluster point g of the sequence $\{g_n: n \in N\}$ is the (pointwise) limit of a suitable subsequence. Denote by G the closure of the range of $\{g_n: n \in N\}$ in Y^X. The assumptions on Y imply that Y has a countable base $\mathfrak{G}$ of open sets. Consider the topology $\mathfrak{T}$ on X generated by $g^{-1}(\mathfrak{G})$ and $\bigcup_n g_n^{-1}(\mathfrak{G})$; $\mathfrak{T}$ has a countable base and is coarser than the given topology of X. The relation R on X defined by "$t \sim s$ if $g_n(t) = g_n(s)$ for all $n \in N$" is a closed equivalence relation, and the quotient

$(X, \mathfrak{X})/R$ is a metrizable, compact space X'. It is clear that g and all g_n define continuous functions g' and g_n', respectively, on X' into Y; moreover, each $h \in G$ defines a function $h' \in Y^{X'}$. Denoting the corresponding subset of $Y^{X'}$ by G', it follows from the first part of the proof that G' is contained in $\mathscr{C}_Y(X')$ and compact (for the topology of simple convergence), and from this it follows that the topology of simple convergence agrees on G' with the topology of simple convergence in any dense subset X_0' of X' (the latter being a coarser Hausdorff topology). Since X' (being compact and metrizable) is separable, there exists a countable dense subset X_0' of X', and this implies that the restriction of the topology of simple convergence (in X') to G' is metrizable. It follows that g' is the limit of a suitable subsequence of $\{g_n': n \in N\}$, and, clearly, this implies the assertion. This completes the proof.

We are now prepared to prove the following theorem on weak compactness, usually referred to as the theorem of Eberlein. The equivalence (a) $\Leftrightarrow$ (b) is essentially Eberlein's result (cf. Corollary 2, above) in a more general setting, (a) $\Leftrightarrow$ (c) is due to Dieudonné [4], (a) $\Leftrightarrow$ (d) to Grothendieck [6]. See also Ptak [2]-[5], Smulian [2].

11.2

Theorem. *Let E be a l.c.s. and let H be a subset of E whose closed, convex hull is complete. The following properties of H are equivalent:*

(a) *H is relatively weakly compact.*
(b) *Each sequence in H has a weak cluster point in E.*
(c) *For each decreasing sequence $\{H_n: n \in N\}$ of closed convex subsets of E such that $H_n \cap H \neq \varnothing$ for all n, $\bigcap_n H_n$ is non-empty.*
(d) *H is bounded and, for each sequence $\{x_m: m \in N\}$ in H and each equicontinuous sequence $\{x_n': n \in N\}$ in E', one has $\lim_n \lim_m \langle x_m, x_n' \rangle = \lim_m \lim_n \langle x_m, x_n' \rangle$ whenever both double limits exist.*

Proof. It is clear that (a) implies (b) and (c). To see that (a) implies (d), we observe that the sequence $\{x_m\}$ has a weak cluster point $x \in E$, since H is relatively weakly compact, and the sequence $\{x_n'\}$ has a weak cluster point $x' \in E'$, since it is equicontinuous. Now if the first double limit exists, it necessarily equals $\lim_n \langle x, x_n' \rangle = \langle x, x' \rangle$; likewise, the second double limit (if it exists) equals $\langle x, x' \rangle$.

To prove the reverse implications, we observe the following: Since the closed, convex hull C of H is complete, C is closed and hence weakly closed in the completion $\tilde{E}$ of E; hence to show that a $\sigma(E'^*, E')$-cluster point x^* of H actually belongs to E it suffices, by Grothendieck's theorem (6.2), to show that the restriction of x^* to X is $\sigma(E', E)$-continuous, where X is an arbitrary $\sigma(E', E)$-closed equicontinuous set in E'. In the following, let X be any such set (which is a compact space under $\sigma(E', E)$). When considering the elements of E (respectively, of E'^*) as elements of $\mathscr{C}_K(X)$ (respectively, of K^X, K the scalar field of E), we actually mean their restrictions to X.

(b) $\Rightarrow$ (a): This is now immediate from (11.1), taking (Y, d) to be K with its usual metric.

(c) $\Rightarrow$ (a): Note first that H is bounded in E; otherwise, there would exist a continuous, real linear form u on E and a sequence $\{x_n\}$ in H such that $u(x_n) > n$ ($n \in N$), and the sets $H_n = \{x \in E: u(x) \geqq n\}$ would satisfy (c) but have empty intersection. Now consider H as a subset of $\mathscr{C}_K(X)$, and denote by $\bar{H}$ the closure of H in K^X; it suffices to show that $\bar{H} \subset \mathscr{C}_K(X)$. On the assumption that some $g \in \bar{H}$ is not continuous on X we construct, as in the first part of the proof of (11.1), a sequence $\{t_0, t_1, t_2, ...\}$ in X and a sequence $\{f_1, f_2, ...\}$ in H such that the relations (1), (2), (3) are satisfied. Now denote by H_n the closed, convex hull of the set $\{f_\nu: \nu \geqq n\}$ in E, and let h be an element of $\bigcap_n H_n$. Then we obtain (4) again, and replacing the element f_m in the proof of (11.1) by a suitable element $f_m' \in H_m$ we arrive at a contradiction as before.

(d) $\Rightarrow$ (b): Since H is bounded, each sequence $\{x_n\}$ in H has a $\sigma(E'', E')$-cluster point $x^* \in E''$, by (5.4); it suffices to show that the restriction of x^* to X is continuous. Suppose it is not; we can assume that $0 \in X$ and that x^* is discontinuous at $0 \in X$. Denoting by U_n ($n \in N$) the weak 0-neighborhood $\{x' \in E': |\langle x_\nu, x' \rangle| < n^{-1}, \nu = 1, ..., n\}$, there exists an $\varepsilon > 0$ and elements $x_n' \in U_n \cap X$ such that $|\langle x^*, x_n' \rangle| > \varepsilon$ for all n. Let $V_m = \{z \in E'': |\langle z, x_\mu' \rangle| < m^{-1}, \mu = 1, ..., m\}$ for each $m \in N$. There exists a subsequence $\{y_m\}$ of $\{x_n\}$ such that $y_m \in x^* + V_m$ for all m, and we can further arrange (by choosing a subsequence of $\{x_n'\}$ if necessary) that $\lim_n \langle x^*, x_n' \rangle$ exists.

Now $\lim_m \langle y_m, x_n' \rangle = \langle x^*, x_n' \rangle$ for each n; hence $\lim_n \lim_m \langle y_m, x_n' \rangle$ exists and is $\geqq \varepsilon$ in absolute value. On the other hand, $\lim_n \langle y_m, x_n' \rangle = 0$ for all m, which implies $\lim_m \lim_n \langle y_m, x_n' \rangle = 0$, contradicting (d). This completes the proof of (11.2).

COROLLARY. *Let E be a l.c.s. which is quasi-complete for $\tau(E, E')$. Then each weakly closed and countably compact subset of E is weakly compact.*

The following result is essential for the theorem of Krein to be proved below. Our proof, which uses the dominated convergence theorem of Lebesgue (see, e.g., Bourbaki [9] chap. IV, §3, theor. 6 or Halmos [1], §26, theor. D), is simple but not elementary; for a combinatorial proof and an enlightening discussion of related questions, we refer to Ptak [7]. Another combinatorial proof was given by Namioka [2].

11.3

Let E be a l.c.s., let B be a compact subset of E, and let C be the closed, convex, circled hull of B. If $\{x_n': n \in N\}$ is a sequence in E', uniformly bounded on B and such that $\lim_n \langle x, x_n' \rangle = 0$ for each $x \in B$, then $\lim_n \langle x, x_n' \rangle = 0$ for each $x \in C$.

Proof. Consider the normed space E_C; without loss in generality we can suppose that (algebraically) $E_C = E$. The dual E' of E can be identified with a subspace of E'_C, and the mapping $x' \to f$ (where $f(x) = \langle x, x' \rangle$, $x \in B$) is a norm isomorphism ψ of the Banach space E'_C into $\mathscr{C}(B)$. It follows from (7.9) that ψ' maps $\mathscr{C}(B)'$ onto the bidual E''_C; in particular, each $x \in E$ is the image under ψ' of a suitable $\mu \in \mathscr{C}(B)'$. On the other hand, if $f_n = \psi(x'_n)$ $(n \in N)$, then, since $\{f_n\}$ is uniformly bounded on B and each $\mu \in \mathscr{C}(B)'$ is a linear combination of positive Radon measures on B, Lebesgue's dominated convergence theorem implies that $\lim_n \mu(f_n) = 0$ for all $\mu \in \mathscr{C}(B)'$; hence $\lim_n \langle x, x'_n \rangle = \lim_n \langle \psi'(\mu), x'_n \rangle = \lim_n \mu(f_n) = 0$, where $x = \psi'(\mu)$, thus completing the proof.

Now we can prove the theorem of Krein (cf. M. G. Krein [1]); our proof follows Kelley-Namioka [1].

11.4

Theorem. *Let B be a weakly compact subset of the l.c.s. E, and denote by C the closed, convex hull of B. Then C is weakly compact if and only if C is complete for the Mackey topology $\tau(E, E')$.*

Proof. The condition is clearly necessary, for if C is weakly compact, it is weakly complete and hence complete for $\tau(E, E')$. The sufficiency will be proved via condition (d) of (11.2) (cf. Ptak [7]). Let $\{x_n\}$ be a sequence in C and let $\{x'_n\}$ be an equicontinuous (with respect to $\tau(E, E')$) sequence in E'; the former has a $\sigma(E'', E')$-cluster point $x^* \in E''$ (cf. (5.4)); the latter has a $\sigma(E', E)$-cluster point $x' \in E'$. By (11.1) there exists a subsequence of $\{x'_n\}$ which converges to x' pointwise on B. Assume hence that $\{x'_n\}$ has this property; it follows then from (11.3), applied to $(E'', \sigma(E'', E'))$, that $\lim_n \langle z, x'_n \rangle = \langle z, x' \rangle$ for each $z \in \bar{C}$, where $\bar{C}$ denotes the $\sigma(E'', E')$-closure of C in E''.

In particular, $\lim_n \langle x_m, x'_n \rangle = \langle x_m, x' \rangle$ for each $m \in N$; if $\lim_m \lim_n \langle x_m, x'_n \rangle$ exists then, since x^* is a $\sigma(E'', E')$-cluster point of $\{x_m\}$, it necessarily equals $\langle x^*, x' \rangle$. On the other hand, if $\lim_m \langle x_m, x'_n \rangle$ exists, it clearly equals $\langle x^*, x'_n \rangle$ and from the preceding it follows that $\lim_n \langle x^*, x'_n \rangle = \langle x^*, x' \rangle$. Hence the double limit condition (d) of (11.2) is satisfied and C, being weakly closed, is weakly compact. The proof is complete.

Since by (I, 5.2) the circled hull of every weakly compact subset of E is weakly compact, (11.4) continues to hold with C the closed, convex, circled hull of B.

If B is a compact subset of E and C the closed, convex (or closed, convex, circled) hull of B, then C is precompact by (II, 4.3) and (I, 5.1); hence C is compact if and only if it is complete. In view of this remark, it is easy to prove from (11.4) the following slightly more general version of Krein's theorem.

11.5

Let B be a compact subset of the l.c.s. E and let C be the closed, convex, circled hull of B. Then C is compact if and only if C is complete for $\tau(E, E')$.

EXERCISES

1. Let F be a vector space over K.

(a) Let Q be any non-empty subset of F^*, and denote by M the linear hull of Q in F^*. If $\mathfrak{T}$ is the topology generated by the semi-norms $x \to |\langle x, y \rangle|$, $y \in Q$, then $(F, \mathfrak{T})' = M$ and M is in duality with the Hausdorff t.v.s. F_0 associated with $(F, \mathfrak{T})$.

(b) If F is a l.c.s. with dual F', then $\sigma(F, F')$ is the $\mathfrak{S}$-topology, $\mathfrak{S}$ denoting the set of all $\sigma(F', F)$-bounded sets, each of which is contained in a suitable finite-dimensional subspace of F'.

2. Denote by E_σ, F_σ two l.c.s. under their respective weak topologies. The space of continuous bilinear forms on $E_\sigma \times F_\sigma$ (equivalently, the dual of the projective tensor product $E_\sigma \otimes F_\sigma$) is the tensor product $E' \otimes F'$.

3. Let F be a l.c.s. over C, let G be a properly real subspace of F (Chapter I, Exercise 16) such that $F_0 = G + iG$, where F_0 is the underlying real space of F, and denote by G' the set of all real linear forms on G that are continuous for the topology induced by F. Show the following assertions to be equivalent:

(α) $F_0 = G + iG$ is a topological direct sum for $\sigma(F_0, F_0')$.

(β) F' is the linear hull (over C) of the linear forms

$$x + iy \to f(x) + if(y), \qquad (x, y \in G),$$

where f runs through G'.

4. Let E be a normed space which is not complete. Then the $\mathfrak{S}$-topology on E', where $\mathfrak{S}$ is the family of all compact subsets of E, is not consistent with $\langle E, E' \rangle$. (Using (6.3), Corollary 1, observe that there exists a compact subset of E whose closed, convex, circled hull is not weakly compact.)

5. (Perfect Spaces): We use the notation of Section 1, Example 4. If λ is a sequence space with elements $x = (x_n)$, we define an order in λ (Chapter V, Section 1) by " $x \leq y$ if $x_n \leq y_n$ for all n "; by $|x|$, we denote the sequence $(|x_n|)$ (which is not necessarily in λ). λ is **perfect** (vollkommen, Köthe [4]) if $\lambda = \lambda^{\times\times}$, **solid** if $x \in \lambda$ and $|y| \leq |x|$ implies $y \in \lambda$. Let λ be given, and denote by P the set of sequences $u \geq 0$ in $\lambda^\times$; the topology $\mathfrak{T}$ generated by the semi-norms $x \to \langle |x|, u \rangle$, $u \in P$, is called the **normal topology** of λ. (In connection with the following problems, the reader might want to consult Köthe [4], [5]; see also Dieudonné [3]. A lattice theoretic characterization of the normal topology can be found in Chapter V, Exercise 20.)

(a) The spaces l^p ($1 \leq p \leq + \infty$) are perfect; each " gestufter Raum " (Chapter III, Exercise 25) is perfect.

(b) The normal topology $\mathfrak{T}$ is consistent with $\langle \lambda, \lambda^\times \rangle$.

(c) For each $u \in \omega(= K_0^N)$, define $\lambda_u = \{x \in \omega : \sum_1^\infty |u_n||x_n| < + \infty\}$. Every space λ_u is isomorphic (as a sequence space) to one of the spaces l^1, ω, or $l^1 \oplus \omega$; deduce that each space λ_u is perfect.

(d) If λ is perfect then λ is solid and $(\lambda, \mathfrak{T})$ is the projective limit $\varprojlim g_{uv}\lambda_v(\mathfrak{T})$, where $u, v \in P$, P being directed ($\leq$), and where g_{uv} is the canonical imbedding of λ_v in λ_u ($u \leq v$).

(e) For a sequence space λ, these propositions are equivalent: (α) λ is perfect, (β) $(\lambda, \mathfrak{T})$ is complete, (γ) $(\lambda, \mathfrak{T})$ is weakly semi-complete. (Use (d).)

6. (Spaces of Minimal Type. Cf. Martineau [1].) A l.c.s. E is said to be **minimal** if its topology is minimal (i.e., if there exists no strictly coarser separated l.c. topology on E).

(a) For a l.c.s. E, these propositions are equivalent: (α) E is minimal; (β) E is isomorphic with the product K_0^d for a suitable cardinal d; (γ) E is weakly complete.

(b) If E is minimal, then $\sigma(E, E') = \beta(E, E')$ and $\tau(E', E) = \beta(E', E)$; $\tau(E', E)$ is the finest l.c. topology on E'. On a minimal space of infinite dimension there exists no continuous norm; an infinite dimensional normed space is never weakly complete.

(c) Each closed subspace and each separated quotient of a minimal space is minimal; the product of any family of minimal spaces is minimal.

(d) Let E, F be minimal and let u be a linear map of E into F. u is continuous if and only if u is closed; if u is continuous, it is a topological homomorphism.

(e) Let E be minimal. If M, N are closed subspaces, then $M + N$ is closed in E. If, in addition, $M \cap N = \{0\}$, then $M + N$ is a topological direct sum; in particular, if E is the algebraic direct sum of the closed subspaces M, N, then $E = M \oplus N$.

7. We use the notation of (4.1).

(a) Let Z denote the unit disk of the complex plane and let F denote the vector space (over C) of continuous complex-valued functions on Z possessing continuous first derivatives in the interior of Z. Denote by $\mu_n (n \in N)$ and μ_0 the Radon measures on Z defined by $\mu_n(f) = n[f(n^{-1}) - f(0)]$ and $\mu_0(f) = (2\pi i)^{-1} \int f(\zeta)\zeta^{-2} d\zeta$ (Cauchy integral over the positively oriented boundary of Z), respectively. Let $B = \{\mu_n: n \in N\}$ and let G be the subspace of F^* generated by μ_0 and by the set Φ of Radon measures of finite support on Z. We define $\mathfrak{S}_1'$ to be the $\sigma(G, F)$-saturated hull of the family $\Phi \cup \{B\}$; let $\mathfrak{T}_1$ be the $\mathfrak{S}_1'$-topology on F. Finally, let M be the subspace of F of functions analytic in the interior of Z, let $\mathfrak{T}_2$ be the topology on M induced by $\mathfrak{T}_1$, and let $\mathfrak{S}_2'$ be the ($\sigma(G/M^\circ, M)$-saturated) family of all $\mathfrak{T}_2$-equicontinuous subsets of G/M°.

In these circumstances, one has $\phi(\mathfrak{S}_1') \neq \mathfrak{S}_2'$. (Prove that μ_0 is not contained in any $S_1 \in \mathfrak{S}_1'$, and observe that $\phi(\mu_0)$ is in the $\sigma(G/M^\circ, M)$-closure of $\phi(B)$.)

(b) Condition (b) of (4.1) implies that $\mathfrak{S}_2'$ is the $\sigma(G/M^\circ, M)$-saturated hull of the family $\phi(\mathfrak{S}_1')$.

(c) Even supposing $M^\bullet$ to be a weakly closed subspace of F, the implication (c) $\Rightarrow$ (d) of (4.1) is false unless $\mathfrak{T}_1'$ is consistent with $\langle F, G \rangle$. (Consider a complete l.c.s. E whose strong dual E_β is not barreled (Exercise 13), and imbed E as a closed subspace M of a product F of Banach spaces, (II, 5.4), Corollary 2. By (4.2), the strong dual F_β' is barreled as a l.c. direct sum of barreled spaces; hence the quotient of $\beta(F'. F)$ is a barreled topology, and therefore distinct from $\beta(F'/M^\circ, M)$.)

8. Let $\{\langle F_\alpha, G_\alpha \rangle : \alpha \in A\}$ be a family of dualities over K and let $F = \prod_\alpha F_\alpha$, $G = \bigoplus_\alpha G_\alpha$. F and G are placed in duality in the canonical way.

(a) $\beta(F, G)$ is the product of the topologies $\beta(F_\alpha, G_\alpha)$, and $\beta(G, F)$ is the l.c. direct sum topology of the topologies $\beta(G_\alpha, F_\alpha)$. (Use (4.2).)

(b) The weak topology $\sigma(G, F)$ is the l.c. direct sum topology of the topologies $\sigma(G_\alpha, F_\alpha)$ if and only if the number of spaces $F_\alpha \neq \{0\}$ is finite.

(c) A fundamental family of convex, circled, weakly compact subsets of F (respectively, G) is obtained by forming arbitrary products of like subsets of the F_α (respectively, by forming arbitrary finite sums of like subsets of the G_α).

9. Construct an example of a l.c.s. F and a closed subspace H such that not every bounded subset of F/H is the canonical image of a bounded subset of F. (Consider $E = l^1$; for each vector $x \geq 0$ of E (for notation, see Exercise 5) let $B(x) = \{y \in E : |y| \leq x\}$. Show that the (norm) topology of E is the finest l.c. topology for which all $B(x)$ are bounded. Represent E as a quotient F/H, where $F = \bigoplus_{x \geq 0} E_{B(x)}$.) Similarly, show that a weakly compact subset of F/H is not necessarily the canonical image of a like set of F. (Apply the preceding method to $E = l^2$.) See also Exercise 20.

10. Show that a closed subspace of a barreled space is not necessarily barreled. (Consider a complete l.c.s. E which is not barreled, and imbed E as a closed subspace of a product of Banach spaces (use (4.3), Corollary 3); for example, it suffices to take for E a sequence space l^p, $p > 1$, under its normal topology (Exercise 5).)

REMARK. Using the theorem of Mackey-Ulam (cf. Chapter II, Section 8), the same method provides an example of a bornological space possessing a non-bornological closed subspace. See also Exercise 20.

11. (Incomplete Quotients.) Let X denote a completely regular topological space, $R(X)$ the l.c.s. (over R) of real-valued continuous functions on X endowed with the topology of compact convergence.

(a) If a function $f \in R^X$ is continuous whenever its restriction to each compact subset of X is continuous, then $R(X)$ is complete.

(b) Let Y be a closed subset of X and let H be the subspace of $R(X)$ whose elements vanish on Y. The quotient space $R(X)/H$ can be identified with the subspace $R_0(Y)$ of $R(Y)$ whose elements have a continuous extension to X. $R_0(Y)$ is dense in $R(Y)$ (use the theorem of Stone-Weierstrass, (V, 8.1)).

(c) If Y is such that $f \in R^Y$ is continuous whenever its restrictions to compact sets are continuous, then $R(X)/H$ is complete if and only if $R_0(Y) = R(Y)$.

(d) Deduce from the preceding an example of a complete l.c.s. $R(X)$ such that for a suitable closed subspace H, $R(X)/H$ is not complete. (Consider, for example, a locally compact space X which is not normal (Bourbaki [5], §4, Exercise 13); there exists a closed subspace Y such that not every continuous real function on Y has a continuous extension to X.)

12. (Topological Complementary Subspaces):

(a) Let E be a non-normable (F)-space on which there exists a

continuous norm (e.g., the space $\mathcal{D}_G$ (Chapter II, Section 6, Example 2));
E contains a closed subspace which has no topological complement.
(Take an increasing fundamental sequence $\{B_n: n \in N\}$ of convex, circled,
bounded subsets of E' such that $E'_{B_n} \neq E'_{B_{n+1}}$, and select $x'_n \in E'_{B_{n+1}} \sim$
E'_{B_n}. Show that the linear hull F of $\{x'_n: n \in N\}$ is a subspace of E' whose
bounded subsets are finite dimensional; deduce from this, using the
theorem of Krein-Šmulian (6.4), that F is a weakly closed subspace of
E'. Since every linear form on F is continuous (use (6.2)), the dual E/F°
of F is weakly complete, hence isomorphic with K_0^N (Exercise 6). If
F° had a topological complement, it would be isomorphic with K_0^N,
which contradicts the fact that there exists no continuous norm on K_0^N
(Exercise 6).)

(b) Let E be a given Banach space, $\{y_\alpha: \alpha \in A\}$ a dense subset of the
unit ball of E. Then E is isomorphic with a quotient space of $l^1(A)$.
(For each $x = (\xi_\alpha) \in l^1(A)$, define $u(x) = \sum_{\alpha \in A} \xi_\alpha y_\alpha$ (Chapter III, Exercise
23); conclude from (III, 2.1) that $x \to u(x)$ is a homomorphism of $l^1(A)$
onto E.) Deduce that whenever A is infinite, $l^1(A)$ contains a closed subspace that has no topological complement.

(c) If E is a complete t.v.s. and H is a closed subspace such that E/H
is not complete (Exercise 11), then H does not have a topological
complement.

For a discussion of the problem of complementary subspaces, see Day
[2] and Köthe [5], §31.

13. A l.c.s. E is called **distinguished** if its strong dual E'_β is barreled.
Let E be the vector space of all numerical double sequences $x = (x_{ij})$
such that for each $n \in N$, $p_n(x) = \sum_{i,j} |a_{ij}^{(n)} x_{ij}| < +\infty$, where $a_{ij}^{(n)} = j$
for $i \leq n$ and all j, $a_{ij}^{(n)} = 1$ for $i > n$ and all j. The semi-norms p_n ($n \in N$)
generate a l.c. topology under which E is an (F)-space which is not dis-
tinguished (Grothendieck [10]). Establish successively the following
partial results:

(a) The dual E' can be identified with the space of double sequences
$u = (u_{ij})$ such that $|u_{ij}| \leq c a_{ij}^{(n)}$ for all i, j and suitable $c > 0$, $n \in N$
(cf. Chapter III, Exercise 25). If B_n is the polar of $U_n = \{x: p_n(x) \leq 1\}$,
then $\{nB_n: n \in N\}$ is a fundamental family of bounded subsets of E'.

(b) Let W denote the convex, circled hull of $\bigcup_n 2^{-n} B_n$; W absorbs all
bounded subsets of E', and W does not contain a $u \in E'$ such that for
each i, there exists j with $|u_{ij}| \geq 2$.

(c) Given a sequence $\rho = (\rho_n)$ of strictly positive numbers, define
elements $u^{(n)} \in E'$ so that $u_{ij}^{(n)} = 0$ for $(i,j) \neq (n, k_n)$, $u_{n,k_n}^{(n)} = 1$, where
k_n is chosen so that $2^{n+1} u^{(n)} \in \rho_n B_n$. For each given ρ, the sequence with
general term $s_N = 2 \sum_1^N u^{(n)}$ is a weak Cauchy sequence in E', hence convergent to $s \in E'$.

(d) For each strong 0-neighborhood B° in E' (B a bounded subset of E)
there exists a sequence $\rho = (\rho_n)$ of numbers > 0 such that $\Gamma_n \rho_n B_n \subset B^\circ$;
if $\{s_N\}$ is a sequence as constructed in (c), then $s_N \in B^\circ$ for all
$N \in N$; hence $s \in B^\circ$ but $s \notin W$. It follows that W contains no B°; hence
E'_β is not bornological and, therefore, not barreled (use (6.6)).

14. If $\langle F, G \rangle$ is a duality, the strong topology $\beta(F, G)$ is, in general, not inherited by closed subspaces or separated quotients. (Concerning subspaces, consider a l.c.s. F as constructed in Exercise 9; then $\beta(H^\circ, F/H)$ is distinct from the topology induced on H° by $\beta(F', F)$. Concerning quotients, consider a non-distinguished complete l.c.s. E (Exercise 13) as a closed subspace of a suitable product G of Banach spaces; then $\beta(G'/E^\circ, E)$ is distinct from the quotient of $\beta(G', G)$.)

If F is a Banach space, then the strong topology is inherited by closed subspaces and separated quotients; the same is true (with respect to $\langle F, F' \rangle$) for the strong dual of F.

15. We use the notation of Section 5. There exist l.c.s. E such that the families $\mathfrak{E}$, $\mathfrak{C}$, $\mathfrak{B}$, $\mathfrak{B}_\sigma$ of subsets of E' are all distinct. (Consider a suitable space F which is not barreled (Chapter II, Exercise 14), and take E to be the space F'_σ.)

16. (Infrabarreled Spaces). The product and l.c. direct sum of any family of infrabarreled spaces is infrabarreled (use Exercise 8). Every separated quotient of an infrabarreled space is infrabarreled (immediate verification), but a closed subspace is not necessarily infrabarreled (Exercise 20).

17. (Theorem of Banach-Mackey). If E is a l.c.s. and B is a bounded, convex, circled subset of E such that the normed space E_B is complete, then B is bounded for $\beta(E, E')$. (Observe that for each barrel $D \subset E$, $D \cap E_B$ is a barrel in the Banach space E_B, and use (1.6).)

18. (Reflexive Spaces). (See also Exercise 20.)

(a) The Banach space $l^p(A)$ is reflexive whenever $1 < p < +\infty$.

(b) If E is a quasi-complete Mackey space such that E'_β is semi-reflexive, then E is reflexive.

(c) If E is a weakly semi-complete l.c.s. whose strong dual E'_β is separable, then E is semi-reflexive. (Use (1.7).)

(d) Give an example of a non-reflexive l.c.s. whose strong dual is reflexive.

(e) If E is a non-reflexive (F)-space, the canonical inclusions $E \subset E''$ $\subset E^{(iv)} \subset \cdots$ and $E' \subset E''' \subset E^{(v)} \subset \cdots$ are all proper. (Conduct the proof indirectly, using (b).)

(f) There exist non-reflexive (B)-spaces E such that the canonical image of E in E'' is of finite codimension (see James [2] and Civin-Yood [1]). Deduce from this that there exist infinite-dimensional Banach spaces E such that E is not isomorphic (as a t.v.s.) with $E \times E$.

19. (Montel Spaces)

(a) Every product and l.c. direct sum of a family of (M)-spaces is an (M)-space; the strict inductive limit of a sequence of (M)-spaces is an (M)-space. By contrast, closed subspaces and separated quotients of (M)-spaces are, in general, not even reflexive (Exercise 20).

(b) Each barreled, quasi-complete nuclear space is an (M)-space (use (III, 7.2), Corollary 2).

(c) A separable (F)-space E is an (M)-space if and only if each $\sigma(E', E)$-convergent sequence in E' is strongly convergent. (For the sufficiency of the condition, use (1.7). Show that each bounded subset of E' is strongly precompact (Chapter I, Exercise 5) (hence relatively compact since E'_β is complete). From (6.2), Corollary 3, conclude that $E = E''$, hence that E and E'_β are reflexive; finally use (5.9).)

(d) Every metrizable (M)-space is separable (Dieudonné [7]). (Select a sequence $\{U_n\}$ of convex, circled 0-neighborhoods forming a base at 0, and imbed E as a subspace of $\prod_n \tilde{E}_{U_n}$. Supposing E to be non-separable, it can be assumed that E_{U_1} is not separable. Denoting by B_1 an uncountable subset of E_{U_1} whose elements have mutual distance $\geq \delta > 0$, let $M_1 = \phi_{U_1}^{-1}(B_1)$. There exists an uncountable proper subset M_2 of M_1 such that $\phi_{U_2}(M_2)$ is bounded in E_{U_2}, etc. If $x_n \in M_n \sim M_{n+1}$ for all $n \in N$, then $\{x_n\}$ is a bounded sequence in E such that $\{\phi_{U_1}(x_n)\}$ contains no Cauchy subsequence, which is contradictory.)

20. Let E be the vector space (over R) of all double sequences $x = (x_{ij})$ such that for each $n \in N$, $p_n(x) = \sum_{i,j} a_{ij}^{(n)} |x_{ij}| < +\infty$, where $a_{ij}^{(n)} = j^n$ for $i < n$ and all j, and $a_{ij}^{(n)} = i^n$ for $i \geq n$ and all j. Under the topology generated by the semi-norms $p_n (n \in N)$, E is an (F)-space and an (M)-space (cf. Exercise 13). The dual E' can be identified with the space of all double sequences $x' = (x'_{ij})$ such that $|x'_{ij}| \leq c a_{ij}^{(n)}$ for all i, j and suitable $c > 0$, $n \in N$ (the canonical bilinear form being $(x, x') \to \langle x, x' \rangle = \sum_{i,j} x_{ij} x'_{ij}$). Each $x \in E$ defines a summable family $\{x_{ij} : (i,j) \in N \times N\}$ (Chapter III, Exercise 23); if one puts $y_j = \sum_i x_{ij}$, then $y = (y_j) \in l^1$, and $x \to y = u(x)$ is a continuous linear map of E onto a dense subspace of l^1. (Köthe [1], Grothendieck [10].)

(a) The adjoint u' is biunivocal and onto a closed subspace $u'(l^\infty)$ of E'. Hence u is a topological homomorphism of E onto l^1 (use (7.7)); thus $E/u^{-1}(0)$ is isomorphic with l^1.

(b) The canonical map $E \to E/u^{-1}(0)$ maps the family of all bounded subsets of E onto the family of all relatively compact subsets of $E/u^{-1}(0)$. (Use that E is an (M)-space and an (F)-space, and (6.3), Corollary 1.) Infer from this that $\beta(E', E)$ induces on $u'(l^\infty)$ the $\mathfrak{S}$-topology, where $\mathfrak{S}$ is the family of all relatively compact subsets of $E/u^{-1}(0)$. $u'(l^\infty)$ is not infrabarreled for this topology.

(c) Conclude from the preceding that (1) a closed subspace or a separated quotient of an (M)-space is not necessarily reflexive; (2) a bounded set in the quotient of an (F)-space E is not necessarily the canonical image of a bounded set in E; (3) a closed subspace of a barreled (respectively, bornological) l.c.s. is not necessarily barreled (respectively, bornological). (The strong dual of the space E, above, is barreled, and bornological by (6.6), Corollary 1.)

21. The Banach-Mackey theorem (Exercise 17) implies that in (II, 8.5) it suffices for the conclusion to assume that B is semi-complete. This permits us to improve several earlier results, for example, to show that every semi-complete bornological space is barreled. The following results show that, in general, the concepts of semi-completeness, quasi-completeness, and completeness should be carefully distinguished.

(a) There exists a semi-complete l.c.s. that is not quasi-complete. (Consider l^1 under its weak topology.)

(b) Let E be a l.c.s. For E'_σ to be complete, it is necessary and sufficient that $\tau(E, E')$ be the finest l.c. topology on E (cf. Exercise 6). Infer from this that if E is metrizable and E'_σ is complete, then E is finite dimensional. (Use Chapter II, Exercise 7.) If E is normable and E_σ complete, then E is finite dimensional. (Apply the preceding result to E'_β.)

(c) In each Hilbert space H, there exists a weakly complete convex set C such that $H = C - C$. (Suppose H to be a space $l^2(A)$ over $\mathbf{R}$ (Chapter II, Section 2, Example 5), and take $C = \{x \in H: \xi_\alpha \geq 0$ for all $\alpha \in A\}$.)

22. Let E be a l.c.s., let E' be its dual, and let $\mathfrak{T}_f$ be the finest topology on E' that agrees with $\sigma(E', E)$ on every equicontinuous subset of E'.

(a) $\mathfrak{T}_f$ is a translation-invariant topology possessing a 0-neighborhood base of radial and circled sets.

(b) In general, $\mathfrak{T}_f$ fails to be locally convex (Collins [1]). (Consider an infinite-dimensional space E on which there exists a topology $\mathfrak{T}_1$ such that $(E, \mathfrak{T}_1)$ is an (F)-space; denote by $\mathfrak{T}_0$ the finest l.c. topology on E (Chapter II, Exercise 7). Then $(E, \mathfrak{T}_1)' = E' \neq E^*$ (Exercise 21). Consider $\mathfrak{T}_f$ on $E^* = (E, \mathfrak{T}_0)'$; then E' is $\mathfrak{T}_f$-closed in E^*. If $\mathfrak{T}_f$ were locally convex, it would be consistent with $\langle E, E^* \rangle$ (observe that $(E, \mathfrak{T}_0)$ is complete, and use (6.2), Corollary 3); hence E' would be closed and dense in $(E^*, \sigma(E^*, E))$, which is contradictory.)

23. Let E be a separable, metrizable l.c.s. and let $\tilde{E}$ be its completion. Then each bounded subset of $\tilde{E}$ is contained in the closure (taken in $\tilde{E}$) of a suitable bounded subset of E. (Observe that $\tilde{E}$ can be identified with the closure of E in the strong bidual E''. If $B \subset \tilde{E}$ is bounded, then B is separable, hence equicontinuous in E'' (use (6.5), Corollary 1); thus $B \subset A^{\circ\circ}$ (bipolar with respect to $\langle E', E'' \rangle$) for a suitable bounded subset A of E.)

24. ((DF)-spaces. See also Grothendieck [10], Köthe [5]).

(a) The strong dual E'_β of a (DF)-space E is an (F)-space. (Use the corollary of (6.7) to show that E'_β is complete.)

(b) If E is a (DF)-space and M is a closed subspace, then $\beta(M^\circ, E/M)$ is the topology induced on M° by $\beta(E', E)$. (Prove that the identity map of $(M^\circ, \beta(E', E))$ onto $(M^\circ, \beta(M^\circ, E/M))$ is continuous by showing that each $\beta(E', E)$-null sequence in M° is equicontinuous in E' (hence $\beta(M^\circ, E/M)$-bounded), using that $\beta(E', E)$ is metrizable (cf. Chapter II, Exercise 17).)

(c) If E is a l.c.s., and M is a subspace which is a (DF)-space, then $\beta(E'/M^\circ, M)$ is the quotient of $\beta(E', E)$. (Employ the same method as in (b), using that each $\beta(E'/M^\circ, M)$-null sequence is equicontinuous and hence by (7.5) the canonical image of an equicontinuous sequence in E'.)

(d) Let E be a (DF)-space. The completion $\tilde{E}$ can (algebraically) be identified with a subspace of E''; moreover, each bounded subset of the strong bidual E'' is contained in the bipolar (with respect to $\langle E', E'' \rangle$) of a suitable bounded subset of E. Each bounded subset of $\tilde{E}$ is contained in the closure (taken in $\tilde{E}$) of a suitable bounded subset of E. (Use (c) to prove that $\beta(E', E) = \beta(E', \tilde{E})$.) Conclude that $\tilde{E}$ is a (DF)-space, and that every quasi-complete (DF)-space is complete.

(e) Each separated quotient of a (DF)-space is a (DF)-space. (Use

(b) to show that each bounded subset of E/M is the canonical image of bounded subset of E.) The l.c. direct sum of a sequence of (DF)-spaces is a (DF)-space, and an inductive limit of a sequence of (DF)-spaces is a (DF)-space. On the other hand, an infinite product of (DF)-spaces (each not reduced to $\{0\}$) is not a (DF)-space; a closed subspace of a (DF)-space is not necessarily a (DF)-space.

(f) Let $E = \lim\limits_{\longrightarrow} h_{nm}E_m$ be an inductive limit of a sequence of reflexive (DF)-spaces. Then the strong dual of E can be identified with the projective limit $\lim\limits_{\longleftarrow} g_{mn}E'_n$ of the strong duals E'_n with respect to the adjoint maps $g_{mn} = h'_{nm}$ (cf. end of Section 4).

(g) There exist complete (DF)-spaces not isomorphic with the strong dual of a metrizable l.c.s. (Consider a non-separable reflexive Banach space E under the topology of uniform convergence on the strongly separable, bounded subsets of E'.)

25. Let E, F be l.c.s. with respective duals E', F'. For any subset $Q \subset \mathscr{L}(E, F)$, denote by Q' the set of adjoints $\{u' : u \in Q\}$. Consider the propositions:

(i) Q is equicontinuous.

(ii) For each equicontinuous set $B \subset E'$, $Q'(B)$ is equicontinuous in F'.

(iii) Q' is equicontinuous in $\mathscr{L}(F'_\beta, E'_\beta)$.

(iv) Q is bounded in $\mathscr{L}_b(E, F)$.

(v) Q' is simply bounded in $L(F', E')$ with respect to $\sigma(E', E)$.

(vi) Q' is simply bounded in $L(F', E')$ with respect to $\beta(E', E)$.

One has the following implications: (a) (i) $\Leftrightarrow$ (ii) and (iii) $\Leftrightarrow$ (iv); (b) If E is infrabarreled, then (i) $\Leftrightarrow$ (iii) $\Leftrightarrow$ (vi); (c) If E is barreled, propositions (i) through (vi) are equivalent.

26. Give an example of two l.c.s. E, F and a $u \in \mathscr{L}(E, F)$ such that u is a topological homomorphism for $\sigma(E, E')$ and $\sigma(F, F')$ but not for $\tau(E, E')$ and $\tau(F, F')$. (Take F to be a Mackey space such that some subspace E is not a Mackey space (cf. Exercise 9), and consider the canonical imbedding $E \to F$.)

Give an example of two (F)-spaces E, F and a topological homomorphism u of E onto F such that u' is not an isomorphism of F'_β into E'_β. (Cf. Exercise 20.)

27. Let E, F be normable spaces and $u \in \mathscr{L}(E, F)$.

(a) Give an example where E is not complete, u is a homomorphism, and u' is a strong homomorphism but not a weak homomorphism. (Take $E = F_0$, where F_0 is a dense subspace $\neq F$ of F.)

(b) Give an example where E is not complete, u is not a homomorphism, and u' is a weak and strong homomorphism.

(c) Give an example where E is complete, F is not complete, and u' is a weak but not a strong homomorphism.

28. Let E, F be l.c.s. and let u be a linear map of E onto F.

(a) The following properties of u are equivalent:

(i) u is nearly open.

(ii) For each subset $A \subset E$, $u(\mathring{A})$ is contained in the interior of $[u(A)]^-$.

(iii) For each convex, circled 0-neighborhood $U \subset E$, $u(U)$ is weakly
       dense in some 0-neighborhood $V \subset F$.

(b) If F is barreled, u is nearly open.

29. (B-Completeness).

(a) Let E be an infinite-dimensional vector space such that there exists
a topology $\mathfrak{T}_1$ under which E is an (F)-space, and denote by $\mathfrak{T}_0$ the
finest l.c. topology on E. Then $(E, \mathfrak{T}_0)$ is complete but not B_r-complete.
(Observe that $(E, \mathfrak{T}_0)$ is not metrizable (Chapter II, Exercise 7) and use
(8.4); cf. Exercise 22.)

(b) Show that products and l.c. direct sums of B-complete spaces are,
in general, not B-complete. (Concerning l.c. direct sums, use (a); concerning products, imbed a complete but not B-complete space into a
product of Banach spaces, and use (8.2).)

(c) The following assertions are equivalent: (i) Every B_r-complete
space is B-complete. (ii) Every separated quotient of a B_r-complete
space is B_r-complete.

(d) Let X be a completely regular topological space, $R(X)$ the
space of real-valued continuous functions on X under the topology
of compact convergence. Of the following propositions, (i) implies (ii):
(i) $R(X)$ is B_r-complete. (ii) If Y is a dense subset of X such that $Y \cap C$
is compact whenever C is compact, then $Y = X$. (Note that the mapping $t \to (f \to f(t))$ is a homeomorphism of X into the weak dual of
$R(X)$.)

30. Let E, F be Banach spaces with respective strong duals E', F'.

(a) The dual of $\mathscr{L}_c(E, F)$ (topology of compact convergence) can be
identified with a quotient of $E \tilde{\otimes} F'$ and, if F is reflexive, with $E \tilde{\otimes} F'$.
(Observe that $\mathscr{L}_b(E, F)$ is canonically isomorphic with a closed sub-
space of $\mathscr{B}_b(E, F')$, the latter space being the strong dual of $E \tilde{\otimes} F'$ by
(9.8). The dual of $\mathscr{L}_s(E, F)$ can be identified with $E \otimes F'$ by (4.3), Corollary 4, and on the bounded subsets of $\mathscr{L}_b(E, F)$ the topologies of simple
and of compact convergence agree (theorem of Banach-Steinhaus).
Finally, use Grothendieck's theorem (6.2).)

(b) E has the approximation property if and only if the canonical
map $\tau: E' \tilde{\otimes} E \to \mathscr{L}(E)$ is biunivocal. (Use (a). Cf. Chapter III, Section 9,
and Grothendieck [13], I, §5, prop. 35.)

(c) If E' possesses the approximation property, then so does E. (Ob-
serve that the canonical imbedding $\rho: E' \tilde{\otimes} E \to E' \tilde{\otimes} E''$ is a topological
isomorphism (in particular, injective). If $u \in E' \tilde{\otimes} E$ and $w = \tau(u)$ is its
canonical image in $\mathscr{L}(E)$, if $v = \rho(u)$, and if $q \in \mathscr{L}(E')$ is the endomorphism defined by v, then $q = w'$. Hence if $w = 0$, conclude from (b) that
$v = 0$ and thus $u = 0$. Use (b) once more.)

31. Show that in the dual of a nuclear space E, each equicontinuous
subset is metrizable for $\sigma(E', E)$ (hence separable). (Use (9.3).) Infer from
this that if d is an uncountable cardinal, then K_0^d is a nuclear space whose
strong dual fails to be nuclear.

32. Let E, F be (DF)-spaces.

(a) The projective tensor product $E \otimes F$ as well as its completion
$E \tilde{\otimes} F$ are (DF)-spaces. (Use (9.8); cf. Exercise 24.)

(b) If E is nuclear and complete, then E is an (M)-space, namely the strong dual of the (F)- and (M)-space E'_τ. (Observe that E is semi-reflexive, and prove that its strong dual $E'_\beta = E'_\tau$ is a reflexive (F)-space; for this, note that each strongly bounded sequence in E' is equicontinuous, and use (11.2). Conclude that E'_τ is an (M)-space (use (III, 4.5)), and thus that each strongly bounded set is separable and hence equicontinuous.)

(c) Suppose that E is nuclear. The strong dual of $E \tilde\otimes F$ can be identified with $E'_\beta \tilde\otimes F'_\beta$, and the strong bidual with $E'' \tilde\otimes F''$. (Assume without loss of generality that E is complete, hence an (M)-space (use (b)); then proceed as in the second part of the proof of (9.9). To obtain the second assertion, apply (9.9) to E'_β and F'_β. Cf. Exercise 33(b).)

(d) (Proof of (9.9), Corollary 3). Suppose that E, F are strong duals of (F)-spaces and that E is nuclear. Then $\mathfrak{B}(E, F) = \mathscr{B}(E, F)$. (Show that a continuous linear map u of E into $(F'', \sigma(F'', F'))$ is continuous for $\sigma(F'', F''')$, using that E is bornological.)

33. (Nuclear (F)- and (DF)-spaces).

(a) If E is an (F)-space whose strong dual is nuclear, then E is nuclear. (Establish this through the following steps, F denoting an arbitrary (B)-space:

1. The canonical map $\psi: E \tilde\otimes F \to \mathfrak{B}(E'_\sigma, F'_\sigma)$ is surjective. [Observe that E is reflexive, and that by (9.3) $E_A \to E$ is nuclear for each closed, convex, circled, bounded subset $A \subset E$. Consider an element of $\mathfrak{B}(E'_\sigma, F'_\sigma)$ as a weakly continuous linear map of F'_β into E.]

2. ψ is injective. [Each $u \in \mathscr{L}(E, F'_\beta)$ can be approximated, uniformly on every compact subset of E, by maps of finite rank; infer that $E' \otimes F'$ is dense in $\mathscr{B}(E, F)$ for the $\mathfrak{S} \times \mathfrak{T}$-topology, $\mathfrak{S}$ and $\mathfrak{T}$ denoting the families of relatively compact subsets of E and F, respectively. Since the canonical imbedding $E' \otimes F' \to \mathscr{B}(E, F)$ can be viewed as the adjoint of ψ, the assertion rests on the fact that the $\mathfrak{S} \times \mathfrak{T}$-topology is consistent with the duality $\langle E \tilde\otimes F, \mathscr{B}(E, F)\rangle$. For this, see Grothendieck [13], I, §4, prop. 21.]

3. From the preceding and Banach's theorem (III, 2.1) it follows that ψ is a topological isomorphism onto $\mathfrak{B}_e(E'_\sigma, F'_\sigma)$. Use (10.7), Corollary 1.)

(b) If E is an (F)-space or a complete (DF)-space, then E is nuclear if and only if its strong dual is nuclear. (Use (a), (9.7) and Exercise 32(b).)

(c) Let E denote a nuclear (DF)-space and let F denote a nuclear (F)-space; then $\mathscr{L}_b(E, F)$ and its strong dual are nuclear.

34. If F is a nuclear space, then its bidual F''_e (natural topology) is nuclear. Use this to show that in (9.7) the assumption that E is semireflexive is dispensable.

35. We use the notation of Section 10. Let A be any non-empty index set and let E be a given l.c.s. For a subset P of $l^1[A, E]'$ (respectively, for a subset $Q \subset l^1(A, E)'$) to be equicontinuous, it is necessary and sufficient that the union of the ranges of all $\mathbf{x}' \in P$ be equicontinuous in E' (respectively, that Q be presentable as described in (9.2)).

36. The notation is as in Section 10.

(a) If E is a l.c.s. such that each summable sequence in E is absolutely summable, then each bounded subset of $l^1(N, E)$ is bounded as a subset of $l^1[N, E]$. (Consider a sequence $\{x_n'\}$ in E' such that $\sum_n |\langle x_n, x_n' \rangle| < +\infty$ for each $\mathbf{x} = (x_n) \in l^1(N, E)$, and show that the function $\mathbf{x} \to \sum_n |\langle x_n, x_n' \rangle|$ is bounded on every bounded subset of $l^1(N, E)$. See Pietsch [5], p. 54.)

(b) If $l^1(N, E)$ is infrabarreled (e.g., if E is metrizable) and each summable sequence in E is absolutely summable, then the canonical map $l^1[N, E] \to l^1(N, E)$ is a topological isomorphism (hence E nuclear). (Use (a).)

(c) Deduce from (b) the following improvement of the theorem of Dvoretzky-Rogers: If E is a normable space in which each summable sequence is absolutely summable, then E is finite dimensional.

37. Let E be the Montel space K_0^d, where d is the cardinality of the continuum. Denote by B the subset $[0, 1]^d$ of E, and by B_0 the subset of B whose elements have not more than countably many non-zero co-ordinates. Show that (under the topology induced by E) B and B_0 are completely regular topological spaces such that B_0 is sequentially compact but not compact, and B is compact but not sequentially compact. Moreover, B_0 is an example of a uniform space which is semi-complete and precompact but not complete.

38. (Weak Countable Compactness). Let $(E, \mathfrak{T})$ be a l.c.s.

(a) If there exists a metrizable l.c. topology on E which is coarser than $\mathfrak{T}$, then each weakly countably compact subset B of E is weakly sequentially compact. (Dieudonné-Schwartz [1]. If $\{x_n\}$ is a sequence in B, the closed linear hull M of its range is separable. Show that the dual M' is weakly separable (cf. (1.7)), and hence that $\{x_n\}$ contains a weakly convergent subsequence.)

(b) Suppose that E' is the union of countably many weakly compact subsets. Then, given a subset $M \subset E$, each point in the weak closure of M is contained in the closure of a suitable countable subset of M (Kaplansky). (Use a method similar to thàt employed in the second part of the proof of (11.1), or see Köthe [5], §24.1.)

(c) Deduce from (a) and (b) extensions of Theorem (11.2) for metrizable l.c.s. and for (LF)-spaces.

The two remaining problems are included for later reference; the results (needed in the Exercises of Chapter V and, to some extent, in the Appendix) are easy to prove if the reader is familiar with Sections 1–3 of this chapter and the basic theory of functions of one complex variable.

39. (Vector Valued Analytic Functions). Let G be a non-empty open subset of the Riemann sphere and let E be a l.c.s. over C. A function $f: G \to E$ is called **holomorphic** at $\zeta_0 \in G$ if there exists a neighborhood Z of ζ_0 such that for each $x' \in E'$, the function $\zeta \to \langle f(\zeta), x' \rangle$ is complex differentiable in the interior Z_0 of Z, and if for each $\zeta \in Z_0$ the linear form

$$x' \to \frac{d}{d\zeta} \langle f(\zeta), x' \rangle$$

is $\sigma(E', E)$-continuous. The unique element $x(\zeta) \in E$ representing this linear form is called the **derivative** of f at ζ, and is usually denoted by $f'(\zeta)$. If $\infty \in G$, f is called holomorphic at ∞ if $\zeta \to f(\zeta^{-1})$ is holomorphic at 0. A function $f: G \to E$ which is holomorphic at each $\zeta \in G$ is called **locally holomorphic** in G. (Note that G need not be connected.)

Let δ denote a rectifiable oriented arc in the complex plane. A function $f: \delta \to E$ is called (Riemann) **integrable** over δ if for each $x' \in E'$, the function $\zeta \to \langle f(\zeta), x' \rangle$ is integrable over δ in the Riemann-Cauchy sense, and if the linear form $x' \to \int_\delta \langle f(\zeta), x' \rangle d\zeta$ is $\sigma(E', E)$-continuous. The unique element of E representing this linear form is called the (Riemann) **integral** of f over δ, and is usually denoted by $\int_\delta f(\zeta) d\zeta$.

In the following we suppose E to be a l.c.s. over C such that the closed, convex hull of each compact subset of E is compact (which is, in particular, the case if E is quasi-complete).

(a) Each continuous function $f: \delta \to E$ is integrable over δ. (Use the fact that the closed, convex, circled hull of $f(\delta)$ is compact in E.) The linear map $f \to \int_\delta f(\zeta) d\zeta$ is continuous on $\mathscr{C}_E(\delta)$, endowed with the topology of uniform convergence, into E.

(b) Each function $f: G \to E$ which is locally holomorphic is continuous. (Consider the difference quotient of f at $\zeta_0 \in G$.)

(c) (Cauchy's theorem). Let G be open and let $\gamma \subset G$ be a positively oriented, closed, rectifiable Jordan curve whose interior G_γ belongs to G. Then $\int_\gamma f(\zeta) d\zeta = 0$ for every function f which is locally holomorphic in G. Deduce from this that if $\zeta_0 \in G_\gamma$, then (f being locally holomorphic in G)

$$f(\zeta_0) = \frac{1}{2\pi i} \int_\gamma \frac{f(\zeta)}{\zeta - \zeta_0} \, d\zeta.$$

(d) Infer from (c) that if f is locally holomorphic in G and $\zeta_0 \in G$, then

$$f(\zeta) = \sum_{n=0}^\infty a_n (\zeta - \zeta_0)^n,$$

where $a_n = (2\pi i)^{-1} \int_\gamma f(\zeta)(\zeta - \zeta_0)^{-n-1} d\zeta$ $(n = 0, 1, \ldots)$ (γ denoting a circle about ζ_0 whose interior is in G) is an expansion valid in a circular neighborhood of ζ_0 (in contrast with ordinary usage, scalars are written to the right of elements of E). The series converges in the interior of γ, with respect to $\tau(E, E')$ and uniformly on compact sets. Moreover, γ can be taken as the largest circle of center ζ_0 to whose interior f has a holomorphic extension $\bar{f}$; the series then converges to $\bar{f}(\zeta)$.

(e) (Liouville's theorem). Every E-valued function f, holomorphic and uniformly bounded on the entire complex plane, is constant (i.e., has a range consisting of a single element of E).

(f) Define for E-valued functions the concepts of pole and isolated singularity, and generalize the classical results on Laurent expansions.

40. (Locally Convex Algebras). Let A be an algebra over K. (Recall that A can be defined as a vector space A_0 over K on which a bilinear associative map, called multiplication and usually denoted by

$(a, b) \rightarrow ab$, is specified; A_0 is then called the underlying vector space of A. If A has a unit, it will be denoted by e; the inverse of an element $a \in A$ is denoted by a^{-1}.)

An algebra A over K ($K = \mathbf{R}$ or $\mathbf{C}$) is called a **locally convex algebra** over K if A_0 is a l.c.s. and if multiplication is separately continuous. A **normed algebra** over K is an algebra A over K such that A_0 is a normed space, with the additional requirement that $\|ab\| \leqq \|a\| \|b\|$ for all $a, b \in A$, and that $\|e\| = 1$ if A has a unit e. A **Banach algebra** is a normed algebra A such that A_0 is a Banach space.

(a) If E is a l.c.s., $\mathfrak{S}$ a total family of bounded subsets of E such that $\mathfrak{S}$ is invariant under each $u \in \mathscr{L}(E)$, then (with respect to the composition of maps) $\mathscr{L}(E)$ is a l.c. algebra under the $\mathfrak{S}$-topology. If $\mathfrak{B}$ denotes the family of all bounded subsets of E, multiplication is right and left $\mathfrak{B}$-hypocontinuous for the topology of bounded convergence. If E is barreled, multiplication is left $\mathfrak{B}$-hypocontinuous for the topology of simple convergence. If $\mathscr{L}_{\mathfrak{S}}(E)$ is an (F)-space, then multiplication is continuous by (III, 5.1).

(b) Let A be a l.c. algebra over $\mathbf{C}$ with unit e. The **spectrum** $\sigma(a)$ of $a \in A$ is the complement of the largest open subset G of the Riemann sphere such that $\lambda \rightarrow (\lambda e - a)^{-1}$ exists and is locally holomorphic in G. (If $\lambda \rightarrow R(\lambda) = (\lambda e - a)^{-1}$ is holomorphic in a neighborhood of ∞, the definition $R(\infty) = 0$ renders R holomorphic at ∞.) Show that $\sigma(a) \neq \varnothing$ for all $a \in A$. (Use Exercise 39(e).)

(c) Let A be a l.c. algebra over $\mathbf{C}$ with unit e and let $a \in A$. For the resolvent $\lambda \rightarrow R(\lambda)$ of a to be holomorphic at $\lambda_0 \in \mathbf{C}$, it is necessary and sufficient that $R(\lambda)$ exists in some neighborhood U of λ_0 such that for each sequence $\{\lambda_n\}$ in U, the sequence $\{R(\lambda_n)\}$ is bounded in A. (Use the resolvent equation $R(\lambda) - R(\mu) = -(\lambda - \mu)R(\lambda)R(\mu)$, which holds whenever $R(\lambda)$, $R(\mu)$ exist.)

(d) If A is a Banach algebra over $\mathbf{C}$ with unit e, then for each $a \in A$, $\sigma(a)$ is a compact subset of $\mathbf{C}$. If $r(a)$ is the radius of the smallest circle of center 0 in $\mathbf{C}$ that contains $\sigma(a)$, we have the relation $r(a) = \lim_n \|a^n\|^{1/n}$. (See, e.g., Hille-Phillips [1].) ($r(a)$ is called the **spectral radius** of $a \in A$ whenever A is a l.c. algebra over $\mathbf{C}$ with unit e.)

Chapter V

ORDER STRUCTURES

The present chapter is devoted to a systematic study of order structures within the framework of topological vector spaces. No attempt has been made to give an account of the extensive literature on Banach lattices, for a survey of which we refer the reader to Day [2], nor is any special emphasis placed on ordered normed spaces. Our efforts are directed towards developing a theory that is in conformity with the modern theory of topological vector spaces, that is to say, a theory in which duality plays the central role. This approach to ordered topological vector spaces is of fairly recent origin, and thus cannot be presented in a form as definite as a mature theory; it is nonetheless hoped that the reader who has encountered parts of it in the literature (e.g., Gordon [1], [2], Kist [1], Namioka [1], Schaefer [1]–[5]) will obtain a certain survey of the methods available and of the results to which they lead. The fact that ordered topological vector spaces abound in analysis is perhaps motivation enough for a systematic study; beyond this, the present chapter is followed by an appendix intended to illustrate some applications to spectral theory. As in the preceding chapters, further information can be found in the exercises.

Section 1 is concerned with algebraic aspects only and supplies, in particular, the basic tools needed in working with vector lattices. For simplicity of exposition we restrict attention to ordered vector spaces over R; Section 2 discusses briefly how these concepts can be applied to vector spaces over C, which is often called for by applications (particularly to measure theory and spectral theory). Section 3 gives the basic results on the duality of convex cones. The concept of normal cone, probably the most important concept of the theory, is introduced and a number of immediate consequences are established. The discussion proceeds covering the real and complex cases simultaneously; the reader who finds this too involved may well assume first that all occurring vector spaces are defined over R. Section 4 introduces ordered topological vector spaces and establishes two more properties of normal cones, among them (Theorem (4.3)) the abstract version of a classical theorem of Dini on monotone convergence. The duality of ordered vector spaces is not discussed there, since such a discussion would have amounted to a direct application of the results of Section 3, which can be left to the reader.

Section 5 is concerned with the induced order structure on spaces of linear mappings; the principal results are Theorem (5.4) on the extension of continuous positive linear forms, and Theorem (5.5) establishing the continuity of a large class of positive linear forms and mappings. The order topology, a locally convex topology accompanying every ordered vector space over R, is studied in some detail in Section 6. The importance of this topology stems in part from the fact that it is the topology of many ordered t.v.s. occurring in analysis. Section 7 treats topological (in particular, locally convex) vector lattices. We obtain results especially on the strong dual of a locally convex vector lattice, and characterizations of vector lattices of minimal type in terms of order convergence and in terms of the evaluation map. (For the continuity of the lattice operations see Exercise 20.) The section concludes with a discussion of weak order units.

Section 8 is concerned with the vector lattice of all continuous real valued functions on a compact space, and with abstract Lebesgue spaces. The Stone-Weierstrass theorem is presented in both its order theoretic and its algebraic form. Further, the dual character of (AM)-spaces with unit and (AL)-spaces is studied as an illuminating example of the duality of topological vector lattices treated in Section 7. (AL)-spaces are represented as bands of Radon measures, characterized by a convergence property, on extremally disconnected compact spaces. The classical representation theorem of Kakutani for (AM)-spaces with unit is established, and an application is made to the representation of a much more general class of locally convex vector lattices.

1. ORDERED VECTOR SPACES OVER THE REAL FIELD

Throughout this section, we consider only vector spaces over the real field R.

Let L be a vector space over R which is endowed with an order structure R defined by a reflexive, transitive, and anti-symmetric binary relation "$\leq$"; L is called an **ordered vector space** over R if the following axioms are satisfied:

$(LO)_1$ $x \leq y$ implies $x + z \leq y + z$ for all $x, y, z \in L$
$(LO)_2$ $x \leq y$ implies $\lambda x \leq \lambda y$ for all $x, y \in L$ and $\lambda > 0$.

$(LO)_1$ expresses that the order of L is translation-invariant, $(LO)_2$ expresses the invariance of the order under homothetic maps $x \to \lambda x$ with ratio $\lambda > 0$. Examples of ordered vector spaces abound; for example, every vector space of real-valued functions f on a set T is naturally ordered by the relation "$f \leq g$ if $f(t) \leq g(t)$ for all $t \in T$"; in this fashion, one obtains a large number of ordered vector spaces from the examples given in Chapter II, Section 2, and Chapter III, Section 8, by considering real-valued functions only and taking $K = R$.

It is immediate from the axioms above that in an ordered vector space L, the subset $C = \{x: x \geq 0\}$ is a convex cone of vertex 0 satisfying $C \cap - C = \{0\}$; a cone in L with these properties is called a **proper cone** in L. The elements $x \in C$ are called **positive**, and C is called the **positive cone** of the ordered vector space L.

Two ordered vector spaces L_1, L_2 are **isomorphic** if there exists a linear biunivocal map u of L_1 onto L_2 such that $x \leq y$ if and only if $u(x) \leq u(y)$ (equivalently, such that u maps the positive cone of L_1 onto the positive cone of L_2).

If L is any vector space over R, a proper cone $H \subset L$ is characterized by the properties

(i) $H + H \subset H$,
(ii) $\lambda H \subset H$ *for all* $\lambda > 0$,
(iii) $H \cap - H = \{0\}$.

It is verified without difficulty that each proper cone $H \subset L$ defines, by virtue of "$x \leq y$ if $y - x \in H$", an order of L under which L is an ordered vector space with positive cone H. Hence for any vector space L, there is a biunivocal correspondence between the family of all proper cones in L and the family of all orderings satisfying $(LO)_1$ and $(LO)_2$. If R_1 and R_2 are two such orderings of L with respective positive cones C_1 and C_2, then the relation "R_1 is finer than R_2" is equivalent with $C_1 \subset C_2$; in particular, if $\{R_\alpha: \alpha \in A\}$ is a family of such orderings of L with respective positive cones C_α, the coarsest ordering R which is finer than all $R_\alpha (\alpha \in A)$ is determined by the proper cone $C = \bigcap_\alpha C_\alpha$. (Cf. Exercise 2.) A cone $H \subset L$ satisfying (i) and (ii) is said to be **generating** if $L = H - H$.

Let L be an ordered vector space. The order of L is called **Archimedean** (or L **Archimedean ordered**) if $x \leq 0$ whenever there exists $y \in L$ such that $nx \leq y$ for all $n \in N$ (in other words, if $x \leq 0$ whenever $\{nx: n \in N\}$ is majorized). For example, if L is a t.v.s. and an ordered vector space whose positive cone is closed, L is Archimedean ordered; on the other hand, R_0^n is not Archimedean ordered for $n \geq 2$ under its lexicographic ordering (see below). An **order interval** in L is a subset of the form $\{z \in L: x \leq z \leq y\}$, where x, y are given; it is convenient to denote this set by $[x, y]$. (There is little danger of confusing this with the inner product notation in pre-Hilbert spaces (Chapter III, Section 2, Example 5) if we avoid using the symbol in different meanings in the same context.) A subset A of L is **order bounded** if A is contained in some order interval. Every order interval is convex, and every order interval of the form $[-x, x]$ is circled. An element $e \in L$ such that $[-e, e]$ is radial is called an **order unit** of L. The set L^b of all linear forms on L that are bounded on each order interval is a subspace of L^*, called the **order bound dual** of L.

Let L be an ordered vector space over R and let M be a subspace of L.

If C is the positive cone of L, then the induced ordering on M is determined by the proper cone $C \cap M$; an ordering of L/M is determined by the canonical image $\hat{C}$ of C in L/M, provided that $\hat{C}$ is a proper cone. (Simple examples, with $L = R_0^2$, show that this is not necessarily the case.) If $\{L_\alpha : \alpha \in A\}$ is a family of ordered vector spaces with respective positive cones C_α, then $C = \prod_\alpha C_\alpha$ is a proper cone in $L = \prod_\alpha L_\alpha$ which determines an ordering of L. The orderings so defined are called the **canonical orderings** of M, L/M (provided $\hat{C}$ is proper), and of $\prod_\alpha L_\alpha$. In particular, the algebraic direct sum $\bigoplus_\alpha L_\alpha$ is canonically ordered as a subspace of $\prod_\alpha L_\alpha$, and if T is any set, then L^T is canonically ordered by the proper cone $\{f : f(t) \in C \text{ for all } t \in T\}$.

Let L be an ordered vector space which is the algebraic direct sum of the subspaces $M_i (i = 1, ..., n)$; L is said to be the **ordered direct sum** of the subspaces M_i if the canonical algebraic isomorphism of L onto $\prod_i M_i$ is an order isomorphism (for the canonical ordering of $\prod_i M_i$).

If L_1, L_2 are ordered vector spaces $\neq \{0\}$ with respective positive cones C_1 and C_2, then $C = \{u : u(C_1) \subset C_2\}$ is a proper cone in the space $L(L_1, L_2)$ of linear mappings of L_1 into L_2, if and only if C_1 is generating in L_1; whenever M is a subspace of $L(L_1, L_2)$ such that $C \cap M$ is a proper cone, the ordering defined by $C \cap M$ is called the **canonical ordering** of M. A special case of importance is the following: A linear form f on an ordered vector space over R is **positive** if $x \geq 0$ implies $f(x) \geq 0$; the set C^* of all positive linear forms on L is a cone which is the polar, with respect to $\langle L, L^* \rangle$, of $-C$. The subspace $L^+ = C^* - C^*$ of L^* is called the **order dual** of L; it is immediate that $L^+ \subset L^b$. However, there exist ordered vector spaces L for which $L^+ \neq L^b$ (see Namioka [1], 6.10).

In order to use the tool of duality successfully in the study of ordered vector spaces L, one needs sufficiently many positive linear forms on L to distinguish points; we shall say that L is **regularly ordered** (or that the order of L is **regular**) if L is Archimedean ordered and L^+ distinguishes points in L (cf. (4.1) below).

As above, the canonical ordering of a subspace $M \subset L^*$ is understood to be the ordering defined by $M \cap C^*$ whenever $M \cap C^*$ is a proper cone in M.

Let us note some simple consequences of $(LO)_1$, L being an ordered vector space. The equality

$$z + \sup(x, y) = \sup(z + x, z + y) \tag{1}$$

is valid for given $x, y \in L$ and all $z \in L$ whenever $\sup(z_0 + x, z_0 + y)$ exists for some $z_0 \in L$. If A, B are subsets $\neq \emptyset$ of L such that $\sup A$ and $\sup B$ exist, then $\sup(A + B)$ exists and

$$\sup(A + B) = \sup A + \sup B. \tag{1'}$$

Also from $(LO)_1$ it follows that

$$\sup(x, y) = -\inf(-x, -y) \tag{2}$$

whenever either $\sup(x, y)$ or $\inf(-x, -y)$ exists; more generally,

$$\sup A = -\inf(-A) \tag{2'}$$

whenever either $\sup A$ or $\inf(-A)$ exists.

A **vector lattice** is defined to be an ordered vector space E over $\mathbf{R}$ such that for each pair $(x, y) \in E \times E$, $\sup(x, y)$ and $\inf(x, y)$ exist. This implies, in particular, that E is directed under the order relation $\leq$ (equivalently, that the positive cone C of E is generating). For each $x \in E$, we define the **absolute** $|x|$ by $|x| = \sup(x, -x)$; two elements x, y of a vector lattice E are **disjoint** if $\inf(|x|, |y|) = 0$; two subsets $A \subset E$ and $B \subset E$ are **lattice disjoint** (or simply **disjoint** if no confusion is likely to result) if $x \in A$, $y \in B$ implies $\inf(|x|, |y|) = 0$. The fact that x, y are disjoint is denoted by $x \perp y$, and if A is a subset of E, $A^{\perp}$ denotes the set of all $y \in E$ such that y is disjoint from each element of A. We record the following simple but important facts on vector lattices.

1.1

Let E be a vector lattice. Then

$$x + y = \sup(x, y) + \inf(x, y) \tag{3}$$

is an identity on $E \times E$. Defining x^+ and x^- by $x^+ = \sup(x, 0)$ and $x^- = \sup(-x, 0)$ for all $x \in E$, we have $x = x^+ - x^-$ and $|x| = x^+ + x^-$; $x = x^+ - x^-$ is the unique representation of x as a difference of disjoint elements ≥ 0. Moreover, we have

$$|\lambda x| = |\lambda| \, |x| \tag{4}$$

$$|x + y| \leq |x| + |y| \tag{5}$$

$$|x^+ - y^+| \leq |x - y| \tag{6}$$

for all $x, y \in E$ and $\lambda \in \mathbf{R}$. Finally, we have

$$[0, x] + [0, y] = [0, x + y] \tag{D}$$

for all $x \geq 0$ and $y \geq 0$.

Proof. To prove (3), consider the more general identity

$$a - \inf(x, y) + b = \sup(a - x + b, a - y + b), \tag{3'}$$

where a, b, x, y are arbitrary elements of E. By (2) we have $-\inf(x, y) = \sup(-x, -y)$, whence (3') follows from (1); from (3') we obtain (3) by the substitution $a = x$, $b = y$. Letting $y = 0$ in (3), we obtain $x = x^+ - x^-$, and since $\inf(x^+, x^-) = x^- + \inf(x, 0) = x^- - \sup(-x, 0) = 0$, x^+ and x^- are disjoint elements; we now obtain via (1), $x^+ + x^- = x + \sup(-2x, 0) = \sup(-x, x) = |x|$. Let $x = y - z$, where $y \geq 0$, $z \geq 0$ are disjoint; we show that $y = x^+$, $z = x^-$. Note first that $x = y - z$ implies $y \geqq x$ hence $y \geqq x^+$ and, therefore, $z \geqq x^-$; it follows that $(y - x^+) \perp (z - x^-)$ which, in view of $y - x^+ = z - x^-$, implies $y = x^+$, $z = x^-$, since clearly 0 is the only element of E disjoint from itself.

If $\lambda \geqq 0$, then from $(LO)_2$ we obtain $(\lambda x)^+ = \lambda x^+$ and $(\lambda x)^- = \lambda x^-$; if $\lambda < 0$, then $(\lambda x)^+ = (-\lambda(-x))^+ = |\lambda| x^-$ and $(\lambda x)^- = |\lambda| x^+$; this proves (4). For (5), note that $\pm x \leqq |x|$, $\pm y \leqq |y|$ implies $|x + y| = \sup(x + y, -x - y) \leqq |x| + |y|$. To prove (6) we conclude from $x = y + (x - y)$ that $x \leqq y^+ + |x - y|$; hence, the right-hand side being $\geqq 0$, that $x^+ \leqq y^+ + |x - y|$; therefore, $x^+ - y^+ \leqq |x - y|$ and interchanging x and y yields $y^+ - x^+ \leqq |x-y|$, hence (6).

Finally, it is clear that $[0, x] + [0, y] \subset [0, x + y]$ whenever $x \geqq 0$ and $y \geqq 0$. Let $z \in [0, x + y]$ and define u, v by $u = \inf(z, x)$ and $v = z - u$; there remains to show that $v \in [0, y]$. But $v = z - \inf(z, x) = z + \sup(-z, -x) = \sup(0, z - x) \leqq \sup(0, x + y - x) = y$ which completes the proof of (1.1).

COROLLARY 1. *In every vector lattice E, the relation $x \leqq y$ is equivalent with "$x^+ \leqq y^+$ and $y^- \leqq x^-$", and the relation $x \perp y$ is equivalent with $\sup(|x|, |y|) = |x| + |y|$. Moreover, if $x \perp y$, then $(x + y)^+ = x^+ + y^+$ and $|x + y| = |x| + |y|$.*

Proof. In fact, if $x^+ \leqq y^+$ and $y^- \leqq x^-$, then $x = x^+ - x^- \leqq y^+ - y^- = y$. Conversely, $x \leqq y$ implies $x^+ \leqq y^+$ and $\inf(x, 0) \leqq \inf(y, 0)$; hence $-x^- \leqq -y^-$ or, equivalently, $y^- \leqq x^-$. The second assertion is immediate from (3) replacing x, y by $|x|, |y|$ respectively. Finally, $x + y = (x^+ + y^+) - (x^- + y^-)$, and $\inf(|x|, |y|) = 0$ expresses that the summands on the right are disjoint; hence $(x + y)^+ = x^+ + y^+$ by the unicity of the representation of $x + y$ as a difference of disjoint elements $\geqq 0$. The last assertion is now immediate.

COROLLARY 2. *Let E be a vector lattice and let $A \subset E$ be a subset for which $\sup A = x_0$ exists. If $B \subset E$ is a subset lattice disjoint from A, then B is lattice disjoint from $\{x_0\}$.*

Proof. We have to show that $z \in B$ implies $z \perp x_0$. Now $x_0^- \leqq x^- \leqq |x|$ for all $x \in A$; hence $z \perp x_0^-$ if $z \in B$. It suffices hence to show that $z \perp x_0^+$. In view of Corollary 1, we have $\sup(|z|, x^+) = |z| + x^+$ for all $x \in A$ by hypothesis, and $x_0^+ = \sup\{x^+ : x \in A\}$; (1') implies that $\sup\{|z| + x^+ : x \in A\} = |z| + x_0^+$. Thus we obtain

$$\sup(|z|, x_0^+) = \sup_{x \in A} \sup(|z|, x^+) = \sup_{x \in A}(|z| + x^+) = |z| + x_0^+,$$

which shows that $|z| \perp x_0^+$ (Corollary 1).

The following observation sometimes simplifies the proof that a given ordered vector space is a vector lattice.

1.2

Let E be an ordered vector space over $\mathbf{R}$ whose positive cone C is generating; if for each pair $(x, y) \in C \times C$ either $\sup(x, y)$ or $\inf(x, y)$ exists, then E is a vector lattice.

The detailed verification is left to the reader; one shows that if $\sup(x, y)$ exists $(x, y \in C)$, then $z = x + y - \sup(x, y)$ proves to be $\inf(x, y)$, and conversely. If x, y are any elements of E, there exists $z \in C$ such that $x + z \in C$ and $y + z \in C$, and the existence of $\sup(x, y)$ and $\inf(x, y)$ is shown via (1).

If $\{E_\alpha : \alpha \in A\}$ is a family of vector lattices, it is quickly verified that $\prod_\alpha E_\alpha$ and $\oplus_\alpha E_\alpha$ are vector lattices under their canonical orderings. A **vector sublattice** M of a vector lattice E is a vector subspace of E such that $x \in M$, $y \in M$ implies that $\sup(x, y) \in M$ where the supremum is formed in E; it follows that M is a vector lattice under its canonical ordering. However, it can happen that a subspace M of E is a vector lattice under its canonical order but not a sublattice of E (Exercise 14).

A subset A of a vector lattice E is called **solid** if $x \in A$ and $|y| \leqq |x|$, $y \in E$, imply that $y \in A$. It is easy to see that a solid subspace of E is necessarily a sublattice of E; for example, the algebraic direct sum $\oplus_\alpha E_\alpha$ of a family $\{E_\alpha : \alpha \in A\}$ of vector lattices is a solid subspace of $\prod_\alpha E_\alpha$ (for the canonical ordering of the product). Also it is easy to see that if M is a solid subspace of E, then E/M is a vector lattice under its canonical order (cf. the examples below).

A subset A of a vector lattice E is called **order complete** if for each nonempty subset $B \subset A$ such that B is order bounded in A, $\sup B$ and $\inf B$ exist and are elements of A; E is **order complete** if it is order complete as a subset of itself. If E is an order complete vector lattice, a subspace M of E which is solid and such that $A \subset M$, $\sup A = x \in E$ implies $x \in M$, is called a **band** in E. E itself is a band, and clearly the intersection of an arbitrary family of bands in E is a band; hence every subset A of E is contained in a smallest band B_A, called the **band generated by** A (in E).

Examples

1. Let T be any set and consider the vector space $\mathbf{R}_0^T$ of all real-valued functions on T under its canonical order, where $\mathbf{R}_0$ is ordered as usual. Obviously $\mathbf{R}_0^T$ is an order complete vector lattice. If A is any subset of $\mathbf{R}_0^T$, denote by T_A the subset $\{t:$ *there exists $f \in A$ such that* $f(t) \neq 0\}$ of T. Then the band generated by A is the subspace $B_A = \{f : f(t) = 0$ whenever $t \notin T_A\}$; the quotient $\mathbf{R}_0^T/B_A$, under its canonical order, is a vector lattice which is isomorphic with $\mathbf{R}_0^{T \sim T_A}$. The canonical ordering of $\mathbf{R}_0^T$ is regular (in particular, Archimedean); in fact, the order dual and the order bound dual coincide with the (ordered) direct sum of card T copies of $\mathbf{R}_0$ (Chapter IV, Section 1, Example 4).

2. Let β be any ordinal number > 0 and let $\mathbf{R}_0^\beta$ denote the vector space of all real valued functions defined on the set of all ordinals

$\alpha < \beta$, and consider the subset H of R_0^β defined by the property "*if
there exists a smallest ordinal $\alpha < \beta$ such that $f(\alpha) \neq 0$ then $f(\alpha) > 0$*".
We verify without difficulty that H is a proper cone in R_0^β; the order determined by H is called the **lexicographical order** of R_0^β. The lexicographical order of R_0^β is not Archimedean (hence not regular) if $\beta > 1$; in fact,
the set of all functions f such that $f(0) = 0$ is majorized by each function
f for which $f(0) > 0$. It is worth noting that the lexicographical order of
R_0^β is a total ordering, since $R_0^\beta = H \cup -H$; thus R_0^β is a vector lattice
under this order which is, however, not order complete if $\beta > 1$. More-
over, (up to a positive scalar factor) $f \to f(0)$ is the only non-trivial
positive linear form, hence the order dual and the order bound dual
(cf. (1.4) below) are of dimension 1.

   3. Let (X, Σ, μ) be a measure space (Chapter II, Section 2, Example 2).
Under the ordering induced by the canonical ordering of R_0^X (Example 1
above), the spaces $\mathscr{L}^p(\mu)$ $(1 \leqq p \leqq +\infty)$ are vector lattices (take the
scalar field $K = R$) which are countably order complete (each majorized
countable family has a supremum) but, in general, not order complete
(Exercise 13). The subspace $\mathscr{N}_\mu$ of μ-null functions is a solid subspace
but, in general, not a band in $\mathscr{L}^p(\mu)$; the quotient spaces $L^p(\mu) =
\mathscr{L}^p(\mu)/\mathscr{N}_\mu$ are order complete vector lattices under their respective
canonical orderings $(1 \leqq p < +\infty)$.

If E is any order complete vector lattice and A a subset of E, the set $A^\perp$ is
a band in E; this is clear in view of Corollary 2 of (1.1). Concerning the
bands B_A and $A^\perp$, we have the following important theorem (F. Riesz [1]).

1.3

Theorem. *Let E be an order complete vector lattice. For any subset
$A \subset E$, E is the ordered direct sum of the band B_A generated by A and of the
band $A^\perp$ of all elements disjoint from A.*

Proof. Since $A^{\perp\perp}$ is a band containing A, it follows that $B_A \subset A^{\perp\perp}$ and
hence that $B_A \cap A^\perp = \{0\}$. Let $x \in E$, $x \geqq 0$, be given; we show that $x = x_1
+ x_2$, where $x_1 \in B_A$, $x_2 \in A^\perp$, and $x_1 \geqq 0$, $x_2 \geqq 0$. Define x_1 by $x_1 = \sup
[0, x] \cap B_A$ and x_2 by $x_2 = x - x_1$; it is clear that x_1, x_2 are positive and
that $x_1 \in B_A$, since B_A is a band in E. Let us show that $x_2 \in B_A^\perp$. For any
$y \in B_A$ let $z = \inf(x_2, |y|)$; then $0 \leqq z \in B_A$, since B_A is solid and $z + x_1
\leqq x_2 + x_1 = x$. This implies, by the definition of x_1 and by virtue of $z + x_1
\in B_A$, that $z + x_1 \leqq x_1$ and hence that $z = 0$. Thus $x_2 \in B_A^\perp$ and a fortiori
$x_2 \in A^\perp$. Since the positive cone of E is generating, it follows that $E = B_A
+ A^\perp$ is the ordered direct sum of the subspaces B_A and $A^\perp$. For, the relations
$x \geqq 0$ and $x = x_1 + x_2$, $x_1 \in B_A$, $x_2 \in A^\perp$ imply $x_1 \geqq 0$, $x_2 \geqq 0$.

   COROLLARY 1. *If A is any subset of E, the band B_A generated by A is the band
$A^{\perp\perp}$.*

Proof. Applying (1.3) to the subset $A^\perp$ of E, we obtain the direct sum
$E = A^{\perp\perp} + A^\perp$; since $E = B_A + A^\perp$ and $B_A \subset A^{\perp\perp}$, it follows that $B_A = A^{\perp\perp}$.

COROLLARY 2. *If* x, y *are disjoint elements of* E *and* B_x, B_y *are the bands generated by* $\{x\}$, $\{y\}$ *respectively, then* B_x *is disjoint from* B_y.

In fact, we have $y \in \{x\}^{\perp}$ and $x \in \{y\}^{\perp}$.

A general example of an order complete vector lattice is furnished by the order dual E^{+} of any vector lattice E; however, E^{+} can be finite dimensional (Example 2 above) or reduced to $\{0\}$ (Exercise 14), even if E is of infinite dimension. We prove the result in the following more general form which shows it to depend essentially on property (D) of (1.1). (Cf. Exercise 16.)

1.4

Let E *be an ordered vector space over* **R** *whose positive cone* C *is generating and has property* (D) *of* (1.1). *Then the order bound dual* E^{b} *of* E *is an order complete vector lattice under its canonical ordering; in particular,* $E^{b} = E^{+}$.

Proof. We show first that for each $f \in E^{b}$, $\sup(f, 0)$ exists; it follows then from (1) that $\sup(f, g) = g + \sup(f - g, 0)$ exists for any pair $(f, g) \in E^{b} \times E^{b}$; hence E^{b} is a vector lattice by (1.2). This implies clearly that $E^{b} = E^{+}$.

Let $f \in E^{b}$ be given; we define a mapping r of C into the real numbers ≥ 0 by

$$r(x) = \sup\{f(y): y \in [0, x]\} \qquad (x \in C).$$

Since $f(0) = 0$ it follows that $r(x) \geq 0$, and clearly $r(\lambda x) = \lambda r(x)$ for all $\lambda \geq 0$. Also, by virtue of (1') and (D),

$$r(x + y) = \sup\{f(z): z \in [0, x] + [0, y]\} = r(x) + r(y).$$

Hence r is positive homogeneous and additive on C. By hypothesis, each $z \in E$ is of the form $z = x - y$ for suitable elements $x, y \in C$, and it is readily seen that the number $r(x) - r(y)$ is independent of the particular decomposition $z = x - y$ of z. A short computation now shows that $z \rightarrow w(z) = r(x) - r(y)$ is a linear form w on E, evidently contained in E^{b}. (We have, in fact, $w(x) = r(x)$ for $x \in C$.) We show that $w = \sup(f, 0)$; indeed, $w(x) \geq \sup(f(x), 0)$ for all $x \in C$, and if $h \geq 0$ is a linear form on E such that $x \in C$ implies $h(x) \geq f(x)$, then $h(x) \geq h(y) \geq f(y)$ for all $y \in [0, x]$, which shows that $h(x) \geq r(x) = w(x)$ whenever $x \in C$.

It remains to prove that $E^{b} = E^{+}$ is order complete; for this it suffices to show that each non-empty, majorized set A of positive linear forms on E has a supremum. Without restriction of generality, we can assume that A is directed under " $\leq$ ". (This can be arranged, if necessary, by considering the set of suprema of arbitrary, non-empty finite subsets of A.) We define a mapping s of C into the real numbers by

$$s(x) = \sup\{f(x): f \in A\} \qquad (x \in C).$$

The supremum is finite for all $x \in C$, since A is majorized. It is clear that $s(\lambda x) = \lambda s(x)$ for all $\lambda \geq 0$ and, since A is directed, that $s(x + y) = s(x) + s(y)$. Hence, as before, s defines a linear form f_0 on E by means of $f_0(z) = s(x)$

$- s(y)$, where $z = x - y$ and $x, y \in C$. It is evident that $f_0 \in E^b$ (since $f_0 \geqq 0$) and that $f_0 = \sup A$.

COROLLARY. *The order dual of every vector lattice is an order complete vector lattice under its canonical ordering.*

From the construction of $f^+ = \sup(f, 0)$ in the proof of (1.4), we obtain the following useful relations; the proof of these is purely computational and will be omitted.

1.5

Let E be a vector lattice and let f, g be order bounded linear forms on E. For each $x \in E$, we have

$$\sup(f, g)(|x|) = \sup\{f(y) + g(z) \colon y \geqq 0, z \geqq 0, y + z = |x|\}$$
$$\inf(f, g)(|x|) = \inf\{f(y) + g(z) \colon y \geqq 0, z \geqq 0, y + z = |x|\} \tag{7}$$

$$|f|(|x|) = \sup\{f(y - z) \colon y \geqq 0, z \geqq 0, y + z = |x|\}$$
$$|f(x)| \leqq |f|(|x|). \tag{8}$$

In particular, two linear forms $f \geqq 0$, $g \geqq 0$ are disjoint if and only if for each $x \geqq 0$ and each real number $\varepsilon > 0$, there exists a decomposition $x = x_1 + x_2$ with $x_1 \geqq 0$, $x_2 \geqq 0$, and such that $f(x_1) + g(x_2) \leqq \varepsilon$.

COROLLARY. *Let E be a vector lattice, and let $\langle E, G \rangle$ be a duality such that G is a sublattice of E^+. Then the polar $A^\circ \subset G$ of each solid subset $A \subset E$ is solid.*

Proof. In fact, if $x \in A$, $y \geqq 0, z \geqq 0$, and $y + z = |x|$, then $y - z \in A$, since $-|x| \leqq y - z \leqq |x|$; hence, if $f \in A^\circ$ and $|g| \leqq |f|$, then from (8) it follows that

$$|g(x)| \leqq |g|(|x|) \leqq |f|(|x|) \leqq 1,$$

which shows that $g \in A^\circ$.

If E is an ordered vector space over $\boldsymbol{R}$ such that the order dual E^+ is an ordered vector space (equivalently, if C^* is a proper cone in E^* where C is the positive cone of E), then the space $(E^+)^+$ is called the **order bidual** of E and denoted by E^{++}. Under the assumptions of (1.4) (in particular, if E is a vector lattice), E^{++} is a vector lattice, and the evaluation (or canonical) map of E into E^{++}, defined by $x \to \tilde{x}$ where $\tilde{x}(f) = f(x)$ ($f \in E^+$), is clearly order preserving. Assuming that E is a vector lattice, let us show that $x \to \tilde{x}$ is an isomorphism onto a sublattice of E^{++} if E is regularly ordered (equivalently, if $x \to \tilde{x}$ is one-to-one). For later use, we prove this result in a somewhat more general form.

1.6

Let E be a vector lattice and let G be a solid subspace of E^+ that separates points in E; the evaluation map $x \to \tilde{x}$, defined by $\tilde{x}(f) = f(x)$ ($f \in G$), is an isomorphism of E onto a sublattice of G^+.

Proof. We must show that for each $x \in E$, the element $\tilde{x}^+$ $(= \sup(0, \tilde{x})$
taken in $G^+)$ is the canonical image of $x^+ \in E$. Denote by P the subset
$\bigcup\{\rho[0, x^+]: \rho \geq 0\}$ of E and define, for each $f \geq 0$ in G, a mapping t_f of the
positive cone C of E into R by

$$t_f(y) = \sup\{f(z): z \in [0, y] \cap P\} \qquad (y \in C).$$

As in the proof of (1.4) it follows that t_f is additive and positive homogeneous, and hence defines a unique linear form $g_f \in C^*$; it is clear that
$g_f \leq f$, hence $g_f \in G$, since G is solid, and that $g_f(x^-) = 0$ because of $[0, x^-]$
$\cap P = \{0\}$. Hence $g_f(x) = g_f(x^+)$, and we obtain $\tilde{x}^+(f) = \sup\{g(x): 0 \leq g \leq f\}$
$\geq g_f(x) = g_f(x^+) = f(x^+)$ for all $f \in C^* \cap G$. This implies $\tilde{x}^+ \geq (x^+)^\sim$;
since it is clear that $(x^+)^\sim \geq \tilde{x}^+$ in G^+, the assertion follows.

We point out that the canonical image of E in G^+ is, in general, not an
order complete sublattice of G^+ even if E is order complete (see the example
following (7.4)). In particular (taking $G = E^+$), a regularly ordered, order
complete vector lattice E need not be mapped onto a band in E^{++} under
evaluation. If E is an order complete, regularly ordered vector lattice whose
canonical image in E^{++} is order complete, E will be called **minimal** (or of
minimal type).

If E, F are vector lattices, a linear map u of E onto F is called a **lattice
homomorphism** provided that u preserves the lattice operations; in view of the
linearity of u, the translation-invariance of the order and the identity (3), this
condition on u is equivalent to each of the following: (i) $u(\sup(x, y))$
$= \sup(u(x), u(y))$ $(x, y \in E)$. (ii) $u(\inf(x, y)) = \inf(u(x), u(y))$ $(x, y \in E)$. (iii)
$u(|x|) = \sup(u(x^+), u(x^-))$ $(x \in E)$. (iv) $\inf(u(x^+), u(x^-)) = 0$ $(x \in E)$. If, in
addition, u is biunivocal, then u is called a **lattice isomorphism** of E onto F.
It is not difficult to show that a linear map u of E onto F is a lattice homomorphism if and only if $u^{-1}(0)$ is a solid sublattice of E and $u(C_1) = C_2$,
where C_1, C_2 denote the respective positive cones of E, F. In particular, if N
is a solid vector sublattice of E, then E/N is a vector lattice under its canonical
order and the canonical map ϕ is a lattice homomorphism of E onto E/N
(Exercise 12).

The linear forms on a vector lattice E that are lattice homomorphisms onto
R have an interesting geometric characterization; let us recall (Chapter II,
Exercise 30) that $\{\lambda x: \lambda \geq 0\}, 0 \neq x \in C$ is called an extreme ray of the cone C
if $x - y \in C$, $y \in C$ imply $y = \rho x$ for some ρ, $0 \leq \rho \leq 1$.

1.7

*Let E be a vector lattice, $f \neq 0$ a linear form on E. The following assertions are
equivalent* :

(a) *f is a lattice homomorphism of E onto R.*
(b) *$\inf(f(x^+), f(x^-)) = 0$ for all $x \in E$.*
(c) *f generates an extreme ray of the cone C^* in E^*.*
(d) *$f \geq 0$ and $f^{-1}(0)$ is a solid hyperplane in E.*

Proof. (a)⇔(b) is clear from the preceding remarks. (b)⇒(d): Since $\inf(f(x^+), f(x^-)) = 0$ for each $x \in E$, it follows that $f \geq 0$, and $f(x) = 0$ implies $f(|x|) = 0$; hence $|y| \leq |x|$ and $f(x) = 0$ imply $|f(y)| \leq f(|y|) \leq f(|x|)$ $= 0$. (d)⇒(c): Suppose $g \in C^*$ is such that $f - g \in C^*$ or, equivalently, that $0 \leq g \leq f$. Then since $f^{-1}(0)$ is solid, $f(x) = 0$ implies $|g(x)| \leq g(|x|) \leq f(|x|)$ $= 0$ and hence $f^{-1}(0) \subset g^{-1}(0)$. Thus (since $f^{-1}(0)$ is a hyperplane) either $g = 0$ or $f^{-1}(0) = g^{-1}(0)$; in any case, $g = \rho f$ for some ρ, $0 \leq \rho \leq 1$. (c)⇒(b): Let f generate an extreme ray of C^*, let $x \in E$ be given, and suppose that $f(x^+) > 0$. Let $P = \bigcup\{\rho[0, x^+]: \rho \geq 0\}$, and define $h \in E^*$ by putting, for $y \geq 0$, $h(y) = \sup\{f(z): z \in [0, y] \cap P\}$ (see proof of (1.6)). It follows that $0 \leq h \leq f$ and hence $h = \rho f$ by the assumption made on f, and since $h(x^+)$ $= f(x^+) > 0$ we must have $\rho = 1$. Thus $h = f$, and since clearly $h(x^-) = 0$, it follows that $f(x^-) = 0$, which completes the proof.

2. ORDERED VECTOR SPACES OVER THE COMPLEX FIELD

It is often useful to have the concept of an ordered vector space over the complex field C. Such is the case, for instance, in spectral theory and in measure theory. It is the purpose of this section to agree on a definite terminology. We define a vector space L over C to be **ordered** if its underlying real space L_0 (Chapter I, Section 7) is an ordered vector space over R; thus by definition, order properties of L are order properties of L_0. The usefulness of this (otherwise trivial) definition lies in the fact that the transition to L_0 does not have to be mentioned continually.

The canonical orderings of products, subspaces, direct sums, quotients, function spaces, and spaces of linear maps are then defined with reference to the respective underlying real spaces; only the term "positive linear form" on L has to be additionally specified when L is an ordered vector space over C. We define $f \in L^*$ to be **positive** if $\operatorname{Re} f(x) \geq 0$ whenever $x \geq 0$ in L; this definition guarantees that whenever the canonical ordering of $(L_0)^*$ is defined, then L^* is ordered, and the canonical isomorphism of (I, 7.2) is an order isomorphism (a corresponding statement holding for subspaces of L^*). The **order bound dual** L^b of an ordered vector space L over C is then defined as the subspace of L^* containing exactly the linear forms bounded on each order interval in L; the **order dual** L^+ is the (complex) subspace of L^* which is the linear hull of the cone C^* of positive linear forms. In accordance with the definition given above, the order of L is called regular if L_0 is regularly ordered; we point out that this is not implied by the fact that C^* separates points in L, and that in general $(L^+)_0$ cannot be identified with $(L_0)^+$ by virtue of (I, 7.2) (Exercise 4).

The term *vector lattice* will not be extended to complex spaces; we shall, however, say that an ordered vector space L over C with positive cone C is **lattice ordered** if the real subspace $C - C$ of L is a vector lattice. For example, the complexification (Chapter I, Section 7) of a vector lattice L is a lattice ordered vector space L_1 over C.

3. DUALITY OF CONVEX CONES

Let L be a vector space (over R or C); by a **cone** in L we shall henceforth understand a convex cone C of vertex 0 and such that $0 \in C$. Let C be a fixed cone in L; for any pair $(x, y) \in L \times L$, we shall write $[x, y] = (x + C) \cap (y - C)$. This notation is consistent with the notation introduced for order intervals in Section 1; if C is the positive cone of an ordering of L, then $(x + C) \cap (y - C)$ is the order interval $\{z: x \leqq z \leqq y\}$. For any subset $A \subset L$, define

$$[A] = (A + C) \cap (A - C) = \bigcup\{[x, y]: x \in A, y \in A\}.$$

A subset $B \subset L$ is called C-**saturated** if $B = [B]$; it is immediate that for any $A \subset L$, $[A]$ is the intersection of all C-saturated subsets containing A, and hence called the C-**saturated hull** of A. It is also quickly verified that $A \to [A]$ is monotone: $A \subset B$ implies $[A] \subset [B]$, that $[A]$ is convex if A is convex, and that $[A]$ is circled with respect to R if A is circled with respect to R. Finally we note that if $\mathfrak{F}$ is a filter (more generally, a filter base) in L, then the family $\{[F]: F \in \mathfrak{F}\}$ is a filter base in L; the corresponding filter will be denoted by $[\mathfrak{F}]$.

Assume now that L is a t.v.s. A cone C in L is said to be **normal** if $\mathfrak{U} = [\mathfrak{U}]$ where $\mathfrak{U}$ is the neighborhood filter of 0. Hence C is a normal cone in the t.v.s. L if and only if there exists a base of C-saturated neighborhoods of 0 (equivalently, if and only if the family of all C-saturated 0-neighborhoods is a base at 0). It will be useful to have a number of alternative characterizations of normal cones.

3.1

Let L be a t.v.s. over K and let C be a cone in L. The following propositions are equivalent:

(a) *C is a normal cone.*
(b) *For every filter $\mathfrak{F}$ in L, $\lim \mathfrak{F} = 0$ implies $\lim[\mathfrak{F}] = 0$.*
(c) *There exists a 0-neighborhood base $\mathfrak{B}$ in L such that $V \in \mathfrak{B}$ implies $[V \cap C] \subset V$.*

If $K = R$ and the topology of L is locally convex, then (a) *is equivalent to each of the following:*

(d) *There exists a 0-neighborhood base consisting of convex, circled, and C-saturated sets.*
(e) *There exists a generating family $\mathscr{P}$ of semi-norms on L such that $p(x) \leqq p(x + y)$ whenever $x \in C$, $y \in C$ and $p \in \mathscr{P}$.*

Proof. Denote by $\mathfrak{U}$ the neighborhood filter of 0 in L. (a) $\Rightarrow$ (b): If $\mathfrak{F}$ is a filter on L which is finer than $\mathfrak{U}$, then $[\mathfrak{F}]$ is finer than $[\mathfrak{U}]$; hence the assertion follows from $\mathfrak{U} = [\mathfrak{U}]$. (b) $\Rightarrow$ (c): (b) implies that $[\mathfrak{U}]$ is the neighborhood

filter of 0 in L; hence $\mathfrak{B} = \{[U]: U \in \mathfrak{U}\}$ is a neighbourhood base of 0 such
that $V \in \mathfrak{B}$ implies $[V \cap C] \subset [V] = V$. (c) $\Rightarrow$ (a): Given $U \in \mathfrak{U}$, it suffices to
show there exists $W \in \mathfrak{U}$ such that $[W] \subset U$. Let $\mathfrak{B}$ be a 0-neighborhood base
as described in (c); select $V \in \mathfrak{B}$ such that $V + V \subset U$ and a circled $W \in \mathfrak{U}$
such that $W + W \subset V$. We obtain

$$[W] = \bigcup_{x, y \in W} [x, y] = \bigcup_{x, y \in W} (x + [0, y - x]) \subset W + [(W + W) \cap C]$$
$$\subset V + [V \cap C] \subset V + V \subset U,$$

which proves the implication (c) $\Rightarrow$ (a).

Assume now that $K = \mathbf{R}$ and the topology of L is locally convex. (a) $\Rightarrow$ (d):
If $\mathfrak{U}_1$ is the family of all convex, circled 0-neighborhoods in L, then $\mathfrak{W}$
$= \{[U]: U \in \mathfrak{U}_1\}$ is a base at 0 consisting of convex, circled, and C-saturated
sets. (d) $\Rightarrow$ (e): If $\mathfrak{W}$ is a 0-neighborhood base as in (d) and p_W is the gauge
function of $W \in \mathfrak{W}$, the family $\{p_W: W \in \mathfrak{W}\}$ is of the desired type. (e) $\Rightarrow$ (c):
If $\mathscr{P}$ is as in (e) then the family of all finite intersections of the sets $V_{p,\varepsilon}$
$= \{x \in L: p(x) \leq \varepsilon\}$ $(p \in \mathscr{P}, \varepsilon > 0)$ is a neighborhood base $\mathfrak{B}$ of 0 having the
property stated in (c). This completes the proof.

COROLLARY 1. *If L is a Hausdorff t.v.s., every normal cone C in L is a proper
cone.*

Proof. In fact, if $x \in C \cap -C$, then $x \in [\{0\}] \subset [U]$ for each 0-neighbor-
hood U, and it follows that $x = 0$.

COROLLARY 2. *If C is a normal cone in L and $B \subset L$ is bounded, then $[B]$ is
bounded; in particular, each set $[x, y]$ is bounded.*

Proof. If B is bounded and U is a 0-neighborhood in L, there exists $\lambda > 0$
such that $B \subset \lambda U$; it follows that $[B] \subset [\lambda U] = \lambda[U]$.

COROLLARY 3. *If the topology of L is locally convex, the closure $\bar{C}$ of a normal
cone is a normal cone.*

Proof. It is immediate that $\bar{C}$ is a cone in L, and $\bar{C}$ is also the closure of C
in the real space L_0; the assertion follows now from proposition (e) of (3.1).

It will become evident from the results in this chapter and the Appendix
that the concept of a normal cone is an important (and perhaps the most
important) notion in the theory of ordered topological vector spaces; for
cones in normed spaces over $\mathbf{R}$ it goes back to M.G. Krein [2]. The original
definition of Krein postulates the existence of a constant $\gamma (\geq 1)$ such that
$\|x\| \leq \gamma \|x + y\|$ for all $x, y \in C$; it follows at once that this definition is
equivalent, for normed spaces $(L, \| \ \|)$ over $\mathbf{R}$, with the one given above, and
(3.1) (e) implies that there exists an equivalent norm on L for which one can
suppose $\gamma = 1$.

If M is a subspace of the t.v.s. L and C is a normal cone in L, it is clear that
$M \cap C$ is a normal cone in M; it is also easy to verify that if $\{L_\alpha: \alpha \in A\}$ is a

family of t.v.s., C_α a cone in L_α, and $L = \prod_\alpha L_\alpha$, then $C = \prod_\alpha C_\alpha$ is a normal cone in L if and only if C_α is normal in $L_\alpha (\alpha \in A)$. Let us record the following result on locally convex direct sums.

3.2

If $\{L_\alpha : \alpha \in A\}$ is a family of l.c.s., C_α a cone in $L_\alpha (\alpha \in A)$, and $L = \bigoplus_\alpha L_\alpha$ the locally convex direct sum of this family, then $C = \bigoplus_\alpha C_\alpha$ is a normal cone in L if and only if C_α is normal in L_α $(\alpha \in A)$.

Proof. The necessity of the condition is immediate, since each L_α can be identified with a subspace of L such that C_α is identified with $L_\alpha \cap C$ $(\alpha \in A)$. To prove that the condition is sufficient assume that $K = R$ (which can be arranged, if necessary, by transition to the underlying real space L_0 of L). Let $\mathfrak{V}_\alpha$ be a neighborhood base of 0 in L_α $(\alpha \in A)$ satisfying (3.1) (d); the family of all sets $V = \Gamma_\alpha V_\alpha$ ($V_\alpha \in \mathfrak{V}_\alpha, \alpha \in A$) is a neighborhood base of 0 in L (Chapter II, Section 6). Now it is clear that $[V \cap C]$ is the convex hull of $\bigcup_\alpha [V_\alpha \cap C_\alpha]$; since $[V_\alpha \cap C_\alpha] \subset V_\alpha$ for all $V_\alpha \in \mathfrak{V}_\alpha$ $(\alpha \in A)$, it follows that $[V \cap C] \subset V$, which proves the assertion in view of (3.1) (c).

It can be shown in a similar fashion that a corresponding result holds for the direct sum topology introduced in Exercise 1, Chapter I (in this case, the spaces L_α need not be supposed to be locally convex). On the other hand, if C is a normal cone in L and M is a subspace of L, then the canonical image $\hat{C}$ of C in L/M is, in general, not a proper cone, let alone normal. (For a condition under which $\hat{C}$ is normal, see Exercise 3.)

Intuitively speaking, normality of a cone C in a t.v.s. L restricts the "width" of C and hence, in a certain sense, is a gauge of the pointedness of C. For example, a normal cone in a Hausdorff space cannot contain a straight line ((3.1), Corollary 1); a cone C in a finite-dimensional Hausdorff space L is normal if its closure $\bar{C}$ is proper (cf. (4.1) below). In dealing with dual pairs of cones, one also needs a tool working in the opposite direction and gauging, in an analogous sense, the bluntness of C. The requirement that $L = C - C$ goes in this direction; in fact, it indicates that every finite subset S of L can be recovered from C in the sense that $S \subset S_0 - S_0$ for a suitable finite subset $S_0 \subset C$. The precise definition of the property we have in mind is as follows.

Let L be a t.v.s., let C be a cone in L, and let $\mathfrak{S}$ be a family of bounded subsets of L (Chapter III, Section 3); for each $S \in \mathfrak{S}$, define S_C to be the subset $S \cap C - S \cap C$ of L. We say that C is an $\mathfrak{S}$-**cone** if the family $\{\bar{S}_C : S \in \mathfrak{S}\}$ is a fundamental subfamily of $\mathfrak{S}$; C is called a **strict** $\mathfrak{S}$-**cone** if $\{S_C : S \in \mathfrak{S}\}$ is fundamental for $\mathfrak{S}$. If L is a l.c.s. over R and $\mathfrak{S}$ is a saturated family, in place of S_C we can use the convex, circled hull of $S \cap C$ in the preceding definitions. A case of particular importance is the case where $\mathfrak{S} = \mathfrak{B}$ is the family of all bounded subsets of L: C is a $\mathfrak{B}$-**cone** in L. The notion of a $\mathfrak{B}$-cone in a normed space $(L, \| \ \|)$ appears to have been first used by Bonsall

[2]; Bonsall defines L to have the decomposition property if each z, $\|z\| \leq 1$, can be approximated with given accuracy by differences $x - y$, where $x \in C$, $y \in C$ and $\|x\| \leq k$, $\|y\| \leq k$ for a fixed constant $k > 0$.

The property of being an $\mathfrak{S}$-cone satisfies certain relations of permanence (Exercise 5); since these are consequences of (3.3), below, the permanence properties of normal cones, and the duality theorems (IV, 4.1) and (IV, 4.3), they will be omitted here. Let us point out that, as the concept of a normal cone, the concept of an $\mathfrak{S}$-cone is independent of the scalar field ($\boldsymbol{R}$ or $\boldsymbol{C}$) over which L is defined.

Examples

1. The set of real-valued, non-negative functions determines a normal cone in each of the Banach spaces enumerated in Chapter II, Examples 1–3. If E is any one of these spaces and C the corresponding cone, then $E = C - C$ if $K = \boldsymbol{R}$; this implies that C is a strict $\mathfrak{B}$-cone in E (see (3.5) below). If the functions (or classes of functions) that constitute E are complex valued, then C and $C + iC$ are normal cones and $C + iC$ is a strict $\mathfrak{B}$-cone.

2. Let C denote the set of all non-negative functions in the space $\mathscr{D}$ of L. Schwartz (Chapter II, Section 6, Example 2). C is not a normal cone in $\mathscr{D}$, but $C + iC$ is a strict $\mathfrak{B}$-cone. The cone C_1 of all distributions T such that $(Tf) \geq 0$ for $f \in C$ (which can be identified with the set of all positive Radon measures on $\boldsymbol{R}^n$ (cf. L. Schwartz [1])) is a normal cone in $\mathscr{D}'$, but $C_1 + iC_1$ is not a $\mathfrak{B}$-cone (Exercise 6).

3. Let E be the space of complex-valued, continuous functions with compact support on a locally compact space X with its usual topology (Chapter II, Section 6, Example 3), and let C be the cone of non-negative functions in E. C is a normal cone in E, $C + iC$ is normal and a strict $\mathfrak{B}$-cone. If C_1 denotes the set of all positive Radon measures on X, $C_1 + iC_1$ is normal and a strict $\mathfrak{B}$-cone in the strong dual E'.

The proofs for these assertions will become clear from the following results and are therefore omitted.

If C is a cone in the t.v.s. E, the **dual cone** C' of C is defined to be the set $\{f \in E' : \operatorname{Re} f(x) \geq 0 \text{ if } x \in C\}$; hence C' is the polar of $-C$ with respect to $\langle E, E' \rangle$. In the following proofs it will often be assumed that the scalar field K of E is $\boldsymbol{R}$; whenever this is done, implicit reference is made to (I, 7.2) (cf. also Section 2). Before proving the principal result of this section, we establish this lemma which is due to M. G. Krein [2].

LEMMA 1. *If C is a normal cone in the normed space E, then $E' = C' - C'$.*

Proof. We can assume that $K = \boldsymbol{R}$. Let $f \in E'$ and define the real function $p \geq 0$ on C by $p(x) = \sup\{f(z) : z \in [0, x]\}$. Then it is clear that $p(\lambda x) = \lambda p(x$ if $\lambda \geq 0$ and that $p(x + y) \geq p(x) + p(y)$, since $[0, x] + [0, y] \subset [0, x + y]$ for) all $x, y \in C$. It follows that the set

$$V = \{(t, x) : 0 \leq t \leq p(x)\}$$

is a cone in the product space $R_0 \times E$. Let $\{x_n: n \in N\}$ be a null sequence in E and suppose that $\{t_n: n \in N\}$ is a sequence of real numbers such that $(t_n, x_n) \in V$ $(n \in N)$. Since C is a normal cone and f is continuous, it follows that $p(x_n) \to 0$ and hence that $t_n \to 0$; this implies that $(1, 0)$ is not in the closure $\bar{V}$ of V in the normable space $R_0 \times E$. By (II, 9.2) there exists a closed hyperplane H strictly separating $\{(1, 0)\}$ and V; it can be arranged that $H = \{(t, x): h(t, x) = -1\}$, where $h(1, 0) = -1$ and h is ≥ 0 on V. By (IV, 4.3) h is of the form $(t, x) \to -t + g(x)$; since $g \in E'$ and $(0, x) \in V$ for each $x \in C$, it follows that $g \in C'$. Now $(p(x), x) \in V$ for all $x \in C$; hence we have $-p(x) + g(x) \geq 0$ if $x \in C$. Since $f(x) \leq p(x) \leq g(x)$ for $x \in C$, we obtain $f = g - (g - f)$, where $g \in C'$, $g - f \in C'$, and the lemma is proved.

3.3

Theorem. *Let E be a l.c.s., let C be a cone in E with dual cone $C' \subset E'$, and let $\mathfrak{S}$ be a saturated family of weakly bounded subsets of E'. If C' is an $\mathfrak{S}$-cone, then C is normal for the $\mathfrak{S}$-topology on E; conversely, if C is normal for an $\mathfrak{S}$-topology consistent with $\langle E, E' \rangle$, then C' is a strict $\mathfrak{S}$-cone in E'.*

Proof. We can assume that $K = R$. If C' is an $\mathfrak{S}$-cone in E', then the saturated hull of the family $\{\Gamma(S \cap C'): S \in \mathfrak{S}\}$ equals $\mathfrak{S}$; hence the $\mathfrak{S}$-topology is generated by the semi-norms

$$x \to p_S(x) = \sup\{|\langle x, x' \rangle|: x' \in S \cap C'\} \qquad (S \in \mathfrak{S})$$

which are readily seen to satisfy proposition (e) of (3.1).

Suppose now that $\mathfrak{T}$ is an $\mathfrak{S}$-topology on E consistent with $\langle E, E' \rangle$ and that C is normal with respect to $\mathfrak{T}$. By (3.1) (d) there exists a 0-neighborhood base $\mathfrak{U}$ in $(E, \mathfrak{T})$ consisting of convex, circled, and C-saturated sets. Since $\{U^\circ: U \in \mathfrak{U}\}$ is a fundamental subfamily of $\mathfrak{S}$, it suffices to show that there exists, for each $U \in \mathfrak{U}$, an integer n_0 such that $U^\circ \subset n_0(U^\circ \cap C' - U^\circ \cap C')$. Let $U \in \mathfrak{U}$ be fixed.

Now the dual of the normed space E_U (for notation, see Chapter III, Section 7) can be identified with E'_{U° and the cone $C' \cap E'_{U^\circ}$ can be identified with the dual cone of $C_U = \phi_U(C)$ where, as usual, ϕ_U is the canonical map $E \to \tilde{E}_U$. Using the fact that U is C-saturated, it is readily seen that C_U is a normal cone in E_U; hence if we define the set $M \subset E'$ by $M = U^\circ \cap C' - U^\circ \cap C'$, Lemma 1 implies that $E'_{U^\circ} = \bigcup_{n \in N} nM$. Now M is $\sigma(E', E)$-compact by (I, 1.1) (iv), hence $\sigma(E', E)$-closed and a fortiori closed in the Banach space E'_{U°; since the latter is a Baire space, it follows that M has an interior point and hence (being convex and circled) is a neighborhood of 0 in E'_{U°; it follows that $U^\circ \subset n_0 M$ for a suitable $n_0 \in N$ and the proof is complete.

COROLLARY 1. *Let C be a cone in the l.c.s. E. The following assertions are equivalent:*

(a) *C is a normal cone in E.*

(b) *For any equicontinuous set $A \subset E'$, there exists an equicontinuous set $B \subset C'$ such that $A \subset B - B$.*

(c) *The topology of E is the topology of uniform convergence on the equicontinuous subsets of C'.*

COROLLARY 2. *If $\mathfrak{S}$ is a saturated family, covering E', of $\sigma(E', E)$-relatively compact sets and if H is an $\mathfrak{S}$-cone in E', then the $\sigma(E', E)$-closure $\bar{H}$ of H is a strict $\mathfrak{S}$-cone.*

Proof. In fact, the cone $C = -H^0$ is normal for the $\mathfrak{S}$-topology which is consistent with $\langle E, E' \rangle$, (IV, 3.2), and $C' = \bar{H}$ by (IV, 1.5).

COROLLARY 3. *If C is a cone in the l.c.s. E, then $E' = C' - C'$ if and only if C is weakly normal; in particular, every normal cone in E is weakly normal.*

We obtain this corollary by taking $\mathfrak{S}$ to be the saturated hull of the family of all finite subsets of E'. Let us point out that if C is a cone in a l.c.s. E over $\mathbf{C}$, it is sometimes of interest to consider the cone $H \subset E'$ of linear forms whose real *and* imaginary parts are $\geqq 0$ on C; we have $H = C' \cap (-iC)'$, and it follows from Corollary 3 above and (IV, 1.5), Corollary 2, that $E' = H - H$ if and only if $C + iC$ (equivalently, $C - iC$) is weakly normal in E (for, $H = (C - iC)'$).

> REMARK. In normed spaces, weak normality and normality of cones are equivalent (see (3.5) below).

The following is an application of (3.3) to the case where $\mathfrak{S}$ is the family of all strongly bounded subsets of the dual of an infrabarreled space E; recall that this class comprises all barreled and all bornological (hence all metrizable l.c.) spaces.

3.4

Let E be an infrabarreled l.c.s., C a cone in E, $\mathfrak{B}$ the family of all strongly bounded subsets of E'. The following assertions are equivalent:

(a) *C is a normal cone in E.*

(b) *The topology of E is the topology of uniform convergence on strongly bounded subsets of C'.*

(c) *C' is a $\mathfrak{B}$-cone in E'.*

(d) *C' is a strict $\mathfrak{B}$-cone in E'.*

The proof is clear from the preceding in view of the fact that $\mathfrak{B}$ is the family of all equicontinuous subsets of E' (Chapter IV, Section 5).

COROLLARY. *If E is a reflexive space, normal cones and $\mathfrak{B}$-cones correspond dually to each other (with respect to $\langle E, E' \rangle$).*

It is an interesting fact that the complete symmetry between normal and $\mathfrak{B}$-cones under the duality $\langle E, E' \rangle$ remains in force, without reflexivity assumptions, when E is a Banach space (cf. Ando [2]). From the proof of this result we isolate the following lemma, which will be needed later, and is of some interest in itself.

LEMMA 2. *Let* $(E, \mathfrak{T})$ *be a metrizable t.v.s. over* **R**, *let* C *be a cone in* E *which is complete, and let* $\{U_n : n \in N\}$ *be a neighborhood base of* 0 *consisting of closed, circled sets such that* $U_{n+1} + U_{n+1} \subset U_n$ $(n \in N)$. *Then the sets*

$$V_n = U_n \cap C - U_n \cap C \qquad (n \in N)$$

form a 0-*neighborhood base for a topology* $\mathfrak{T}_1$ *on* $E_1 = C - C$ *such that* $(E_1, \mathfrak{T}_1)$ *is a complete (metrizable) t.v.s. over* **R**.

Proof. It is clear that each set V_n is radial and circled in E_1, and obviously $V_{n+1} + V_{n+1} \subset V_n$ for all $n \in N$. It follows from (I, 1.2) that $\{V_n : n \in N\}$ is a 0-neighborhood base for a (unique translation invariant) topology $\mathfrak{T}_1$ on E_1 under which E_1 is a t.v.s. Of course $(E_1, \mathfrak{T}_1)$ is metrizable, and there remains to prove that $(E_1, \mathfrak{T}_1)$ is complete. In fact, given a Cauchy sequence in $(E_1, \mathfrak{T}_1)$, there exists a subsequence $\{z_n\}$ such that $z_{n+1} - z_n \in V_n (n \in N)$; we have, consequently, $z_{n+1} - z_n = x_n - y_n$, where x_n and y_n are elements of $U_n \cap C$, and it is evidently sufficient to show that the series $\sum\limits_{n=1}^{\infty} x_n$ and $\sum\limits_{n=1}^{\infty} y_n$ converge in $(E_1, \mathfrak{T}_1)$. Let us show this for $\sum\limits_{n=1}^{\infty} x_n$. Letting $u_n = \sum\limits_{v=1}^{n} x_v$ $(n \in N)$, we obtain

$$u_{n+p} - u_n \in (U_{n+1} + \cdots + U_{n+p}) \cap C \subset (U_n \cap C) \subset V_n$$

for all $p \in N$ and $n \in N$. Since C is complete in $(E, \mathfrak{T})$, $\{u_n\}$ converges for $\mathfrak{T}$ to some $u \in C$ and we have $u - u_{n+1} \in U_n \cap C \subset V_n$, since $U_n \cap C$ is closed in $(E, \mathfrak{T})$. Now the last relation shows that $u_n \to u$ in $(E_1, \mathfrak{T}_1)$, and the proof is complete.

3.5

Theorem. *Let* E *be a Banach space and let* C *be a closed cone in* E. *Then* C *is normal (respectively, a strict* $\mathfrak{B}$-*cone) if and only if* C' *is a strict* $\mathfrak{B}$-*cone (respectively, normal) in* E'_β.

Proof. The assertion concerning normal cones $C \subset E$ is a special case of (3.4), and if C is a $\mathfrak{B}$-cone, then C' is normal in E'_β by (3.3). Hence suppose that C' is normal in E'_β and denote by U the unit ball of E. The bipolar of $U \cap C$ (with respect to $\langle E', E'' \rangle$) is $U^{\circ\circ} \cap C''$ and by (3.3) C'' is a strict $\mathfrak{B}$-cone in the strong bidual E''; hence $U^{\circ\circ} \cap C'' - U^{\circ\circ} \cap C''$ is a 0-neighborhood in E''. It follows that $V = U \cap C - U \cap C$ is dense in a 0-neighborhood V_1 in E; if $E_1 = C - C$ and $\mathfrak{T}_1$ is the topology on E_1 defined in Lemma 2, this means precisely that the imbedding ψ of $(E_1, \mathfrak{T}_1)$ into E is nearly open and continuous, with dense range. Consequently, Banach's homomorphism theorem (III, 2.1) implies that ψ is a topological isomorphism of $(E_1, \mathfrak{T}_1)$ onto E, and hence C is a strict $\mathfrak{B}$-cone in E.

COROLLARY. *Let E be a Banach space and let C be a cone in E with closure $\bar{C}$. The following assertions are equivalent:*

(a) *C is a $\mathfrak{B}$-cone in E.*
(b) *$E = \bar{C} - \bar{C}$.*
(c) *$\bar{C}$ is a strict $\mathfrak{B}$-cone in E.*

Proof. (a) $\Rightarrow$ (c) is clear from the preceding since C' is normal in E' whenever C is a $\mathfrak{B}$-cone, by (3.3). (c) $\Rightarrow$ (b) is trivial. (b) $\Rightarrow$ (a): Let M denote the closure of $U \cap C - U \cap C$, where U is the unit ball of E. Then M is convex, circled (over R), and such that $E = \bigcup_{1}^{\infty} nM$; since E is a Baire space, it follows that M is a 0-neighborhood in E and hence C is a $\mathfrak{B}$-cone.

4. ORDERED TOPOLOGICAL VECTOR SPACES

Let L be a t.v.s. (over R or C) and an ordered vector space; we say that L is an **ordered topological vector space** if the following axiom is satisfied:

(*LTO*)   *The positive cone $C = \{x: x \geqq 0\}$ is closed in L.*

Recall that an Archimedean ordered vector space is called regularly ordered if the real bilinear form $(x, x^*) \to \operatorname{Re} \langle x, x^* \rangle$ places L_0 and L_0^+ in duality, where L_0 is the real underlying space of L (Chapter I, Section 7). In order to prove some alternative characterizations, we need the following lemma which is of interest in itself. (Cf. Exercise 21.)

LEMMA. *Let E be an ordered vector space of finite dimension over R. The order of E is Archimedean if and only if the positive cone C is closed for the unique topology under which E is a Hausdorff t.v.s.*

Proof. If C is closed, then clearly the order of E is Archimedean. Conversely, suppose that E is Archimedean ordered; without restriction of generality we can assume that $E = C - C$. If the dimension of E is n ($\geqq 1$), then C contains n linearly independent elements $x_1, \ldots, x_n$ and hence the n-dimensional simplex with vertices $0, x_1, \ldots, x_n$; since the latter has non-empty interior, so does C. Now let $x \in \bar{C}$ and let y be interior to C; by (II, 1.1) $n^{-1}y + x$ is interior to C ($n \in N$), and hence we have $-x \leqq n^{-1}y$ for all n. This implies $-x \leqq 0$ or, equivalently, $x \in C$.

4.1

If L is an ordered vector space over R with positive cone C, the following propositions are equivalent:

(a) *The order of L is regular.*
(b) *C is sequentially closed for some Hausdorff l.c. topology on L, and L^+ distinguishes points in L.*

(c) *The order of L is Archimedean, and C is normal for some Hausdorff l.c. topology on L.*

Proof. (a) $\Rightarrow$ (b): It suffices to show that the intersection of C with every finite dimensional subspace M is closed (Chapter II, Exercise 7), and this is immediate from the preceding lemma, since the order of L is Archimedean, and hence the canonical order of each subspace $M \subset L$ is Archimedean. (b) $\Rightarrow$ (c): If $\mathfrak{T}$ is a Hausdorff l.c. topology under which C is sequentially closed, then, clearly, L is Archimedean ordered. Moreover, since L^+ separates points in L, the canonical bilinear form on $L \times L^*$ places L and $L^+ = C^* - C^*$ in duality, and by (3.3) C is normal for the Hausdorff l.c. topology $\sigma(L, L^+)$. (c) $\Rightarrow$ (a): It suffices to show that L^+ separates points in L. If $\mathfrak{T}$ is a Hausdorff l.c. topology for which C is normal, then $(L, \mathfrak{T})' = C' - C'$ by (3.3); hence $C' - C'$, and a fortiori $L^+ = C^* - C^*$ separates points in L. This completes the proof.

COROLLARY 1. *The canonical orderings of subspaces, products, and direct sums of regularly ordered vector spaces are regular.*

COROLLARY 2. *Every ordered locally convex space is regularly ordered.*

Proof. If $(E, \mathfrak{T})$ is an ordered l.c.s., then C is closed by definition, and the bipolar theorem (IV, 1.5) shows $-C$ to be the polar of C' with respect to $\langle E, E' \rangle$. Since $C \cap -C = \{0\}$, it follows that $C' - C'$ is weakly dense in $(E, \mathfrak{T})'$, hence $L^+ = C^* - C^*$ separates points in L.

If A is an ordered set and $S \subset A$ is a subset ($\neq \varnothing$) directed for $\leq$, recall that the section filter $\mathfrak{F}(S)$ is the filter on A determined by the base $\{S_x : x \in S\}$, where $S_x = \{y \in S : y \geq x\}$, and S_x is called a section of S. In particular, if S is a monotone sequence in A, then $\mathfrak{F}(S)$ is the filter usually associated with S.

4.2

Let L be an ordered t.v.s. and let S be a subset of L directed for $\leq$. If the section filter $\mathfrak{F}(S)$ converges to $x_0 \in L$, then $x_0 = \sup S$.

Proof. Let $x \in S$ and let z be any element of L majorizing S; we have $x \leq y \leq z$ for all $y \in S_x$, and from $x_0 \in \bar{S}_x$ it follows that $x \leq x_0 \leq z$, since the positive cone is closed in L. This proves that $x_0 = \sup S$.

A deeper result is the following monotone convergence theorem, which can be viewed as an abstract version of a classical theorem of Dini. Although it can be derived from Dini's theorem, using (4.4) below (Exercise 9), we give a direct proof based on the Hahn-Banach theorem.

4.3

Theorem. *Let E be an ordered l.c.s. whose positive cone C is normal, and suppose that S is a subset of L directed for $\leq$. If the section filter $\mathfrak{F}(S)$ converges for $\sigma(E, E')$, then it converges in E.*

Proof. Without loss in generality we can suppose that S is directed for $\geqq$, and that $\lim \mathfrak{F}(S) = 0$ for $\sigma(E, E')$; it follows now from (4.2) that $S \subset C$. Assume that the assertion is false; then there exists a 0-neighborhood U in E that contains no section of S, and since C is normal we can suppose that U is convex and C-saturated. Since $x \in S \cap U$ implies $S_x \subset U$, it follows that $S \cap U = \varnothing$; moreover, $(S + C) \cap U = \varnothing$, since U is C-saturated, and $S + C$ is convex, since it is the union of the family $\{x + C : x \in S\}$ of convex sets which is directed under inclusion. Hence by (II, 9.2) U and S can be separated by a closed real hyperplane in E, and this contradicts the weak convergence of $\mathfrak{F}(S)$ to 0.

COROLLARY 1. *Let S be a directed $(\leqq)$ subset of E such that $x_0 = \sup S$, where E is an ordered l.c.s. with normal positive cone. If for every real linear form f which is positive and continuous on E one has $f(x_0) = \sup\{f(x) : x \in S\}$, then $\lim_{x \in S} g(x) = g(x_0)$ $(g \in E')$ uniformly on each equicontinuous subset of E'.*

Proof. In fact, in view of (3.3), Corollary 3, the weak convergence of $\mathfrak{F}(S)$ to x_0 is equivalent to the relation $f(x_0) = \sup\{f(x) : x \in S\}$ for every real linear form f on E which is positive and continuous (cf. (I, 7.2)).

The reader will note that the preceding corollary is equivalent with (4.3). The following result can be viewed as a partial converse of (4.2).

COROLLARY 2. *Let E be a semi-reflexive, ordered l.c.s. whose positive cone is normal. If S is a directed $(\leqq)$ subset of E which is majorized or (topologically) bounded, then $x_0 = \sup S$ exists and $\mathfrak{F}(S)$ converges to x_0.*

Proof. Let S_x be any fixed section of S; it suffices to show that $\sup S_x$ exists in E. If S is majorized by some $z \in E$, then $S_x \subset [x, z]$ and hence S_x is bounded in E by (3.1), Corollary 2; hence assume that S_x is bounded in E. The weak normality of C implies that $E' = C' - C'$, and hence that the section filter $\mathfrak{F}(S_x)$ is a weak Cauchy filter in E which is bounded. It follows from (IV, 5.5) that $\mathfrak{F}(S_x)$ converges to some $x_0 \in E$, and (4.2) implies that $x_0 = \sup S_x$, since C is closed and hence (being convex) weakly closed in E.

The following result is an imbedding (or representation) theorem for ordered l.c.s. over $\mathbf{R}$; let us denote by X a (separated) locally compact space, and by $R(X)$ the space of all real-valued continuous functions on X under the topology of compact convergence and endowed with its canonical order (Section 1).

4.4

Let E be an ordered l.c.s. over $\mathbf{R}$. If (and only if) the positive cone C of E is normal, there exists a locally compact space X such that E is isomorphic (as an ordered t.v.s.) with a subspace of $R(X)$.

Proof. The condition is clearly necessary, for the positive cone of $R(X)$ (and hence of every subspace of $R(X)$) is normal. To show that the condition is sufficient, we note first from (3.3), Corollary 1, that the topology of E is the topology of uniform convergence on the equicontinuous subsets of the dual cone $C' \subset E'$. Let $\{B_\alpha : \alpha \in A\}$ be a fundamental family of $\sigma(E', E)$-closed equicontinuous subsets of C'; under the topology induced by $\sigma(E', E)$, each B_α is a compact space. We define X as follows: Endow A with the discrete topology, C' with the topology induced by $\sigma(E', E)$, and let X_α be the subspace $\{\alpha\} \times B_\alpha$ of the topological product $A \times C'$; then X is defined to be the subspace $\bigcup_\alpha X_\alpha$ of $A \times C'$. The space X is the topological sum of the family $\{B_\alpha : \alpha \in A\}$; clearly, X is a locally compact space in which every X_α is open and compact, and hence every compact subset of X is contained in the union of finitely many sets X_α. For each $x \in E$, we define an element $f_x \in R(X)$ by putting $f_x(t) = \langle x, x' \rangle$ for every $t = (\alpha, x') \in X$; it is clear that $x \to f_x$ is an algebraic and order isomorphism of E into $R(X)$. Finally, since a closed subset of X is compact if and only if it is contained in a finite union $\bigcup_{\alpha \in H} X_\alpha$, it is also evident that $x \to f_x$ is a homeomorphism.

REMARKS. It is easy to see that $R(X)$ is complete; hence the image of E under $x \to f_x$ is closed in $R(X)$ if and only if E is complete. Moreover, if E is metrizable, then the family $\{B_\alpha : \alpha \in A\}$ can be assumed to be countable, and hence X countable at infinity; if E is normable, one can take $X = U^\circ \cap C'$ (under $\sigma(E', E)$), where U is any bounded neighborhood of 0 in E (in particular, the unit ball if E is normed). If E is a separable normed space, $U^\circ \cap C'$ is a compact metrizable space for $\sigma(E', E)$ by (IV, 1.7) and hence a continuous image of the Cantor set (middle third set) in $[0, 1] \subset R$; in this case X can be taken to be the Cantor set itself, or $[0, 1]$ (for details, see Banach [1], chap. XI, § 8, theor. 9).

Finally, proposition (4.4) can be specialized to the case $C = \{0\}$; we obtain thus a representation of an arbitrary l.c.s. E over R as a subspace of a suitable space $R(X)$; it is immediate that in this particular case, the restriction to the scalar field R can be dropped.

5. POSITIVE LINEAR FORMS AND MAPPINGS

The present section is concerned with special properties of linear maps $u \in L(E, F)$ which map the positive cone C of E into the positive cone D of F, where E, F are ordered vector spaces (respectively, ordered t.v.s.); these mappings are called **positive**. It is clear that the set H of all positive maps is a cone in $L(E, F)$; whenever M is a subspace of $L(E, F)$ such that $H \cap M$ is a proper cone, $H \cap M$ defines the canonical ordering of M (Section 1). Recall also (Section 2) that a linear form f on an ordered vector space E is called positive if $\operatorname{Re} f(x) \geqq 0$ for each x in the positive cone C of E.

We begin our investigation with some simple but useful observations concerning the properties of the cone $\mathscr{H} \subset \mathscr{L}(E, F)$ of continuous positive maps, where E, F are supposed to be ordered t.v.s. over K. We point out that

in view of the agreements made in Section 2, it suffices in general to consider
the case $K = R$.

5.1

*Let E, F be ordered t.v.s. and let $\mathfrak{S}$ be a family of bounded subsets of E that
covers E. Then the positive cone $\mathscr{H} \subset \mathscr{L}(E, F)$ is closed for the $\mathfrak{S}$-topology.
For $\mathscr{H}$ to be a proper cone, it is sufficient (and, if E is a l.c.s. and $F \neq \{0\}$, necessary) that the positive cone C of E be total in E.*

Proof. In fact, by definition of the $\mathfrak{S}$-topology (Chapter III, Section 3) the
bilinear map $(u, x) \to u(x)$ is separately continuous on $\mathscr{L}_{\mathfrak{S}}(E, F) \times E$ into F;
hence the partial map $f_x: u \to u(x)$ is continuous for each $x \in E$. Since $\mathscr{H}$
$= \bigcap \{f_x^{-1}(D): x \in C\}$ and the positive cone D of F is closed, $\mathscr{H}$ is closed in
$\mathscr{L}_{\mathfrak{S}}(E, F)$. Further, since D is proper, $u \in \mathscr{H} \cap -\mathscr{H}$ implies that $u(x) = 0$ for
$x \in C$; hence $u = 0$ if C is total in E. Finally, if E is a l.c.s. and C is not total
in E, there exists an $f \in E'$ such that $f \neq 0$ but $f(C) = \{0\}$, by virtue of the
Hahn-Banach theorem; if y is any element $\neq 0$ of F, the mapping $u = f \otimes y$
(defined by $x \to f(x)y$) satisfies $u \in \mathscr{H} \cap -\mathscr{H}$.

COROLLARY. *If C is total in E and if F is a l.c.s., the (canonical) ordering of
$\mathscr{L}(E, F)$ defined by $\mathscr{H}$ is regular.*

Proof. In fact, $\mathscr{H}$ is a closed proper cone for the topology of simple convergence which is a Hausdorff l.c. topology by (III, 3.1), Corollary; the
assertion follows from (4.1), Corollary 2.

5.2

*Let E, F be ordered l.c.s. with respective positive cones C, D and let $\mathfrak{S}$ be a
family of bounded subsets of E. If C is an $\mathfrak{S}$-cone in E and D is normal in F, the
positive cone $\mathscr{H} \subset \mathscr{L}(E, F)$ is normal for the $\mathfrak{S}$-topology.*

Proof. Since D is normal in F, there exists, by (3.1), a family $\{q_\alpha: \alpha \in A\}$ of
real semi-norms on F that generate the topology of F, and which are mono-
tone (for the order of F) on D. Since C is an $\mathfrak{S}$-cone in E, it follows that the
real semi-norms

$$u \to p_{\alpha, S}(u) = \sup\{q_\alpha(ux): x \in S \cap C\} \qquad (\alpha \in A, S \in \mathfrak{S})$$

generate the $\mathfrak{S}$-topology on $\mathscr{L}(E, F)$. Now, evidently, each $p_{\alpha, S}$ is monotone
on $\mathscr{H}$ (for the canonical order of $\mathscr{L}(E, F)$); hence $\mathscr{H}$ is a normal cone in
$\mathscr{L}_{\mathfrak{S}}(E, F)$, as asserted.

On the other hand, there are apparently no simple conditions guaranteeing
that $\mathscr{H}$ is a $\mathfrak{T}$-cone in $\mathscr{L}_{\mathfrak{S}}(E, F)$, even for the most frequent types of families
$\mathfrak{T}$ of bounded subsets of $\mathscr{L}_{\mathfrak{S}}(E, F)$, except in every special cases (cf. Exercise
7). At any rate, the following result holds where E, F are ordered l.c.s. with
respective positive cones C, D, and $\mathscr{L}_s(E, F)$ denotes $\mathscr{L}(E, F)$ under the
topology of simple convergence.

5.3

If C is weakly normal in E and if $F = D - D$, then $\mathscr{H} - \mathscr{H}$ is dense in $\mathscr{L}_s(E, F)$.

Proof. Since the weak normality of C is equivalent with $E' = C' - C'$ by (3.3), Corollary 3, the assumptions imply that $\mathscr{H} - \mathscr{H}$ contains the subspace $E' \otimes F$ of $\mathscr{L}(E, F)$. On the other hand, the dual of $\mathscr{L}_s(E, F)$ can be identified with $E \otimes F'$ by (IV, 4.3), Corollary 4, and it is known that (under the duality between $\mathscr{L}(E, F)$ and $E \otimes F'$) $E' \otimes F$ separates points in $E \otimes F'$ (Chapter IV, Section 1, Example 3); it follows from (IV, 1.3) that $E' \otimes F$ is weakly dense in $\mathscr{L}_s(E, F)$ and hence (being convex) dense in $\mathscr{L}_s(E, F)$.

We turn to the question of extending a continuous positive linear form, defined on a subspace of an ordered t.v.s. E, to the entire space E. The following extension theorem is due to H. Bauer [1], [2] and, independently, to Namioka [1].

5.4

Theorem. *Let E be an ordered t.v.s. with positive cone C and let M be a subspace of E. For a linear form f_0 on M to have an extension f to E which is a continuous positive linear form, it is necessary and sufficient that $\operatorname{Re} f_0$ be bounded above on $M \cap (U - C)$, where U is a suitable convex 0-neighborhood in E.*

Proof. It suffices to consider the case $K = \mathbf{R}$. If f is a linear extension of f_0 to E which is positive and continuous and if $U = \{x: f(x)\} < 1$, it is clear that $f_0(x) < 1$ whenever $x \in M \cap (U - C)$; hence the condition is necessary. Conversely, suppose that U is an open convex 0-neighborhood such that $x \in M \cap (U - C)$ implies $f_0(x) < \gamma$ for some $\gamma \in \mathbf{R}$. Then $\gamma > 0$ and $N = \{x \in M: f_0(x) = \gamma\}$ is a linear manifold in E not intersecting the open convex set $U - C$. By the Hahn-Banach theorem (II, 3.1) there exists a closed hyperplane H containing N and not intersecting $U - C$, which, consequently, can be assumed to be of the form $H = \{x: f(x) = \gamma\}$; clearly, f is a continuous extension of f_0. Furthermore, since $0 \in U - C$ it follows that $f(x) < \gamma$ when $x \in U - C$ and hence when $x \in -C$; thus $x \in C$ implies $f(x) \geqq 0$.

COROLLARY 1. *Let f_0 be a linear form defined on the subspace M of an ordered vector space L. f_0 can be extended to a positive linear form f on L if and only if $\operatorname{Re} f_0$ is bounded above on $M \cap (W - C)$, where W is a suitable convex radial subset of L.*

In fact, it suffices to endow L with its finest locally convex topology for which W is a neighborhood of 0, and to apply (5.4). The same specialization can be made in the following result which is due to Krein-Rutman [1].

COROLLARY 2. *Let E be an ordered t.v.s. with positive cone C, and suppose that M is a subspace of E such that $C \cap M$ contains an interior point of C. Then every continuous, positive linear form on M can be extended to E under preservation of these properties.*

Proof. If f_0 is the linear form in question and $x_0 \in M$ is an interior point of C, choose a convex 0-neighborhood U in E such that $x_0 + U \subset 2x_0 - C$. Then Re f_0 is bounded above on $M \cap (U - C)$, for we have $M \cap (U - C) \subset (x_0 - C) \cap M$.

REMARK. The condition of Corollary 2 can, in general, not be replaced by the assumption that $C \cap M$ possesses an interior point (Exercise 14). For another condition guaranteeing that every linear form f_0, defined and positive on a subspace M of an ordered vector space L, can be extended to a positive linear form f on L see Exercise 11.

There is a comparatively large class of ordered t.v.s. on which every positive linear form is necessarily continuous; we shall see (Section 7 below) that this class includes all bornological vector lattices that are at least sequentially complete (semi-complete). It is plausible that in spaces with this property, the positive cone must be sufficiently " wide " (cf. the discussion following (3.2)). More precisely, one has the following result (condition (ii) is due to Klee [2], condition (iii) to the author [2]).

5.5

Theorem. *Let E be an ordered t.v.s. with positive cone C. Each of the following conditions is sufficient to ensure the continuity of every positive linear form on E:*

(i) *C has non-empty interior.*
(ii) *E is metrizable and complete, and $E = C - C$.*
(iii) *E is bornological, and C is a semi-complete strict $\mathfrak{B}$-cone.*

Proof. It is again sufficient to consider real linear forms on E. The sufficiency of condition (i) is nearly trivial, for if f is positive, then $f^{-1}(0)$ is a hyperplane in E lying on one side of the convex body C, and hence closed which is equivalent with the continuity of f by (I, 4.2). Concerning condition (ii), we use Lemma 2 of Section 3: The topology $\mathfrak{T}_1$ on E, determined by the neighborhood base of 0, $\{V_n : n \in N\}$, where $V_n = U_n \cap C - U_n \cap C$, is evidently finer than the given topology $\mathfrak{T}$ of E, and hence we have $\mathfrak{T} = \mathfrak{T}_1$ by Banach's theorem (III, 2.1), Corollary 2. Now if f is a positive, real linear form on E which is not continuous, then f is unbounded on each set $U_n \cap C$ hence there exists $x_n \in U_n \cap C$ such that $f(x_n) > 1$ $(n \in N)$. On the other hand, since $U_{n+1} + U_{n+1} \subset U_n$ for all n, the sequence $\{x_n\}$ is summable in E with sum $\sum_{n \in N} x_n = z \in C$ (C being closed), and from $z \geqq \sum_{n=1}^{p} x_n$ we obtain $f(z) > p$ for each $p \in N$, which is contradictory. Finally, concerning condition (iii) we observe that since E is bornological and C is a strict $\mathfrak{B}$-cone, a linear form on E which is bounded on the bounded subsets of C is necessarily continuous by (II, 8.3); now if f is a positive, real linear form on E which is not continuous, there exists a bounded sequence $\{x_n\}$ in C such that $f(x_n) > n$ ($n \in N$). Since E is locally convex by definition, we conclude that $\{n^{-2}x_n: n \in N\}$ is a summable sequence in C with sum $z \in C$, say, and it follows that

$$f(z) \geqq \sum_{n=1}^{p} n^{-2}f(x_n) > \sum_{n=1}^{p} n^{-1} \quad \text{for all } p,$$

which is impossible. The proof is complete.

COROLLARY. *Let E be an ordered l.c.s. which is the inductive limit of a family $\{E_\alpha: \alpha \in A\}$ of ordered (F)-spaces with respect to a family of positive linear maps, and suppose that $E_\alpha = C_\alpha - C_\alpha$ ($\alpha \in A$). Then each positive linear form on E is continuous.*

This is immediate in view of (II, 6.1). For locally convex spaces, an important consequence of (5.5) is the automatic continuity of rather extensive classes of positive linear maps.

5.6

Let E, F be ordered l.c.s. with respective positive cones C, D. Suppose that E is a Mackey space on which every positive linear form is continuous, and assume that D is a weakly normal cone in F. Then every positive linear map of E into F is continuous.

Proof. Let u be a linear map of E into F such that $u(C) \subset D$, and consider the algebraic adjoint u^* of u (Chapter IV, Section 2). For each $y' \in D'$, $x \to \langle x, u^*y' \rangle$ is a positive linear form on E, hence continuous by assumption; since $F' = D' - D'$ by (3.3), Corollary 3, it follows that $u^*(F') \subset E'$; hence u is weakly continuous by (IV, 2.1). Thus $u \in \mathscr{L}(E, F)$ by (IV, 7.4).

We conclude this section with an application of several of the preceding results to the convergence of directed families of continuous linear maps.

5.7

Let E be an ordered barreled space such that $E = C - C$, and let F be an ordered semi-reflexive space whose positive cone D is normal. Suppose that $\mathscr{U}$ is a subset of $\mathscr{L}(E, F)$ which is directed upward for the canonical order of $\mathscr{L}(E, F)$, and either majorized or simply bounded. Then $u_0 = \sup \mathscr{U}$ exists, and the section filter $\mathfrak{F}(\mathscr{U})$ converges to u_0 uniformly on every precompact subset of E.

Proof. In fact, (5.1) and (5.2) show that $\mathscr{H}$ is a closed normal cone in $\mathscr{L}_s(E, F)$ and hence is the positive cone for the canonical order of $\mathscr{L}_s(E, F)$. For each $x \in C$, the family $\{u(x): u \in \mathscr{U}\}$ satisfies the hypotheses of (4.3), Corollary 2, and hence of (4.3), so $\mathfrak{F}(\mathscr{U})$ converges simply to a linear map $u_0 \in \mathscr{L}(E, F)$. By (III, 4.6) u_0 is continuous, and the convergence of $\mathfrak{F}(\mathscr{U})$ is uniform on every precompact subset of E. Since $\mathscr{H}$ is closed in $\mathscr{L}_s(E, F)$, (4.2) implies that $u_0 = \sup \mathscr{U}$.

6. THE ORDER TOPOLOGY

If S is an ordered set, the order of S gives rise to various topologies on S (cf. Birkhoff [1]); however, in general, the topologies so defined do not satisfy axioms $(LT)_1$ and $(LT)_2$ (Chapter I, Section 1) if S is a vector space, even if (LTO) holds (cf. Exercise 17). On the other hand, if L is an ordered vector space over R, there is a natural locally convex topology which, as will be seen below, is the topology of many (if not all) ordered vector spaces occuring in analysis. The present section is devoted to a study of the principal properties of this topology. (See also Gordon [2].)

Let L be an ordered vector space over R; we define the **order topology** $\mathfrak{T}_o$ of L to be the finest locally convex topology on L for which every order interval is bounded. The family of locally convex topologies on L having this property is not empty, since it contains the coarsest topology on L, and $\mathfrak{T}_o$ is the upper bound of this family (Chapter II, Section 5); a subset $W \subset L$ is a 0-neighborhood for $\mathfrak{T}_o$ if and only if W is convex and absorbs every order interval $[x, y] \subset L$. (W is necessarily radial, since $\{x\} = [x, x]$ for each $x \in L$.) Although $\mathfrak{T}_o$ is a priori defined for ordered vector spaces over R only, it can happen (cf. the corollaries of (6.2) and (6.4) below) that $(L, \mathfrak{T})$ is an ordered vector space over C such that $(L_0, \mathfrak{T}) = (L_0, \mathfrak{T}_0)$, where L_0 is the underlying real space of L. We begin with the following simple result.

6.1

The dual of $(L, \mathfrak{T}_0)$ is the order bound dual L^b of L. If L^b separates points in L (in particular, if the order of L is regular), $(L, \mathfrak{T}_0)$ is a bornological l.c.s. If L, M are ordered vector spaces, each positive linear map of L into M is continuous for the respective order topologies.

Proof. It is clear from the definition of $\mathfrak{T}_0$ that each order interval is bounded for $\mathfrak{T}_0$; hence if $f \in (L, \mathfrak{T}_0)'$ then $f \in L^b$. Conversely, if $f \in L^b$, then $f^{-1}([-1, 1])$ is convex and absorbs each order interval, and hence is a 0-neighborhood for $\mathfrak{T}_0$. $\mathfrak{T}_0$ is a Hausdorff topology if and only if L^b distinguishes points in L. Let W be a convex subset of L that absorbs each bounded subset of $(L, \mathfrak{T}_0)$; since W a fortiori absorbs all order intervals in L, W is a $\mathfrak{T}_0$-neighborhood of 0. Hence $(L, \mathfrak{T}_0)$ is bornological if (and only if) $\mathfrak{T}_0$ is a Hausdorff topology. Finally, if u is a positive linear map of L into M, then $u([x, y]) \subset [u(x), u(y)]$ for each order interval in L; hence if V is convex and absorbs order intervals in M, $u^{-1}(V)$ has the same properties in L and thus u is continuous for the order topologies. (Cf. Exercise 12.)

COROLLARY. *Let L_i $(i = 1, ..., n)$ be a finite family of ordered vector spaces, and endow $L = \prod_i L_i$ with its canonical order. Then the order topology of L is the product of the respective order topologies of the L_i.*

Proof. We show that the projection p_i of $(L, \mathfrak{T}_0)$ onto $(L_i, \mathfrak{T}_0)$ $(i = 1, ..., n)$ is a topological homomorphism. In fact, p_i is continuous by (6.1); if I_i is an order interval in L_i then $I_i \times \{0\}$ is an order interval in L, hence if W is a convex 0-neighborhood in $(L, \mathfrak{T}_o)$ then $p_i(W)$ is convex and absorbs I_i which proves the assertion.

The order topology is most easily analyzed when L is an Archimedean ordered vector space with an order unit e. For convenience of expression, let us introduce the following terminology: A sequence $\{x_n: n \in N\}$ of elements ≥ 0 of an ordered vector space L is **order summable** if $\sup_n u_n$ exists in L, where $u_n = \sum_{p=1}^n x_p$. We shall say that a positive sequence $\{x_n: n \in N\}$ is *of type* l^1 if there exists an $a \geq 0$ in L and a sequence $(\lambda_n) \in l^1$ such that $(0 \leq)x_n \leq \lambda_n a$ $(n \in N)$.

6.2

Let L be an Archimedean ordered vector space over **R**, *possessing an order unit e. Then* $(L, \mathfrak{T}_o)$ *is an ordered t.v.s. which is normable,* $\mathfrak{T}_o$ *is the finest locally convex topology on L for which the positive cone C is normal, and the following assertions are equivalent*:

(a) $(L, \mathfrak{T}_o)$ *is complete.*
(b) *Each positive sequence of type* l^1 *in L is order summable.*

Proof. The order interval $[-e, e]$ is convex, circled, and (by the definition of order unit) radial in L; since L is Archimedean ordered, the gauge p_e of $[-e, e]$ is a norm on L. The topology generated by p_e is finer than $\mathfrak{T}_o$ since $[-e, e]$ is $\mathfrak{T}_o$-bounded, and it is coarser than $\mathfrak{T}_o$, since it is locally convex and $[-e, e]$ absorbs order intervals; hence p_e generates $\mathfrak{T}_o$. To see that C is closed in $(L, \mathfrak{T}_o)$, note that e is an interior point of C; the fact that C is closed follows then, as in the proof of the lemma preceding (4.1), from the hypothesis that L is Archimedean ordered. Moreover, since by (3.1), Corollary 2, $\mathfrak{T}_o$ is finer than any l.c. topology on L for which C is normal, the second assertion follows from the fact that the family $\{\varepsilon[-e, e]: \varepsilon > 0\}$ is a 0-neighborhood base for $\mathfrak{T}_o$ that consists of C-saturated sets.

Further, it is clear that (a) $\Rightarrow$ (b), since every positive sequence of type l^1 in L is of type l^1 with respect to $a = e$, and hence even absolutely summable in $(L, \mathfrak{T}_o)$; the assertion follows from (4.2). (b) $\Rightarrow$ (a): We have to show that $(L, \mathfrak{T}_o)$ is complete. Given a Cauchy sequence in $(L, \mathfrak{T}_o)$, there exists a subsequence $\{x_n: n \in N\}$ such that for all n, $p_e(x_{n+1} - x_n) < \lambda_n$, where $(\lambda_n) \in l^1$; hence $x_{n+1} - x_n \in \lambda_n[-e, e]$ and we have $x_{n+1} - x_n = u_n - v_n$, where $u_n = \lambda_n e + (x_{n+1} - x_n)$ and $v_n = \lambda_n e$ $(n \in N)$. To show that $\{x_n\}$ converges, it suffices to show that $\sum_{n=1}^\infty u_n$ converges. Now $0 \leq u_n \leq 2\lambda_n e$; hence $\{u_n\}$ is of type l^1 and $\sup_n \sum_{p=1}^n u_p = u \in C$ exists by hypothesis. Since for all n

$$0 \leq u - \sum_{p=1}^n u_p = \sup_k \sum_{p=n+1}^{n+k} u_p \leq 2\left(\sum_{p=n+1}^\infty \lambda_p\right)e,$$

it follows that $\sum_{n=1}^\infty u_n = u$ for $\mathfrak{T}_o$ and hence $(L, \mathfrak{T}_o)$ is complete.

COROLLARY 1. *If L is Archimedean ordered and has an order unit, the order of L is regular and we have $L^b = L^+$.*

This is immediate in view of (6.1) and (3.3).

COROLLARY 2. *Let $(E, \mathfrak{X})$ be an ordered Banach space possessing an order unit. Then $\mathfrak{X} = \mathfrak{X}_0$ if and only if the positive cone C of E is normal in $(E, \mathfrak{X})$.*

Proof. In fact, the order of E is Archimedean, since C is closed in $(E, \mathfrak{X})$; if $\mathfrak{X} = \mathfrak{X}_0$, then C is normal by (6.2). Conversely, if C is normal, then $\mathfrak{X}$ is coarser than $\mathfrak{X}_0$; since $[-e, e]$ is a barrel in $(E, \mathfrak{X})$ (we can suppose that $K = R$), it follows that $\mathfrak{X} = \mathfrak{X}_0$.

Examples to which the preceding corollary applies are furnished by the spaces $\mathscr{C}(X)$ (X compact) and $L^\infty(\mu)$ (Chapter II, Section 2, Examples 1 and 2) and, more generally, by every ordered Banach space whose positive cone is normal and has non-empty interior. It is readily verified that each interior point of the positive cone C of an ordered t.v.s. L is an order unit, and each order unit is interior to C for $\mathfrak{X}_0$.

However, most of the ordered vector spaces occurring in analysis do not have order units, so that the description of $\mathfrak{X}_0$ given in (6.2) does not apply. Let L be an Archimedean ordered vector space over R and denote, for each $a \geqq 0$, by L_a the ordered subspace $L_a = \bigcup_{n=1}^{\infty} n[-a, a]$ endowed with its order topology; L_a is a normable space. The family $\{L_a: a \geqq 0\}$ is evidently directed under inclusion $\subset$, and if $L_a \subset L_b$, the imbedding map $h_{b,a}$ of L_a into L_b is continuous.

6.3

Let L be a regularly ordered vector space over R, and denote by H any subset of the positive cone C of L which is cofinal with C for $\leqq$. Then $(L, \mathfrak{X}_0)$ is the inductive limit $\lim\limits_{\longrightarrow} h_{b,a} L_a \ (a, b \in H)$.

Proof. By (6.1), the assumption on L implies that $\mathfrak{X}_0$ is Hausdorff. In view of the preceding remarks and the definition of inductive limit (Chapter II, Section 6), it suffices to show that $\mathfrak{X}_0$ is the finest l.c. topology on L for which each of the imbedding maps $f_a: L_a \to L \ (a \in H)$ is continuous. Since H is cofinal with C, each order interval $[x, y] \subset L$ is contained in a translate of some $[-a, a]$ where $a \in H$, and hence $[x, y]$ is bounded for the topology $\mathfrak{X}$ of the inductive limit; hence $\mathfrak{X}_0$ is finer than $\mathfrak{X}$. On the other hand, if W is a convex 0-neighborhood in $(L, \mathfrak{X}_0)$, then W absorbs all order intervals in L, which implies that $f_a^{-1}(W)$ is a 0-neighborhood in $L_a \ (a \in H)$, and hence $\mathfrak{X}$ is finer than $\mathfrak{X}_0$.

COROLLARY 1. *If the order of L is regular and each positive sequence of type l^1 in L is order summable, $(L, \mathfrak{X}_0)$ is barreled.*

Proof. In fact, the assumption implies by (6.2) that each of the spaces L_a $(a \in H)$ is normable and complete and hence barreled; the result follows from (II, 7.2).

COROLLARY 2. *If the order of L is regular and the positive cone C satisfies condition* (D) *of* (1.1), *then C is normal for* $\mathfrak{T}_0$ *(hence the dual of* $(L, \mathfrak{T}_0)$ *is* L^+).

Proof. By definition of the topology of inductive limit, a 0-neighborhood base for $\mathfrak{T}_0$ is given by the family of all convex radial subsets U of L such that $V = U \cap (C - C)$ is of the form $V = \Gamma'\{\rho_a[-a, a]: a \in H\}$ where $a \to \rho_a$ is any mapping of H into the set of real numbers > 0. We prove the normality of C via (3.1) (c) by showing that $x \in U$ and $y \in [0, x]$ imply $y \in U$. If $x \in U$ and $x \geqq 0$, then x is of the form $x = \sum_{i=1}^{n} \lambda_i z_i$, where $\sum_{i=1}^{n} |\lambda_i| \leqq 1$ $(\lambda_i \in \mathbf{R})$ and $z_i \in \rho_{a_i}[-a_i, a_i]$ $(i = 1, ..., n)$. If $y \in [0, x]$ it follows that $y \leqq \sum_{i=1}^{n} |\lambda_i| \rho_{a_i} a_i$; by repeated application of (D), we obtain $y = \sum_{i=1}^{n} |\lambda_i| y_i$, where $y_i \in \rho_{a_i}[0, a_i]$ $(i = 1, ..., n)$. Hence $y \in V \subset U$ as was to be shown.

REMARK. Since $L_a \subset L_b$ (where $a, b \in C$) is equivalent with $a \leqq \lambda b$ for a suitable scalar $\lambda > 0$, it suffices in (6.3) to require that the set of all positive scalar multiples of the elements $a \in H$ be cofinal with C (for $\leqq$); in particular, if L has an order unit e, it suffices to take $H = \{e\}$. Let us note also that the inductive limit of (6.3) is in general not strict (the topology induced by L_b on L_a $(b > a)$ is, in general, not the order topology of L_a). For example, if L is the space $L^2(\mu)$ and a, b are the respective equivalence classes of two functions f, g such that $0 \leqq f \leqq g$, f is bounded and g μ-essentially unbounded, then the topology of L_a is strictly finer than the topology induced on L_a by L_b. (Cf. Exercise 12.)

We apply the preceding description of $\mathfrak{T}_0$ to the case where L is a vector lattice; the lattice structure compensates in part for the lack of an order unit and one obtains a characterization of $\mathfrak{T}_0$ that can be compared with (6.2).

6.4

Let L be a vector lattice whose order is regular and let $\mathfrak{T}$ be a locally convex topology on L. These assertions are equivalent:

(a) $\mathfrak{T}$ *is the order topology* $\mathfrak{T}_0$.
(b) $\mathfrak{T}$ *is the finest l.c. topology on L for which C is normal.*
(c) $\mathfrak{T}$ *is the Mackey topology with respect to* $\langle L, L^+ \rangle$.

Proof. Let us note first that by (1.4), $L^b = L^+$ and that since the order of L is assumed to be regular, $\langle L, L^+ \rangle$ is a duality. (a) $\Leftrightarrow$ (b): Since the positive cone of a vector lattice satisfies (D) of (1.1), this follows from the fact that by (3.1), Corollary 2, $\mathfrak{T}_0$ is finer than any l.c. topology for which C is normal, in view of (6.3), Corollary 2. (a) $\Leftrightarrow$ (c): Since $(L, \mathfrak{T}_0)' = L^b = L^+$, $\mathfrak{T}_0$ is consistent with the duality $\langle L, L^+ \rangle$; since $(L, \mathfrak{T}_0)$ is bornological, $\mathfrak{T}_0$ is necessarily the Mackey topology with respect to $\langle L, L^+ \rangle$.

The following corollary is now a substitute for (6.2), Corollary 2.

COROLLARY. *Let $(E, \mathfrak{T})$ be an ordered* (F)-*space* (*over* **R**) *which is a vector lattice. Then $\mathfrak{T} = \mathfrak{T}_0$ if and only if the positive cone C of E is normal in $(E, \mathfrak{T})$.*

Proof. If $\mathfrak{T} = \mathfrak{T}_0$, then C is normal by (6.4). Conversely, if C is normal, then $E' = C' - C'$ by (3.3), and $C' - C' = E^+$ by (5.5) (for C is closed in $(E, \mathfrak{T})$ and $E = C - C$); since E is a Mackey space by (IV, 3.4), the assertion follows from (6.4) (c).

7. TOPOLOGICAL VECTOR LATTICES

Let L be a t.v.s. over **R** and a vector lattice, and consider the maps $x \to |x|$, $x \to x^+$, $x \to x^-$ of L into itself, and the maps $(x, y) \to \sup(x, y)$ and $(x, y) \to \inf(x, y)$ of $L \times L$ into L. By utilizing the identities (1), (2) and (3) of Section 1, it is not difficult to prove that the continuity of one of these maps implies the continuity (in fact, the uniform continuity) of all of them; in this case, we say that "the lattice operations are continuous" in L. Recall that a subset A of L is called solid if $x \in A$ and $|y| \le |x|$ imply that $y \in A$; we call L **locally solid** if the t.v.s. L possesses a 0-neighborhood base of solid sets.

7.1

*Let L be a t.v.s. over **R** and a vector lattice. The following assertions are equivalent:*

(a) *L is locally solid.*

(b) *The positive cone of L is normal, and the lattice operations are continuous.*

Proof. (a) $\Rightarrow$ (b): Let $\mathfrak{U}$ be a 0-neighborhood base in L consisting of solid sets; if $x \in U \in \mathfrak{U}$ and $0 \le y \le x$, then $y \in U$ and hence the positive cone C of L is normal by (3.1) (c). Moreover, if $x - x_0 \in U$, we conclude from (6) of (1.1) that $x^+ - x_0^+ \in U$ $(U \in \mathfrak{U})$ and hence the lattice operations are continuous.

(b) $\Rightarrow$ (a): Suppose that C is normal and the lattice operations are continuous. Let $\mathfrak{U}$ be a 0-neighborhood base in L consisting of circled C-saturated sets (Section 3). For a given $U \in \mathfrak{U}$, choose $V \in \mathfrak{U}$, $W \in \mathfrak{U}$ so that $V + V \subset U$ and that $x \in W$ implies $x^+ \in V$. Now if $x \in W$, then $-x \in W$, since W is circled; hence x^+ and $x^- = (-x)^+$ are in V and $|x| = x^+ + x^- \in U$. If $|y| \le |x|$, then $y \in [-|x|, |x|]$; hence, since U is C-saturated, it follows that $y \in U$. Therefore, the set $\{y: \text{ there exists } x \in W \text{ such that } |y| \le |x|\}$ is a 0-neighborhood contained in U, and is obviously solid.

It is plausible that for t.v.s. that are vector lattices, just as for more general types of ordered t.v.s., the axiom (*LTO*) (closedness of the positive cone) by itself is too weak to produce useful results. We define a **topological vector lattice** to be a vector lattice and a Hausdorff t.v.s. over R that is locally solid; it will be seen from (7.2) below that in these circumstances, the positive cone of L is automatically closed, and hence every topological vector lattice is an ordered t.v.s. over R. A **locally convex vector lattice** (abbreviated **l.c.v.l.**) is a topological vector lattice whose topology is locally convex. Every solid set is circled (with respect to R, cf. (4) of (1.1)); hence a topological vector lattice possesses a base of circled solid 0-neighborhoods. Since the convex hull of a solid set is solid (hence also circled), a l.c.v.l. possesses a 0-neighborhood base of convex solid sets. The gauge function p of a radial, convex solid set is characterized by being a semi-norm such that $|y| \leq |x|$ implies $p(y) \leq p(x)$, and is called a **lattice semi-norm** on L. Therefore, the topology of a l.c.v.l. can be generated by a family of lattice semi-norms (for example, by the family of all continuous lattice semi-norms). A **Fréchet lattice** is a l.c.v.l. which is an (F)-space; a **normed lattice** is a normed space (over R) whose unit ball $\{x: \|x\| \leq 1\}$ is solid. By utilizing (I, 1.5) and the uniform continuity of the lattice operations, it is easy to see that with respect to the continuous extension of the lattice operations, the completion of a topological vector lattice is a topological vector lattice; in particular, the completion of a normed lattice is a complete normed lattice with respect to the continuous extension of its norm. A complete normed lattice is called a **Banach lattice**. Let us record the following elementary consequences of the definition of a topological vector lattice.

7.2

In every topological vector lattice L, the positive cone C is closed, normal, and a strict $\mathfrak{B}$-cone; if L is order complete, every band is closed in L.

Proof. C is normal cone by (7.1) and, since $C = \{x: x^- = 0\}$, C is closed, since the topology of L is Hausdorff and $x \to x^-$ is continuous. To show that C is a strict $\mathfrak{B}$-cone, recall that if B is a circled, bounded set, then $B^+ = B^-$, and hence $B \subset B^+ - B^+$. It suffices, therefore, to show that B^+ is bounded if B is bounded. If B is bounded and U is a given solid 0-neighborhood in L, there exists $\lambda > 0$ such that $B \subset \lambda U$; since λU is evidently solid, it follows that $B^+ \subset \lambda U$ hence B^+ is bounded. Finally, if A is a band in L, then $A = A^{\perp\perp}$ by (1.3), Corollary 1. Now each set $\{a\}^\perp = \{x \in L: \inf(|x|, |a|) = 0\}$ is closed, since L is Hausdorff and $x \to \inf(|x|, |a|)$ is continuous, and we have $A = \bigcap \{\{a\}^\perp: a \in A^\perp\}$.

Examples

1. The Banach spaces (over R) $L^p(\mu)$ (Chapter II, Section 2, Example 2) are Banach lattices under their canonical orderings; it will be seen below that these are order complete for $p < \infty$, and the spaces $L^1(\mu)$ and $L^\infty(\mu)$ will be important concrete examples for the discussion in Section

8. The corresponding spaces over C can be included in the discussion, as they are complexifications (Chapter I, Section 7) of their real counterparts.

2. Let λ be a subspace of ω_d such that $\lambda = \lambda^{\times\times}$ (Chapter IV, Section 1, Example 4); λ is a perfect space in the sense of Köthe [5]). Under the normal topology (Köthe [5], Peressini [2]) λ is a l.c.v.l. when endowed with its canonical ordering as a subspace of ω_d. The normal topology is the topology of uniform convergence on all order intervals of $\lambda^{\times}$, and the coarsest topology consistent with $\langle \lambda, \lambda^{\times} \rangle$ such that the lattice operations are continuous. (Cf. Exercise 20.)

3. Let X be a locally compact (Hausdorff) space and let E be the space of all real-valued functions with compact support in X, endowed with its inductive limit topology (Chapter II, Section 6, Example 3). The topology of E is the order topology $\mathfrak{T}_0$ (Section 6), so that E is a locally convex vector lattice (see (7.3)); E is, in general, not order complete. The dual of $(E, \mathfrak{T}_0)$ is the order dual E^+ of E (the space of all real Radon measures on X); under its canonical order, E^+ is an order complete vector lattice by (1.4), Corollary, and a l.c.v.l. for its strong topology $\beta(E^+, E)$ ((7.4) below). Of particular interest are the spaces $E = \mathscr{C}(X)$ when X is compact (Section 8).

We now supplement the results on the order topology $\mathfrak{T}_0$ obtained in the previous section.

7.3

Let E be a regularly ordered vector lattice. Then the order topology $\mathfrak{T}_0$ is the finest topology $\mathfrak{T}$ on E such that $(E, \mathfrak{T})$ is a l.c.v.l. Moreover, if E is order complete, then $(E, \mathfrak{T}_0)$ is barreled, and every band decomposition of E is a topological direct sum for $\mathfrak{T}_0$.

Proof. In view of (6.1) (and $E^+ = E^b$, (1.4)), the regularity of the order of E is sufficient (and necessary, cf. Exercise 19) for $\mathfrak{T}_0$ to be a Hausdorff topology. By (6.3), $(E, \mathfrak{T}_0)$ is the inductive limit of the normed spaces L_a ($a \geqq 0$) that are normed lattices in the present circumstances; one shows, as in the proof of (6.3), Corollary 2, that the convex circled hull of any family $\{\rho_a[-a, a]: a \geqq 0\}$ is solid, and hence that $(E, \mathfrak{T}_0)$ is locally solid. The fact that $\mathfrak{T}_0$ is the finest topology $\mathfrak{T}$ such that $(E, \mathfrak{T})$ is a l.c.v.l. then follows from (6.4)(b), since the positive cone is normal for all these topologies, (7.1). If E is order complete, then clearly every positive sequence of type l^1 is order summable; hence $(E, \mathfrak{T}_0)$ is barreled by (6.3), Corollary 1. The last assertion is clear from the corollary of (6.1), since $\mathfrak{T}_0$ induces on each band $B \subset E$ the order topology of B. (Exercise 12.)

COROLLARY 1. *If the order of the vector lattice E is regular, then $(E, \mathfrak{T}_0)$ is a l.c.v.l. whose topology is generated by the family of all lattice semi-norms on E.*

From the corollary of (6.4), we obtain:

COROLLARY 2. *If E is a vector lattice and an ordered* (F)-*space in which the positive cone is normal, the lattice operations are continuous in E.*

It is interesting that the strong dual of a l.c.v.l. E reflects the properties of E in a strengthened form; in addition, E'_β is complete when E is barreled. (As has been pointed out in Chapter IV, Section 6, the strong dual of a barreled l.c.s. is in general not complete.)

7.4

Theorem. *Let E be a l.c.v.l. Then the strong dual E'_β is an order complete l.c.v.l. under its canonical order, and a solid subspace of E^+; moreover, if E is barreled, then E' is a band in E^+, and E'_β is a complete l.c.s.*

Proof. Since the positive cone C of E is normal in $(E, \mathfrak{T})$ by (7.2), it follows from (3.3) that $E' = C' - C' \subset C^* - C^* = E^+$.

It follows from the corollary of (1.5) that the polar U° of every solid 0-neighborhood U in E is a solid subset of E^+. Since E' is the union of these polars, as U runs through a base of solid 0-neighborhoods, E' is a solid subspace and therefore a sublattice of E^+. In particular, it follows that E' is an order complete sublattice of E^+. To see that E' is a l.c.v.l. for the strong topology $\beta(E', E)$, it suffices to observe that the family of all solid bounded subsets of E is a fundamental family of bounded sets; by the corollary of (1.5) the polars B° (with respect to $\langle E, E' \rangle$) of these sets B form a 0-neighborhood base for $\beta(E', E)$ that consists of solid subsets of F'.

If $(E, \mathfrak{T})$ is barreled and S is a directed $(\leq)$ subset of the dual cone C' such that S is majorized in E^+, then each section of S is bounded for $\sigma(E^+, E)$, hence for $\sigma(E', E)$ and, consequently, $\sigma(E', E)$-relatively compact, (IV, 5.2). Thus the section filter of S converges weakly to some $f \in C'$, and it is clear from the definition of the order of E' that $f = \sup S$ (cf. (4.2), which is, however, not needed for the conclusion). Since we have shown before that E' is a solid sublattice of E^+, it is now clear that E' is a band in E^+.

There remains to show that if $(E, \mathfrak{T})$ is barreled, then $(E', \beta(E'E))$ is complete. Let us note first that E^+, which is the dual of $(E, \mathfrak{T}_0)$ by (6.1) (note that $E^+ = E^b$ by (1.4)), is complete under $\beta(E^+, E)$ by (IV, 6.1), for it is the strong dual of a bornological space. Hence by the preceding results and (7.3), $(E^+, \beta(E^+, E))$ is a l.c.v.l. (7.2) shows that E', being a band in E^+, is closed in $(E^+, \beta(E^+, E))$ and hence complete for the topology induced by $\beta(E^+, E)$. On the other hand, this latter topology is coarser than $\beta(E', E)$, since $\mathfrak{T}$ is coarser than $\mathfrak{T}_0$; hence if $\mathfrak{F}$ is a $\beta(E', E)$-Cauchy filter in E', $\mathfrak{F}$ has a unique $\beta(E^+, E)$-limit $g \in E'$. Clearly, $\mathfrak{F}$ converges to g pointwise on E, and (since $\mathfrak{F}$ is a Cauchy filter for $\beta(E', E)$), it follows from a simple argument that $\lim \mathfrak{F} = g$ for $\beta(E', E)$. This completes the proof.

COROLLARY 1. *Every reflexive locally convex vector lattice is order complete, and a complete l.c.s.*

In fact, the strong dual of E is a l.c.v.l. which is reflexive by (IV, 5.6), Corollary 1, and hence barreled, and E can be identified (under evaluation) with the strong dual of E_β'. More generally, if E is a l.c.v.l. that is semi-reflexive, then E is order complete and $(E, \beta(E, E'))$ is complete (cf. Corollary 2 of (7.5) below).

COROLLARY 2. *If E is a normed lattice, its strong dual E' is a Banach lattice with respect to dual norm and canonical order. If, in addition, E is a Banach space then $E' = E^+$.*

Proof. The first assertion is clear, since the unit ball of E' is solid by the corollary of (1.5). The second assertion is a consequence of (5.5) and (7.2).

The following result is the topological counterpart of (1.6).

COROLLARY 3. *If E is an infrabarreled l.c.v.l., then E can be identified, under evaluation, with a topological vector sublattice of its strong bidual E'' (which is an order complete l.c.v.l. under its canonical order).*

Proof. The assumption that E is infrabarreled (Chapter IV, Section 5) means precisely that the evaluation map $x \to \tilde{x}$ is a homeomorphism of E into E''; the remainder follows from (1.6), since E' is a solid subspace of E^+.

It would, however, be a grave error to infer from the foregoing corollary that for an *infinite* subset $S \subset E$ such that $x = \sup S$ exists in E, one has necessarily $\tilde{x} = \sup \tilde{S}$. Thus even if E is order complete, E can, in general, not be identified (under evaluation) with an order complete sublattice of E''. For example, let $E = l^\infty$ be endowed with its usual norm and order; E is an order complete Banach lattice (in fact, E can be identified with the strong dual of the Banach lattice l^1). Denote by x_n $(n \in N)$ the vector in E whose n first coordinates are 1, the remaining ones being 0; $\{x_n: n \in N\}$ is a monotone sequence in E such that $\sup_n x_n = e$, where $e = (1, 1, 1, ...)$. Let $z = \sup_n \tilde{x}_n$ in E'' $(= E^{++}$ by virtue of (5.5)); we assert that $z \neq \tilde{e}$. In view of $E' = C' - C'$, $\{x_n\}$ is a weak Cauchy sequence in E and $z(f) = \sup_n f(x_n)$ for each $f \in C'$; if we had $z = \tilde{e}$, the sequence $\{x_n\}$ would be weakly convergent to e in E, and hence norm convergent by (4.3). On the other hand, one has $\|x_{n+p} - x_n\| = 1$ for all $n \in N$, $p \in N$, which is contradictory, and it follows that $z < \tilde{e}$.

Our next objective is a characterization of those l.c. vector lattices that can be identified (under evaluation) with order complete sublattices of their bidual E''; this will yield, in particular, a characterization of order complete vector lattices of minimal type (Section 1). A filter $\mathfrak{F}$ in an order complete vector lattice is called **order convergent** if $\mathfrak{F}$ contains an order bounded set Y (hence an order interval), and if

$$\sup_Y (\inf Y) = \inf_Y (\sup Y),$$

where Y runs through all order bounded sets $Y \in \mathfrak{F}$. The common value of the right- and left-hand terms is called the **order limit** of $\mathfrak{F}$. Let us note also that if E is a l.c.v.l., then the bidual E'' of E is a l.c.v.l. under its natural topo-logy (the topology of uniform convergence on the equicontinuous subsets of E', Chapter IV, Section 5); in fact, the polar of every solid 0-neighborhood in E is a solid subset of E' by the corollary of (1.5), and hence the family of all solid equicontinuous subsets of E' is a fundamental family of equicon-tinuous sets. Hence their respective polars (in E'') form a 0-neighborhood base for the natural topology, consisting of solid sets.

7.5

Let $(E, \mathfrak{T})$ be an order complete l.c.v.l., and let E'' be endowed with its natural topology and canonical order (under which it is an order complete l.c.v.l.). The following assertions are equivalent:

(a) *Under evaluation, E is isomorphic with an order complete sublattice of E''.*
(b) *For every majorized, directed ($\leq$) subset S of E, the section filter of S $\mathfrak{T}$-converges to sup S.*
(c) *Every order convergent filter in E $\mathfrak{T}$-converges to its order limit.*

REMARK. The equivalences remain valid when "to sup S" and "to its order limit" are dropped in (b) and (c), respectively; if the cor-responding filters converge for $\mathfrak{T}$, they converge automatically to the limits indicated, by (4.2).

Proof of (7.5). (a) $\Rightarrow$ (b): Let S be a directed ($\leq$) subset of E such that $x_0 = \sup S$; identifying E with its canonical image in E'', we obtain (by definition of the canonical order of E'') $f(x_0) = \sup\{f(x): x \in S\}$ for every continuous, positive linear form on E. It follows that the section filter of S converges weakly to x_0, and hence for $\mathfrak{T}$ by (4.3), since the positive cone C is normal in E.

(b) $\Rightarrow$ (c): Let $\mathfrak{T}$ be an order convergent filter in E with order limit x_0 and let $\mathfrak{G}$ be the base of $\mathfrak{F}$ consisting of all order bounded subsets $Y \in \mathfrak{F}$. Let $a(Y) = \inf Y$ ($Y \in \mathfrak{G}$); the family $\{a(Y): Y \in \mathfrak{G}\}$ is directed ($\leq$) with least upper bound x_0; hence by hypothesis its section filter converges to x_0 for $\mathfrak{T}$. Likewise, if $b(Y) = \sup Y$, the family $\{b(Y): Y \in \mathfrak{G}\}$ is directed ($\geq$) with greatest lower bound x_0, and hence its section filter $\mathfrak{T}$-converges to x_0. Let U be any C-saturated 0-neighborhood in E; there exists a set $Y_0 \in \mathfrak{G}$ such that $a(Y_0) \in x_0 + U$ and $b(Y_0) \in x_0 + U$, and this implies that $Y_0 \subset x_0 + U$. Since (C being normal) the family of all C-saturated 0-neighborhoods is a base at 0, it follows that $\mathfrak{F}$ converges to x_0 for $\mathfrak{T}$.

(c) $\Rightarrow$ (a): Let S be a directed ($\leq$) subset of E such that $x_0 = \sup S$. It is clear that the section filter of S is order convergent with order limit x_0, and hence it $\mathfrak{T}$-converges to x_0 by assumption. It follows that $f(x_0) = \sup\{f(x): x \in S\}$ for every $f \in C'$; hence from the definition of order in E'' it follows that $\tilde{x}_0 = \sup \tilde{S}$, where $x \to \tilde{x}$ is the evaluation map of E into E''. The proof is complete.

COROLLARY 1. *Let E be an order complete vector lattice whose order is regular. The following assertions are equivalent*:

(a) *E is of minimal type.*
(b) *For every majorized, directed* ($\leq$) *subset S of E, the section filter converges to* sup *S for* $\mathfrak{T}_0$.
(c) *Every order convergent filter in E converges for* $\mathfrak{T}_0$.

Moreover, if E is minimal, then $\mathfrak{T}_0$ *is the finest l.c. topology on E for which every order convergent filter converges.*

Proof. Applying (7.5) to $(E, \mathfrak{T}_0)$ we see that $E' = E^+$, and $E'' = (E^+, \beta(E^+, E))'$ is a solid subspace of E^{++} by (7.4). Hence E is minimal (that is, isomorphic with an order complete sublattice of E^{++} under evaluation) if and only if E is isomorphic with an order complete sublattice of E'', which proves the first assertion.

For the second assertion there remains, in view of (c), only to show that every l.c. topology $\mathfrak{T}$ on E for which every order convergent filter converges is coarser than $\mathfrak{T}_0$. Hence let $\mathfrak{T}$ be such a topology, and let $a \in C$ be fixed. Now $\{\varepsilon[-a, a]: \varepsilon > 0\}$ ($\varepsilon \in R$) is a filter base in E, and it is immediate that the corresponding filter is order convergent with order limit 0 (E is regular, hence Archimedean ordered); thus if U is a convex 0-neighborhood for $\mathfrak{T}$, it follows that there exists $\varepsilon > 0$ such that $\varepsilon[-a, a] \subset U$. Therefore, U absorbs arbitrary order intervals in E, which shows that $\mathfrak{T}$ is coarser than $\mathfrak{T}_0$.

COROLLARY 2. *Let E be a l.c.v.l. which is semi-reflexive; then E is order complete. If, in addition, every positive linear form on E is continuous, then E is of minimal type,* $\tau(E, E') = \mathfrak{T}_0$, *and* $(E, \mathfrak{T}_0)$ *is reflexive.*

Proof. The first assertion follows at once from (4.3), Corollary 2. If every positive linear form on E is continuous, then $E' = E^+$, and the equality $\tau(E, E') = \mathfrak{T}_0$ follows from (6.4) in view of the fact that the order of E is regular, (4.1), Corollary 2. Hence E is minimal by Corollary 1, for the section filter of every majorized, directed ($\leq$) subset S of E converges weakly to sup S, so it converges for $\mathfrak{T}_0$ by (4.3). Finally, $(E, \mathfrak{T}_0)$ is reflexive, since it is semi-reflexive and (by (7.3)) barreled.

Examples

4. Each of the Banach lattices $L^p(\mu)$, $1 < p < +\infty$ (Chapter II, Section 2, Example 2; take $K = R$) is order complete and of minimal type; in particular, the norm topology is the finest l.c. topology for which every order convergent filter converges.

5. The Banach lattice $L^1(\mu)$ is order complete and of minimal type. In fact, if S is a directed ($\leq$) subset of the positive cone C and majorized by h, then for any subset $\{f_1, ..., f_n\}$ of S such that $f_1 \leq \cdots \leq f_n$ one obtains

$$\|h - f_1\| = \|h - f_n\| + \|f_n - f_{n-1}\| + \cdots + \|f_2 - f_1\|,$$

since the norm of $L^1(\mu)$ is additive on C; this shows that the section filter of S is a Cauchy filter for the norm topology, and hence convergent. Since the latter topology is $\mathfrak{X}_0$, it follows that $L^1(\mu)$ is of minimal type. Obviously these conclusions apply to any Banach lattice whose norm is additive on the positive cone; these lattices are called abstract (L)-spaces (cf. Kakutani [1] and Section 8 below).

6. Suppose μ to be totally σ-finite. As strong duals of $L^1(\mu)$, the spaces $L^\infty(\mu)$ are order complete Banach lattices by (7.4); in general, these spaces are not of minimal type as the example preceding (7.5) shows and hence (in contrast with $L^1(\mu)$), in general, not bands in their respective order biduals.

7. Each perfect space (Example 2 above) is order complete and, if each order interval is $\sigma(\lambda, \lambda^+)$-compact, of minimal type.

As we have observed earlier, ordered vector spaces possessing an order unit are comparatively rare; it will be shown in Section 8 below that every Banach lattice with an order unit is isomorphic (as an ordered t.v.s.) with $\mathscr{C}_R(X)$ for a suitable compact space X. A weaker notion that can act as a substitute was introduced by Freudenthal [1]; an element $x \ge 0$ of a vector lattice L is called a **weak order unit** if $\inf(x, |y|) = 0$ implies $y = 0$ for each $y \in L$. A corresponding topological notion is the following: If L is an ordered t.v.s., an element $x \ge 0$ is called a **quasi-interior point** of the positive cone C of L if the order interval $[0, x]$ is a total subset of L. The remainder of this section is devoted to some results on weak order units and their relationship with quasi-interior points of C.

7.6

Let E be an ordered l.c.s. over R which is metrizable and separable, and suppose that the positive cone C of E is a complete, total subset of E. Then the set Q of quasi-interior points of C is dense in C.

Proof. Since C is separable, there exists a subset $\{x_n: n \in N\}$ which is dense in C; denote by $\{p_n: n \in N\}$ an increasing sequence of semi-norms that generate the topology of E. Since C is complete, $x_0 = \sum_1^\infty 2^{-n} x_n / p_n(x_n)$ is an element of C. Now the linear hull of $[0, x_0]$ contains each x_n $(n \in N)$, and hence is dense in $C - C$ and, therefore, in E; that is, $x_0 \in Q$. It is obvious that $C_1 = \{0\} \cup Q$ is a subcone of C, and that $\bar{Q} = \bar{C}_1$. Suppose that $\bar{Q} \ne C$. There exists, by (II, 9.2), a linear form $f \in E'$ such that $f(x) \ge 0$ when $x \in \bar{Q}$, and a point $y \in C$ such that $f(y) = -1$. Consequently, there exists $\lambda > 0$ such that $f(x_0 + \lambda y) < 0$, which conflicts with $x_0 + \lambda y \in Q$.

COROLLARY. *Let E be a Fréchet lattice which is separable. Then the set of weak order units is dense in the positive cone of E.*

Proof. It suffices to show that each quasi-interior point of C is a weak order unit. But if x is quasi-interior to C, then $y \perp x$ implies that y is disjoint from the linear hull of $[0, x]$ which is dense in E; hence $y = 0$ since the lattice operations are continuous.

REMARK. The assumptions that E be metrizable and separable are not dispensable in (7.6); if E is, for example, either the l.c. direct sum of infinitely many copies of R_0 or the Hilbert direct sum of uncountably many copies of l^2 (under their respective canonical orderings), then the set of quasi-interior points of C (equivalently, by (7.7), the set of weak order units) is empty.

7.7

Let E be an order complete vector lattice of minimal type. For each $x > 0$, the following assertions are equivalent:

(a) *x is weak order unit.*
(b) *For each positive linear form $f \neq 0$ on E, $f(x) > 0$.*
(c) *For each topology $\mathfrak{T}$ on E such that $(E, \mathfrak{T})$ is a l.c.v.l., x is a quasi-interior point of the positive cone.*

Proof. If B_x denotes the band in E generated by $\{x\}$, then x is a weak order unit if and only if $B_x = E$, by virtue of (1.3). Now if f is a positive linear form on E, then, since E is minimal, $f(x) = 0$ is equivalent with $f(B_x) = \{0\}$; this shows that (a) $\Leftrightarrow$ (b). Moreover, E being minimal, $B_x = E$ is equivalent with the assertion that the linear hull of $[0, x]$ is dense in $(E, \mathfrak{T}_0)$ (for the closure of each solid subspace G in $(E, \mathfrak{T}_0)$ contains the band generated by G); hence (a) $\Rightarrow$ (c), since the topologies mentioned in (c) are necessarily coarser than $\mathfrak{T}_0$ by (7.3). (c) $\Rightarrow$ (a) is clear in view of the continuity of the lattice operations in $(E, \mathfrak{T})$ (cf. proof of (7.6), Corollary).

For example, in the spaces $L^p(\mu)$ $(1 \leq p < +\infty)$ the weak order units ($=$ quasi-interior points of C) are those classes containing a function which is > 0 a.e. (μ). By contrast, a point in $L^\infty(\mu)$ is quasi-interior to C exactly when it is interior to C; the classes containing a function which is > 0 a.e. (μ) are weak order units, but not necessarily quasi-interior to C. Hence the minimality assumption is not dispensable in (7.7).

8. CONTINUOUS FUNCTIONS ON A COMPACT SPACE. THEOREMS OF STONE-WEIERSTRASS AND KAKUTANI

This final section is devoted to several theorems on Banach lattices of type $\mathscr{C}(X)$, where X is a compact space, in particular, the order theoretic and algebraic versions of the Stone-Weierstrass theorem and representation theorems for (AM)-spaces with unit and for (AL)-spaces. For a detailed account of this circle of ideas, which is closely related to the Krein-Milman theorem, we refer to Day [2]; the present section is mainly intended to serve as an illustration for the general theory of ordered vector spaces and lattices developed earlier. Let us point out that with only minor modifications most of the following results are applicable to spaces $\mathscr{C}_0(X)$ (continuous functions on a locally compact space X that vanish at infinity); for $\mathscr{C}_0(X)$ can be viewed as a solid sublattice of codimension 1 in $\mathscr{C}(\dot{X})$, where $\dot{X}$ denotes the one-point compactification of X.

With one exception (see (8.3) below) we consider in this section only vector spaces over the real field $\mathbf{R}$; *mutatis mutandis*, many of the results can be generalized without difficulty to the complex case, since $\mathscr{C}(X)$ over $\mathbf{C}$ is the complexification (Chapter I, Section 7) of $\mathscr{C}(X)$ over $\mathbf{R}$; (8.3) is an example for this type of generalization. To avoid ambiguity we shall denote by $\mathscr{C}_{\mathbf{R}}(X)$ the Banach lattice of real-valued continuous functions on X, and by $\mathscr{C}_{\mathbf{C}}(X)$ the (B)-space of complex-valued continuous functions on X.

Let us recall some elementary facts on the Banach lattice $\mathscr{C}_{\mathbf{R}}(X)$, where $X \neq \varnothing$ is any compact space. $\mathscr{C}_{\mathbf{R}}(X)$ possesses order units; $f \in \mathscr{C}_{\mathbf{R}}(X)$ is an order unit if and only if $\inf\{f(t): t \in X\} > 0$. Thus the order units of $\mathscr{C}_{\mathbf{R}}(X)$ are exactly the functions f that are interior to the positive cone C. Distinguished among these is the constantly-one function e; in fact, the norm $f \to \|f\| = \sup\{|f(t)|: t \in X\}$ is the gauge function p_e of $[-e, e]$ and, of course, the topology of $\mathscr{C}_{\mathbf{R}}(X)$ is the order topology $\mathfrak{T}_O$ (Section 6). We begin with the following classical result, the order theoretic form of the Stone-Weierstrass theorem.

8.1

Theorem. *If F is a vector sublattice of $\mathscr{C}_{\mathbf{R}}(X)$ that contains e and separates points in X, then F is dense in $\mathscr{C}_{\mathbf{R}}(X)$.*

REMARK. The subsequent proof will show that a subset $F \subset \mathscr{C}_{\mathbf{R}}(X)$ is dense if it satisfies the following condition: F is a (not necessarily linear) sublattice of the lattice $\mathscr{C}_{\mathbf{R}}(X)$, and for every $\varepsilon > 0$ and quadruple $(s, t; \alpha, \beta) \in X^2 \times \mathbf{R}^2$ such that $\alpha = \beta$ whenever $s = t$, there exists $f \in F$ satisfying $|f(s) - \alpha| < \varepsilon$ and $|f(t) - \beta| < \varepsilon$.

Proof of (8.1). Let s, t be given points of X and α, β given real numbers such that $\alpha = \beta$ if $s = t$; the hypothesis implies the existence of $f \in F$ such that $f(s) = \alpha, f(t) = \beta$. This is clear if $s = t$, since $e \in F$; if $s \neq t$, there exists $g \in F$ such that $g(s) \neq g(t)$, and a suitable linear combination of e and g will satisfy the requirement.

Now let $h \in \mathscr{C}_{\mathbf{R}}(X)$ and $\varepsilon > 0$ be preassigned and let s be any fixed element of X. Then for each $t \in X$, there exists an $f_t \in F$ such that $f_t(s) = h(s)$ and $f_t(t) = h(t)$. The set $U_t = \{r \in X : f_t(r) > h(r) - \varepsilon\}$ is open and contains t; hence $X = \bigcup_{t \in X} U_t$, and the compactness of X implies the existence of a finite set $\{t_1, \ldots, t_n\}$ such that $X = \bigcup_{v=1}^{n} U_{t_v}$. Using the lattice property of F, form the function $g_s = \sup\{f_{t_1}, \ldots, f_{t_n}\}$; it is clear that $g_s(t) > h(t) - \varepsilon$ for all $t \in X$, since each t is contained in at least one U_{t_v}. Moreover, $g_s(s) = h(s)$.

Now consider this procedure applied to each $s \in X$; we obtain a family

$\{g_s: s \in X\}$ in F such that $g_s(s) = h(s)$ for all $s \in X$, and $g_s(t) > h(t) - \varepsilon$ for all $t \in X$ and $s \in X$. The set $V_s = \{r \in X: g_s(r) < h(r) + \varepsilon\}$ is open and contains s; hence $X = \bigcup_{s \in X} V_s$, and the compactness of X implies the existence of a finite set $\{s_1, \ldots, s_m\}$ such that $X = \bigcup_{\mu=1}^{m} V_{s_\mu}$. Let $g = \inf\{g_{s_1}, \ldots, g_{s_m}\}$; then $g \in F$ and $h(r) - \varepsilon < g(r) < h(r) + \varepsilon$ for all $r \in X$; hence $\|h - g\| < \varepsilon$, and the proof is complete.

The algebraic form of the Stone-Weierstrass theorem replaces the hypothesis that F be a sublattice of $\mathscr{C}_R(X)$ by assuming that F be a subalgebra (that is, a subspace of $\mathscr{C}_R(X)$ invariant under multiplication). Our proof follows de Branges [1], but does not involve Borel measures. The proof is an interesting application of the Krein-Milman theorem and provides an opportunity to apply the concept of Radon measure that has been utilized earlier (Chapter IV, Sections 9 and 10).

The space $\mathscr{M}_R(X)$ of (real) Radon measures on X is, by definition, the dual of $\mathscr{C}_R(X)$ (Chapter II, Section 2, Example 3); since E is a Banach lattice, $\mathscr{M}_R(X)$ is a Banach lattice under its dual norm and canonical order by (7.4), Corollary 2. Thus $\|\mu\| = \||\mu|\|$ for each μ; if $\mu \geq 0$, then $\|\mu\| = \sup\{\mu(f): f \in [-e, e]\} = \mu(e)$, and this implies that $\|\mu\| = \mu^+(e) + \mu^-(e)$ for all $\mu \in \mathscr{M}_R(X)$. If $g \in \mathscr{C}_R(X)$ is fixed, then $f \to gf$ (pointwise multiplication) is a continuous linear map u of $\mathscr{C}_R(X)$ into itself; the image of $\mu \in \mathscr{M}_R(X)$ under the adjoint u' is a Radon measure denoted by $g \cdot \mu$. Obviously $|g \cdot \mu| \leq \|g\||\mu|$ and hence u' leaves each band in $\mathscr{M}_R(X)$ invariant; in particular, if $g \geq 0$, then $g \cdot \mu = g \cdot \mu^+ - g \cdot \mu^-$, where $\inf(g \cdot \mu^+, g \cdot \mu^-) = 0$. It follows from (1.1) that $(g \cdot \mu)^+ = g \cdot \mu^+$, $(g \cdot \mu)^- = g \cdot \mu^-$ in this case, and that $|g \cdot \mu| = g \cdot |\mu|$.

The **support** of $f \in \mathscr{C}_R(X)$ is the closure S_f of $\{t \in X: f(t) \neq 0\}$ in X; we define the **support** S_μ of $\mu \in \mathscr{M}_R(X)$ to be the complement (in X) of the largest open set U such that $S_f \subset U$ implies $\mu(f) = 0$ (equivalently, such that $S_f \subset U$ implies $|\mu|(f) = 0$). An application of Urysohn's theorem (cf. Prerequisites) shows that if $f \geq 0$ and $\mu \geq 0$, then $\mu(f) = 0$ if and only if $f(t) = 0$ whenever $t \in S_\mu$. Notice a particular consequence of this: if μ is such that $S_\mu = \{t_0\}$, then μ is of the form $\mu(f) = \mu(e)f(t_0)$ (hence, up to a factor $\mu(e) \neq 0$, evaluation at t_0). For $S_\mu = \{t_0\}$ implies that $|\mu(f - f(t_0)e)| \leq |\mu|(|f - f(t_0)e|) = 0$, which is the assertion. Finally, $\mu = 0$ if and only if $S_\mu = \varnothing$. The following lemma is now the key to the proof of (8.2).

LEMMA. *Let F be a subspace of $\mathscr{C}_R(X)$, and suppose that the Radon measure μ is an extreme point of $F^\circ \cap [-e, e]^\circ \subset \mathscr{M}_R(X)$. If $g \in \mathscr{C}_R(X)$ is such that $g \cdot \mu \in F^\circ$, then g is constant on S_μ.*

Proof. If $\mu = 0$, there is nothing to prove. Otherwise, it can be arranged (by adding a suitable scalar multiple of e and subsequent normalization) that $g \geq 0$ and $|\mu|(g) = 1$. Suppose, for the moment, that $g \leq e$; since μ is an extreme point of $F^\circ \cap [-e, e]^\circ$, we have $\|\mu\| = 1$ and it follows that $|\mu|(e - g) = \|\mu\| - |\mu|(g) = 0$, which implies that $1 = e(t) = g(t)$ for all $t \in S_\mu$, in view of the remarks preceding the lemma. We complete the proof by showing that $\|g\| > 1$ is impossible. In fact, assume that $\|g\| > 1$, let $\beta = \|g\|^{-1}$, and define the Radon measures μ_1, μ_2 by $\mu_1 = g_1 \cdot \mu$ and $\mu_2 = g \cdot \mu$, where $g_1 = (e - \beta g)/(1 - \beta)$. We observe that $\mu_1 \in F^\circ$, $\mu_2 \in F^\circ$ and that $|\mu_2| = g \cdot |\mu|$; hence $\|\mu_2\| = g \cdot |\mu|(e) = 1$. Moreover, $\mu_1^+ = g_1 \cdot \mu^+$ and $\mu_1^- = g_1 \cdot \mu^-$, since $g_1 \geqq 0$, and in view of $\|\mu_1\| = \mu_1^+(e) + \mu_1^-(e)$, it follows from a short computation that $\|\mu_1\| = 1$. On the other hand, it is easy to see that $\mu = (1 - \beta)\mu_1 + \beta\mu_2$, which conflicts with the hypothesis that μ be an extreme point of $F^\circ \cap [-e, e]^\circ$.

The following is the algebraic form of the Stone-Weierstrass theorem.

8.2

Theorem. *If F is a subalgebra of $\mathscr{C}_R(X)$ that contains e and separates points in X, then F is dense in $\mathscr{C}_R(X)$.*

Proof. The set $F^\circ \cap [-e, e]^\circ$ is a convex, circled, weakly compact subset of $\mathscr{M}_R(X)$; hence by the Krein-Milman theorem (II, 10.4) there exists an extreme point μ of $F^\circ \cap [-e, e]^\circ$. Since F is a subalgebra of $\mathscr{C}_R(X)$, each $f \in F$ satisfies the hypothesis of the lemma with respect to μ; hence each $f \in F$ is constant on the support S_μ of μ. This is clearly impossible if S_μ contains at least two points, since F separates points in X; on the other hand, if $S_\mu = \{t_0\}$, then $\mu(f) = \mu(e)f(t_0)$, and it follows that each $f \in F$ vanishes at t_0, which is impossible since $e \in F$. Hence S_μ is empty which implies $\mu = 0$ and, therefore, $F^\circ = \{0\}$; consequently, F is dense in $\mathscr{C}_R(X)$ by the bipolar theorem (IV, 1.5).

The preceding theorem is essentially a theorem on real algebras $\mathscr{C}(X)$; for instance, if X is the unit disk in the complex plane and F is the algebra of all complex polynomials (restricted to X), then F separates points in X and $e \in F$, but F is not dense in $\mathscr{C}_C(X)$ (for each $f \in \bar{F}$ is holomorphic in the interior of X). One can, nevertheless, derive results for the complex case from (8.1) and (8.2) by making appeal to the fact that $\mathscr{C}_C(X)$ is the complexification of $\mathscr{C}_R(X)$; we say that a subset F of the complex algebra $\mathscr{C}_C(X)$ is **conjugation-invariant** if $f \in F$ implies $f^* \in F$ (where $f^*(t) = f(t)^*$, $t \in X$). We consider $\mathscr{C}_C(X)$ as ordered by the cone of real functions $\geqq 0$ (Section 2).

8.3

COMPLEX STONE-WEIERSTRASS THEOREM. *Let F be a vector subspace of the complex Banach space $\mathscr{C}_C(X)$ such that $e \in F$ and F separates points in X and is conjugation-invariant. Then either of the following assumptions implies that F is dense in $\mathscr{C}_C(X)$:*

(i) *F is lattice ordered* (Section 2)
(ii) *F is a subalgebra of $\mathscr{C}_C(X)$.*

Proof. If F_1 denotes the subset of F whose elements are the real-valued functions contained in F, then $F = F_1 + iF_1$ by the conjugation-invariance of the subspace F; clearly, $e \in F_1$ and F_1 separates points in X, since F does. Thus if F is lattice ordered, then F_1 is a vector lattice (Section 2), and (8.1) shows that F_1 is dense in $\mathscr{C}_R(X)$; by (8.2) the same conclusion holds if F is a subalgebra of $\mathscr{C}_C(X)$, for then F_1 is a subalgebra of $\mathscr{C}_R(X)$. This completes the proof.

It is customary to call a Banach lattice E an (AL)-**space** (abstract L-space) if the norm of E is additive on the positive cone C. The reason for this terminology is that every $L^1(\mu)$ (over R) possesses this property and that, conversely, every (AL)-space is isomorphic (as a Banach lattice) with a suitable space $L^1(\mu)$ (Kakutani [1]; cf. Exercise 22). A Banach lattice E is called an (AM)-**space** (abstract (m)-space) if the norm of E satisfies $\|\sup(x, y)\| = \sup$ $(\|x\|, \|y\|)$ for all x, y in the positive cone C; E is called an (AM)-**space with unit** u if, in addition, there exists $u \in C$ such that $[-u, u]$ is the unit ball of E. (Clearly, such u is unique and an order unit of E.) It is immediate that every Banach lattice $\mathscr{C}_R(X)$ is an (AM)-space with unit (the unit being the constantly-one function e); we will show that this property characterizes the spaces $\mathscr{C}(X)$ over R among Banach lattices. More generally, every (AM)-space is isomorphic with a closed vector sublattice of a suitable $\mathscr{C}_R(X)$ (Kakutani [2]). Let us record first the following elementary facts on (AL)- and (AM)-spaces; by the strong dual of a Banach lattice E, we understand the dual E' $(= E^+)$ under its natural norm and canonical order.

8.4

The strong dual of an (AM)-*space with unit is an* (AL)-*space, and the strong dual of an* (AL)-*space is an* (AM)-*space with unit. Moreover, if E is an Archimedean ordered vector lattice, u an order unit of E, and p_u the gauge function of $[-u, u]$, then the completion of (E, p_u) is an* (AM)-*space with unit u.*

Proof. Let E be an (AM)-space with unit u; the strong dual E' is a Banach lattice by (7.4), Corollary 2. If $x' \in C'$ then $\|x'\| = \sup\{|\langle x, x'\rangle|: x \in [-u, u]\}$ $= \langle u, x'\rangle$; hence the norm of E' is additive on the dual cone C'.

If F is an (AL)-space, the norm of F is an additive, positive homogeneous real function on C, and hence defines a (unique) linear form f_0 on F such that $f_0(x) = \|x\|$ for all $x \in C$; evidently we have $0 \leq f_0 \in F'$. It follows that $g \in F'$ satisfies $\|g\| \leq 1$ if and only if $g \in [-f_0, f_0]$, and hence the norm of the strong dual F' is the gauge function of $[-f_0, f_0]$. Now if $g \geq 0$, $h \geq 0$ are elements of F' such that $\|g\| = \lambda_1$, $\|h\| = \lambda_2$, then $g \leq \lambda_1 f_0$ and $h \leq \lambda_2 f_0$, since the order of F' is Archimedean. Consequently, $\|\sup(g, h)\| \leq \sup(\lambda_1, \lambda_2)$, and here equality must hold or else both the relations $\|g\| = \lambda_1$, $\|h\| = \lambda_2$ could not be valid. Therefore, under its canonical order, F' is an (AM)-space with unit f_0.

To prove the third assertion, we observe that if E is an Archimedean ordered vector lattice and u is an order unit of E, then p_u is a norm on E, and even a lattice norm, since $[-u, u]$ is clearly solid. The completion $(\tilde{E}, p_u)$

of (E, p_u) is a Banach lattice (with respect to the continuous extension of the lattice operations) whose unit ball is the set $\{x \in \tilde{E}: -f(u) \leq f(x) \leq f(u), f \in C'\}$, and hence the order interval $[-u, u]$ in $\tilde{E}$. As in the preceding paragraph, it follows that $(\tilde{E}, p_u)$ is an (AM)-space with unit u. This ends the proof.

Let $E \neq \{0\}$ be an (AM)-space with unit u; the intersection of the hyperplane $H = \{x': \langle u, x' \rangle = 1\}$ with the dual cone C' is a convex, $\sigma(E', E)$-closed subset H_0 of the dual unit ball $[-u, u]^\circ$. It follows that H_0, which is called the **positive face** of $[-u, u]^\circ$, is $\sigma(E', E)$-compact; hence C' is a cone with weakly compact base, and $t \in H_0$ is an extreme point of H_0 if and only if $\{\lambda t: \lambda \geq 0\}$ is an extreme ray of C' (Chapter II, Exercise 30).

Now we can prove the representation theorem of Kakutani [2] for (AM)-spaces with unit.

8.5

Theorem. *Let $E \neq \{0\}$ be an (AM)-space with unit and let X be the set of extreme points of the positive face of the dual unit ball. Then X is non-empty and $\sigma(E', E)$-compact, and the evaluation map $x \to f$ (where $f(t) = \langle x, t \rangle$, $t \in X$) is an isomorphism of the (AM)-space E onto $\mathscr{C}_{\mathbf{R}}(X)$.*

Proof. Let u be the unit of E. Since the positive face H_0 of $[-u, u]^\circ$ is convex and $\sigma(E', E)$-compact, the Krein-Milman theorem (II, 10.4) implies that the set X of extreme points of H_0 is non-empty. Since H_0 is a base of C', it follows from (1.7) that $t \in X$ if and only if t is a lattice homomorphism of E onto $\mathbf{R}$ such that $t(u) = 1$. It is clear from this that X is closed, hence compact for $\sigma(E', E)$. The mapping $x \to f$ is clearly a linear map of E into $\mathscr{C}_{\mathbf{R}}(X)$ that preserves the lattice operations, since each $t \in X$ is a lattice homomorphism; to show that $x \to f$ is a norm isomorphism, it suffices (since E and $\mathscr{C}_{\mathbf{R}}(X)$ are Banach lattices) that $\|f\| = \|x\|$ when $x \geq 0$. For $x \geq 0$ we have $\|x\| = \sup\{\langle x, x' \rangle: \|x'\| \leq 1\} = \sup\{\langle x, x' \rangle: x' \in H_0\}$; since H_0 is the $\sigma(E', E)$-closed convex hull of X and each $x \in E$ is linear and $\sigma(E', E)$-continuous, it follows that $(x \geq 0) \sup\{\langle x, x' \rangle: x' \in H_0\} = \sup\{\langle x, t \rangle: t \in X\} = \|f\|$. Thus $x \to f$ is an isomorphism of E onto a vector sublattice F of $\mathscr{C}_{\mathbf{R}}(X)$ that is complete and contains e (the image of u); since E separates points in E' and a fortiori in X, it follows from (8.1) that $F = \mathscr{C}_{\mathbf{R}}(X)$, which completes the proof.

We conclude this section with two applications of the preceding result; the first of these gives us some more information on the structure of (AL)-spaces, the second on more general locally convex vector lattices.

From (8.4) we know that the strong dual $E'(= E^+)$ of an (AL)-space E is an (AM)-space with unit; hence by (8.5), E' can be identified with a space $\mathscr{C}_{\mathbf{R}}(X)$, where X is the set of extreme points of the positive face of the unit ball in E''. By (7.4), the Banach lattice E' is order complete, which has the interesting consequence that X is extremally disconnected (that is, the closure of every open set in X is open). In fact, let $G \subset X$ be open and denote by S

the family of all $f \in \mathscr{C}_{\mathbf{R}}(X)$ such that $f \in [0, e]$ and the support S_f is contained in G. S is directed ($\leqq$) and majorized by e; hence $f_0 = \sup S$ exists. Since G is open, it follows from Urysohn's theorem that $f_0(s) = 1$ whenever $s \in G$, and that $f_0(t) = 0$ whenever $t \notin \bar{G}$. Thus f_0 is necessarily the characteristic function of $\bar{G}$ since f_0 is continuous, and this implies that $\bar{G}$ is open.

Therefore, if E is an (AL)-space, then E' can be identified with a space $\mathscr{C}_{\mathbf{R}}(X)$, where X is compact and extremally disconnected, and it follows that E itself can be identified with a closed subspace of the Banach lattice $\mathscr{M}_{\mathbf{R}}(X)$ which is the strong bidual of E. For a characterization of E within $\mathscr{M}_{\mathbf{R}}(X)$, let us consider the subset $B \subset \mathscr{M}_{\mathbf{R}}(X)$ such that $\mu \in B$ if and only if for each directed ($\leqq$), majorized subset $S \subset \mathscr{C}_{\mathbf{R}}(X)$ it is true that $\lim \mu(f) = \mu(\sup S)$, the limit being taken along the section filter of S. It is not difficult to verify that B is a vector sublattice of $\mathscr{M}_{\mathbf{R}}(X)$; in fact, if S is directed ($\leqq$) and $f_0 = \sup S$, and if $f_0 \geqq 0$ (which is no restriction of generality) then there exists, for given $\mu \in B$ and $\varepsilon > 0$, a decomposition $f_0 = g_0 + h_0 (g_0 \geqq 0, h_0 \geqq 0)$ such that $\mu^+(h_0) < \varepsilon$ and $\mu^-(g_0) < \varepsilon$ ((1.5), formula (7)). Using that $\mu \in B$, we obtain after a short computation that $\mu^+(f_0) < \sup\{\mu^+(f): f \in S\} + 3\varepsilon$, which proves that $\mu^+ \in B$. Thus B is a sublattice of $\mathscr{M}_{\mathbf{R}}(X)$ which is clearly solid; it is another straightforward matter to prove that B is a band in $\mathscr{M}_{\mathbf{R}}(X)$. The only assertion in the following representation theorem that remains to be proved is the assertion that $B = E$.

8.6

Theorem. *Let E be an (AL)-space. The Banach lattice E' ($= E^+$) can be identified with $\mathscr{C}_{\mathbf{R}}(X)$, where X is a compact, extremally disconnected space. Moreover, under evaluation, E is isomorphic with the band of all (real) Radon measures μ on X such that*

$$\lim_{f \in S} \mu(f) = \mu(\sup S)$$

for every majorized, directed ($\leqq$) subset S of $\mathscr{C}_{\mathbf{R}}(X)$.

Proof. It is easy to see that (identifying E' with $\mathscr{C}_{\mathbf{R}}(X)$ and E with its canonical image in $E'' = \mathscr{M}_{\mathbf{R}}(X)$) we have $E \subset B$ (see the preceding paragraph for notation). For if S is majorized and directed ($\leqq$), every section of S is $\sigma(E', E)$-bounded and hence the section filter $\sigma(E', E)$-converges to $\sup S$; the assertion follows since $\mu \in E$ is $\sigma(E', E)$-continuous.

To prove the reverse inclusion, let $0 \leqq v \in B$ and let $\mu_0 = \sup [0, v] \cap E$. Then the section filter of $[0, v] \cap E$ is a Cauchy filter for the norm topology, since E is an (AL)-space (Section 7, Example 5), and hence $\mu_0 \in E$, since E is norm complete. Now $\mu_1 = v - \mu_0$ is an element of B lattice disjoint from E; it will be shown that this implies $\mu_1 = 0$, and hence $B = E$ by (1.3).

Denote by T_1 the support of μ_1; if $T = X \sim T_1$, then T is open and $\mu_1(f) = 0$ for each f whose support T_f is contained in T. The family of all $f \in [0, e]$ such that $T_f \subset T$ is directed, and its least upper bound f_0 is necessarily the characteristic function of the closure $\bar{T}$. Since $\mu_1 \in B$, it follows that $\mu_1(f_0) = 0$; hence $\bar{T} \cap T_1 = \varnothing$, which shows that T_1 is open and closed. If $T_1 = \varnothing$ the proof is complete; hence assume that T_1 is non-empty. Since T_1 is open, there exist elements $\mu \in E$ whose support intersects T_1 (otherwise E would not distinguish points in $E' = \mathscr{C}_{\mathbf{R}}(X)$). There exists, consequently, a positive $\mu \in E$ such that $\|\mu\| = 1$ and whose support is contained in T_1 (it suffices to take a positive $\lambda \in E$ for which $\lambda(g_0) > 0$, where $g_0 = e - f_0$, and to consider $g_0 . \lambda$). The proof will now be completed by showing that this last statement is false.

Let $\varepsilon_n = 2^{-n} \ (n \in N)$. By formula (7) of (1.5) there exist (since $\inf(\mu, \mu_1) = 0$) decompositions $g_0 = f_n + f_n'$, where $f_n \geqq 0$, $f_n' \geqq 0$ and such that $\mu_1(f_n) < \varepsilon_n^2$, $\mu(f_n') < \varepsilon_n^2$, so that $\mu(f_n) > 1 - \varepsilon_n^2 \ (n \in N)$. Let $G_n = \{t : f_n(t) > \varepsilon_n\}$ for all n, then G_n is open and $\bar{G}_n$ is closed and open. If we write $\mu(A)$ in place of $\mu(\chi_A)$ whenever $A \subset X$ is a subset whose characteristic function χ_A is continuous, we obtain $\mu_1(\bar{G}_n) < \varepsilon_n$; in fact, $\mu_1(\bar{G}_n) \geqq \varepsilon_n$ would imply that $\mu_1(f_n) \geqq \varepsilon_n \mu_1(\bar{G}_n) \geqq \varepsilon_n^2$, which is contradictory. Now let $H_k = \bigcup \{G_n : n \geqq k + 1\}$; then $\bar{H}_k$ is closed and open, and it follows from $\mu_1 \in B$ that $\mu_1(\bar{H}_k) < \varepsilon_k$, since the characteristic function of $\bar{H}_k$ is the least upper bound of the characteristic functions of the sets $\bar{G}_n (n \geqq k + 1)$. Now define g_n by $g_n = \sup\{f_v : v \geqq n\}$ $(n \in N)$; then $\{g_n : n \in N\}$ is a monotone $(\geqq)$ sequence; let $h = \inf\{g_n : n \in N\}$. In the complement of $\bar{H}_k$ one has $f_v(t) \leqq \varepsilon_v$ whenever $v \geqq k + 1$, and hence $g_n(t) \leqq \varepsilon_n$ whenever $n \geqq k + 1$; in view of $\mu_1(\bar{H}_k) < \varepsilon_k$, it is clear that $\mu_1(h) \leqq \varepsilon_k$. This implies $\mu_1(h) = 0$ and thus $h = 0$, for the support of h is contained in the support T_1 of μ_1. On the other hand, since $\mu \in E \subset B$, we have $\lim_n \mu(g_n) = \mu(h) = 0$, which conflicts with $\mu(f_n) > 1 - \varepsilon_n^2$, since $0 \leqq f_n \leqq g_n$ for all n. This completes the proof of (8.6).

COROLLARY 1. *In an* (AL)-*space E each order interval is weakly compact.*

Proof. Since E is a band in $E''(= E^{++})$, E is a solid subspace of E''; thus if $x, y \in E$, we have $[x, y] = (x + C) \cap (y - C) = (x + C'') \cap (y - C'')$ where C, C'' denote the positive cones of E, E'' respectively. Since C'' is $\sigma(E'', E')$-closed, it follows that $[x, y]$ is $\sigma(E'', E')$-closed and hence $\sigma(E'', E')$-compact.

COROLLARY 2. *Every* (AL)-*space E is an order complete vector lattice of minimal type; by contrast, its order dual E^+ is not of minimal type, unless E is of finite dimension.*

Proof. Since E can be identified with a band in $E'' = E^{++}$, it is clearly of minimal type (Section 7). If, on the other hand, E^+ (which can be identified with $\mathscr{C}_{\mathbf{R}}(X)$) is of minimal type, then by (7.5), Corollary 1, the section filter of each directed $(\leqq)$, majorized set S converges to sup S pointwise (even uniformly) on X, which implies that each open subset of X is closed, and hence that the topology of X is discrete. Since X is compact, X is finite, and hence E^+ and E are finite dimensional.

Our second application of (8.5) is the following result.

8.7

Let $(E, \mathfrak{T})$ be a l.c.v.l. which is bornological and sequentially complete. There exists a family of compact spaces $X_\alpha (\alpha \in A)$ and a family of vector lattice isomorphisms f_α of $\mathscr{C}_R(X_\alpha)$ into E ($\alpha \in A$) such that $\mathfrak{T}$ is the finest l.c. topology on E for which each f_α is continuous.

Proof. In view of (5.5) and (6.4), the assumption that $(E, \mathfrak{T})$ be bornological implies that $\mathfrak{T}$ is the order topology $\mathfrak{T}_0$. Hence by (6.3), $(E, \mathfrak{T})$ is the inductive limit of the subspaces (E_α, p_α) ($\alpha \in A$) where $E_\alpha = \bigcup_{n=1}^{\infty} n[-a_\alpha, a_\alpha]$, p_α is the gauge of $[-a_\alpha, a_\alpha]$ on E_α, and $\{a_\alpha : \alpha \in A\}$ is a directed subset of the positive cone of E such that $\bigcup_\alpha E_\alpha = E$. By (6.2) each (p_α, E_α) is a Banach lattice, and by (8.4) even an (AM)-space with unit a_α. Hence by (8.5), (E_α, p_α) can be identified with $\mathscr{C}_R(X_\alpha)$ for a suitable compact space X_α, and the assertion follows from the definition of inductive topologies (Chapter II, Section 6).

EXERCISES

1. A reflexive, transitive binary relation " $\prec$ " on a set S is called a **pre-order** on S. A pre-order on a vector space L over R is said to be compatible (with the vector structure of L) if $x \prec y$ implies $x + z \prec y + z$ and $\lambda x \prec \lambda y$ for all $z \in L$ and all scalars $\lambda > 0$.
   (a) If (X, Σ, μ) is a measure space (Chapter II, Section 2, Example 2), the relation " $f(t) \leq g(t)$ almost everywhere (μ) " defines a compatible pre-order on the vector space (over R) of all real-valued Σ-measurable functions on X.
   (b) If " $\prec$ " is a compatible pre-order, the relation " $x \prec y$ and $y \prec x$ " is an equivalence relation on L, the subset N of elements equivalent to 0 is a subspace of L, and L/N is an ordered vector space under the relation " $\hat{x} \leq \hat{y}$ if there exist elements $x \in \hat{x}$, $y \in \hat{y}$ satisfying $x \prec y$ ".
   (c) The family of all compatible pre-orders of a vector space L over R is in one–to–one correspondence with the family of all convex cones in L that contain their vertex 0.
2. The family of all total vector orderings (total orderings satisfying $(LO)_1$ and $(LO)_2$, Section 1) of a vector space L is in one–to–one correspondence with the family of all proper cones that are maximal (under set inclusion). Deduce from this that for each vector ordering R of L, there exists a total vector ordering of L that is coarser than R. (Use Zorn's lemma.) Show that a total vector ordering cannot be Archimedean if the real dimension of L is > 1.
3. Let L be an ordered vector space with positive cone C. Let N be a subspace of L, and denote by $\hat{C}$ the canonical image of C in L/N.
   (a) If N is C-saturated, then $\hat{C}$ defines the canonical order of L/N.
   (b) If L is a t.v.s. and if for each 0-neighborhood V in L there exists a 0-neighborhood U such that $[(U + N) \cap C] \subset V + N$, then $\hat{C}$ is normal for the quotient topology. (Compare the proof of (3.1).)

(c) If L is a topological vector lattice and N is a closed solid sublattice, then L/N is a topological vector lattice with respect to quotient topology and canonical order (cf. Exercise 12 below). (Use (b) above, and (7.1).)

4. Consider the order of the complex space $L = \boldsymbol{C}^N$ defined by the cone C, where $x = (x_n) \in C$ if and only if either $x = 0$, or Re $x_n \geq 0$ $(n \in N)$ and Im $x_1 > 0$. Show that the dual cone $C^* \subset L^*$ separates points in L, but there exists no Hausdorff l.c. topology on L for which C is normal.

5. Let $\{E_\alpha : \alpha \in A\}$ be a family of l.c.s., let $\mathfrak{S}_\alpha$ be a saturated family of bounded subsets of E_α, and let C_α be an $\mathfrak{S}_\alpha$-cone in $E_\alpha(\alpha \in A)$. Show that $\prod_\alpha C_\alpha$ is an $\mathfrak{S}_1$-cone in $\prod_\alpha E_\alpha$ and that $\oplus_\alpha C_\alpha$ is an $\mathfrak{S}_2$-cone in $\oplus_\alpha E_\alpha$, where $\mathfrak{S}_1$ and $\mathfrak{S}_2$ denote the families $\prod_\alpha \mathfrak{S}_\alpha$ and $\oplus_\alpha \mathfrak{S}_\alpha$ respectively (Chapter IV, Section 4). Derive an analogous result for families of strict $\mathfrak{S}_\alpha$-cones, and discuss the permanence properties of $\mathfrak{S}$-cones under the transition to subspaces and quotient spaces. (Use (IV, 4.1), (3.3), and Exercise 3 above.)

6. Let $\mathcal{D}$ be the space of infinitely differentiable real functions on $\boldsymbol{R}^k$ with compact support, and let C be the cone of non-negative functions in $\mathcal{D}$. (Section 3, Example 2.)

(a) Show that C is a strict $\mathfrak{B}$-cone which is not normal.

(b) Each positive linear form on $\mathcal{D}$ is continuous, and has a unique extension which is a positive linear form, to the space of continuous functions on $\boldsymbol{R}^k$ with compact support. Deduce from this that each positive distribution defines a unique positive Radon measure on $\boldsymbol{R}^k$.

(c) If S is a directed $(\leq)$ set of distributions which is majorized, then $f_0 = \sup S$ exists and $\lim f = f_0$ uniformly on every bounded subset of $\mathcal{D}$.

7. Let E, F be ordered l.c.s. with respective positive cones C, D and suppose that F is a quasi-complete; moreover, assume that C is normal and that D is a strict $\mathfrak{B}$-cone. Every nuclear map $u \in \mathcal{L}(E, F)$ is of the form $u = u_1 - u_2$, where u_1, u_2 are positive nuclear maps. Apply this to the case where E is nuclear and F is a Banach space.

8. Denote by E a separable Banach space and suppose that $\{x_n : n \in N\}$ is a maximal topologically free subset of E. If C is the set of all linear combinations of elements $x_n (n \in N)$ with coefficients ≥ 0, the following assertions are equivalent;

(i) C is a normal $\mathfrak{B}$-cone in E.

(ii) $\{x_n : n \in N\}$ is an unconditional basis of E (Chapter III, Section 9). (Use (3.5); see also Schaefer [2].)

9. (Dini's theorem). Let X be a locally compact space, and denote by $R(X)$ the vector space of all real-valued continuous functions on X under the topology of compact convergence. If S is a directed $(\leq)$ subset of $R(X)$ such that the numerical least upper bound f_0 of S is finite and continuous on X, then the section filter of S converges to f_0 in $R(X)$. Deduce from this another proof of (4.3) by utilizing (4.4).

10. Let L be an ordered vector space over $\boldsymbol{R}$ with positive cone C.

(a) L possesses order units if and only if there exists a l.c. topology on L for which C has non-empty interior; if so, each interior point of C is an order unit of L.

(b) If $(L, \mathfrak{T})$ is a t.v.s. such that C has interior points, then each quasi-interior point of C is interior to C.

(c) If $(L, \mathfrak{T})$ is a non-normable Hausdorff t.v.s. in which C is normal, then C possesses no interior points.

11. Let L be an ordered vector space over R with positive cone C. If M is a subspace of L such that $M + C = M - C$, then every linear form f_0 on M, positive for the canonical order of M, can be extended to a positive linear form on L. (Day [2], §6 Theorem 1.)

12. Let L_1, L_2 be vector lattices.

(a) If N is a solid subspace of L_1, the canonical image of the positive cone of L_1 in L_1/N defines the canonical order of L_1/N under which L_1/N is a vector lattice, and the canonical map $L_1 \rightarrow L_1/N$ is a lattice homomorphism. If L is order complete and N is a band in L_1, then L_1/N is isomorphic with $N^\perp$.

(b) If u is a linear map of L_1 onto L_2, then u is a lattice homomorphism if and only if $u(C_1) = C_2$ and $N = u^{-1}(0)$ is solid in L_1, where C_1, C_2 denote the respective positive cones. In these circumstances, the biunivocal map u_0 associated with u is a lattice isomorphism of L_1/N onto L_2.

(c) If u is a vector lattice homomorphism of L_1 onto L_2, then u is a topological homomorphism for the respective order topologies of L_1 and L_2. In particular, if N is a solid subspace of L_1, then the order topology of L_1/N is the quotient of the order topology of L_1.

(d) If L_1 is order complete, if N is a band in L_1, and if $\mathfrak{T}_o$ is the order topology of L_1, then the topology induced by $\mathfrak{T}_o$ is the order topology of N.

(e) Give an example of an order complete vector lattice L and a solid subspace M of L such that $\mathfrak{T}_o$ (of L) does not induce the order topology of M (cf. Remark preceding (6.4)).

13. Let (X, Σ, μ) be a measure space (Chapter II, Section 2, Example 2) such that Σ contains all singletons but not all subsets of X, and μ is bounded. Under their canonical order, the real spaces $\mathscr{L}^p(\mu)$ are vector lattices that are countably order complete (each countable majorized subset has a least upper bound), but not order complete $(1 \leqq p \leqq + \infty)$.

14. Denote by μ Lebesgue measure on the real interval $[0, 1]$ and let L be the real space $L^p(\mu)$, where p is fixed, $0 < p < 1$.

(a) Under its canonical order, L is an order complete topological vector lattice such that $L^b = L^+ = \{0\}$. (Use Chapter I, Exercise 6 and Chapter V, (5.5).)

(b) Infer from (a) that the order of L is Archimedean but not regular, that the positive cone is dense in L for every l.c. topology on L, and that $\mathfrak{T}_o$ is the coarsest topology on L.

(c) There exist vector sublattices $M \subset L$ such that $M \cap C$ has non-empty interior (in M), but no positive linear form on M can be extended to a positive linear form on L.

(d) The two-dimensional subspace of L determined by the functions $t \rightarrow at + b$ ($a, b \in R$) is a vector lattice under the induced order, but not a sublattice of L.

15. Let $X \neq \varnothing$ be a set, let Σ be a σ-algebra of subsets of X, and let E denote the vector space (over R) of all real-valued bounded Σ-measurable functions (that is, functions f such that $f^{-1}(A) \in \Sigma$ for each Borel set $A \subset R$). Under the sup-norm and the canonical order, E is an (AM)-space. Every positive linear form on E defines a real-valued, finitely additive non-negative set function on Σ, and conversely; hence E^+ can be identified with the vector lattice of all real-valued, finitely additive set functions on Σ that are differences of non-negative functions of the same type.

(a) E^+ is order complete, and the set functions $\mu \in E^+$ that are countably additive on Σ form a band M in E^+. Deduce from this that every finitely additive, non-negative set function v on Σ has a unique representation $v = v_1 + v_2$, where v_1, v_2 are ≥ 0, v_1 is countably additive and v_2 is not countably additive unless $v_2 = 0$.

(b) Suppose that Σ contains all finite (hence all countable) subsets of X, and call the elements $\mu \in M$ briefly measures on X. The measures μ such that $\mu(\{t\}) = 0$ for every $t \in X$ are called diffuse; the set of all diffuse measures is a band M_d in M. The elements of the complementary band M_a in M are called atomic. Show that each atomic measure on X is the sum (for $\sigma(E^+, E)$) of a summable family $\{\alpha_n \mu_n : n \in N\}$ (Chapter III, Exercise 23), where $(\alpha_n) \in l^1$ and each μ_n is a point measure on X (that is, $\mu(\{t_0\}) = 1$ for a suitable $t_0 \in X$ and $\mu(A) = 0$ for each $A \in \Sigma$ such that $t_0 \notin A$).

(c) Illustrate the preceding by considering the case, where X is compact and Σ is the family of all Baire subsets of X.

16. Let L be an ordered vector space over R. A **monotone** (non-decreasing) **transfinite sequence** is a mapping $\alpha \to a_\alpha$ of the set of all ordinals $\alpha < \beta$ (where β is an ordinal ≥ 1) into L such that $\alpha_1 < \alpha_2$ $(<\beta)$ implies $a_{\alpha_1} \leq a_{\alpha_2}$. Suppose that the positive cone C generates L, and that C satisfies condition (D) of (1.1).

(a) If each majorized monotone transfinite sequence in L possesses a least upper bound, then L is an order complete vector lattice.

(b) If there exists a linear form f on L such that $x > 0$ implies $f(x) > 0$, for L to be an order complete vector lattice it suffices that each ordinary monotone sequence which is majorized, has a least upper bound. (Schaefer [4].)

17. Let L be a vector lattice which is order complete, and denote by $\mathfrak{T}$ the finest topology on L such that every order convergent filter in L $\mathfrak{T}$-converges to its order limit. Show that $\mathfrak{T}$ is a translation-invariant topology that possesses a 0-neighborhood base of radial and circled sets but fails, in general, to satisfy axiom $(LT)_1$ (Chapter I, Section 1). (Show that every one-point set is closed, but $\mathfrak{T}$ is not necessarily Hausdorff.)

18. Let E, F be vector lattices and suppose that F is order complete. Denote by $H \subset L(E, F)$ the cone of all positive linear maps of E into F. The subspace $M = H - H$ of $L(E, F)$ is an order complete vector lattice under its canonical order, containing exactly those linear maps that map all order intervals of E into order intervals of F.

19. Recall that the order of a vector space L is called Archimedean if $x \leqq 0$ whenever $x \leqq n^{-1}y$ for all $n \in N$ and some $y \in L$. The order of L is called **almost Archimedean** if $x = 0$ whenever $-n^{-1}y \leqq x \leqq n^{-1}y$ for all $n \in N$ and some $y \in L$.

(a) If L is almost Archimedean ordered and possesses an order unit, then $(L, \mathfrak{T}_o)$ is normable and $(L, \mathfrak{T}_o)' = L^b = L^+$.

(b) If L is almost Archimedean ordered and L^b distinguishes points in L, $(L, \mathfrak{T}_o)$ can be characterized as an inductive limit in analogy to (6.3). If, in addition, L is a vector lattice, then $(L, \mathfrak{T}_o)$ is a l.c.v.l.

(c) If L is an almost Archimedean ordered vector lattice such that L^+ distinguishes points in L, then the order of L is Archimedean (hence regular). (Use (b), observing that $\mathfrak{T}_o$ is a Hausdorff topology for which the positive cone is closed.)

(d) Let T be a set containing at least two elements and let E be the vector space (over R) of all bounded real functions on T, ordered by the relation "$g \leqq f$ if either $f = g$ or $\inf\{f(t) - g(t): t \in T\} > 0$". The order of E is almost Archimedean, but not Archimedean.

(e) Every order complete vector lattice is Archimedean ordered.

20. (Continuity of the Lattice Operations; cf. Gordon [1], Peressini [1]). Let E be an ordered l.c.s. over R whose positive cone is generating, and endow E' with its canonical order. Denote by $o(E, E')$ the topology of uniform convergence on all order intervals in E'.

(a) If E' is a C^*-saturated subspace of E^+, then $o(E, E')$ is consistent with $\langle E, E' \rangle$. (Observe that each order interval in E' is $\sigma(E', E)$-compact.)

(b) If E is a l.c. vector lattice, then $o(E, E')$ is the coarsest translation-invariant topology on E finer than $\sigma(E, E')$ and for which the lattice operations are continuous. Deduce from this that if E is an (AL)-space, the norm topology is the only topology consistent with $\langle E, E' \rangle$ and such that the lattice operations are continuous.

(c) If E is a normed lattice, then the lattice operations are weakly continuous if and only if E is finite dimensional. (Use (b) to infer that every order interval of E' must be contained in a finite dimensional subspace of E', and show that this is absurd unless E' is of finite dimension.)

(d) If E is a l.c.v.l., the completion of $(E', o(E', E))$ can be identified with the band in E^+ generated by E'. Deduce that if E is barreled, then E' is complete for $o(E', E)$.

(e) Let L be a vector lattice and let P be a non-empty set of positive linear forms on L. The semi-norms $x \to f(|x|)$ ($f \in P$) generate a l.c. convex topology $\mathfrak{T}$ on L for which the lattice operations are continuous. More precisely, $(L, \mathfrak{T})'$ is the smallest solid subspace M of L^+ that contains P, and if $(E, \mathfrak{T}_1)$ is the Hausdorff t.v.s. associated with $(L, \mathfrak{T})$, then $(E, \mathfrak{T}_1)$ is a l.c.v.l. whose dual E' can be identified with M, and such that $\mathfrak{T}_1 = o(E, E')$.

21. Let L be a vector lattice of finite dimension n.

(a) If L is Archimedean ordered, then L is isomorphic with R^n under its canonical order. (Observe that the positive cone C of L has non-empty interior (proof of the lemma preceding (4.1)), and use (8.4) and

(8.5) to show that L is isomorphic with $\mathscr{C}_R(X)$, where X contains exactly n points.)

(b) If L is not Archimedean ordered, there exist integers k, m such that $2 \leqq k \leqq n$, $m \geqq 0$, $k + m = n$ and such that L is isomorphic with $R_0^k \times R_0^m$, where the order of R_0^k is lexicographic and the order of R_0^m is canonical. (Birkhoff [1], Chapter XV, Theorem 1.)

22. Let E be an (AL)-space.

(a) If (and only if) E is separable, there exists a compact metrizable space X such that E is isomorphic (as a Banach lattice) with $L^1(\mu)$ where μ is a suitable regular Borel measure on X. (Observe that E possesses a weak order unit x_0 and note that E is the band generated by $\{x_0\}$; utilize (8.6) and the Radon-Nikodym theorem.)

(b) If, in the spirit of (8.6), E'' is identified with the Banach lattice of all bounded, signed, regular Borel measures on X (Chapter II, Section 2, Example 3), deduce from (8.6) that the measures in E are exactly those vanishing on each subset of first category in X. (Kelley-Namioka [1].)

23. If X is a compact space, the Banach lattice $\mathscr{C}_R(X)$ is order complete exactly when X is extremally disconnected (a **Stonian space**). Infer from this that $\mathscr{C}_R(X)$ cannot be a dual Banach space, unless X is Stonian, and not reflexive, unless X is finite.

24. Let A be an algebra over R with unit e and denote by A_0 the underlying vector space of A (Chapter IV, Exercise 40). A is called an **ordered algebra** if A_0 is an Archimedean ordered vector space such that $e \geqq 0$ and such that $a \geqq 0$, $b \geqq 0$ imply $ab \geqq 0$.

(a) Suppose that e is an order unit of A_0, and denote by C^* the cone (in A_0^*) of all positive linear forms. Each linear form f generating an extreme ray of C^* and satisfying $f(e) = 1$ is multiplicative: $f(ba) = f(b)f(a)$ for all $a, b \in A$. (Show that for fixed $b \geqq 0$, $a \to f(ba)$ is a linear form g such that $g = \lambda_b f$, and that $\lambda_b = f(b)$.)

(b) (Stone's Algebra Theorem). Let A be an ordered algebra such that the unit e of A is an order unit. Under evaluation, A is isomorphic with a dense subalgebra of $\mathscr{C}_R(X)$, where X is the $\sigma(A_0^*, A_0)$-compact set of multiplicative, positive linear forms f satisfying $f(e) = 1$. (Use (a) and the Krein-Milman theorem, and apply (8.2).) Infer that A is commutative and show that evaluation on X is a norm isomorphism of (A, p_e) into $\mathscr{C}_R(X)$, where p_e denotes the gauge of $[-e, e]$ in A.)

(c) If, in addition to the hypothesis of (b), every positive sequence of type l^1 in A is order summable, then (A, p_e) is isomorphic with the Banach algebra $\mathscr{C}_R(X)$ (Chapter IV, Exercise 40). Conclude that in these circumstances, A_0 is necessarily a vector lattice.

25. (Spectral Measures and Algebras). Let X be a compact space, A a locally convex algebra over K (Chapter IV, Exercise 40). A continuous map μ of $\mathscr{C}_K(X)$ onto A which is an algebraic homomorphism, is called a **spectral measure** on X with range A. The range of a spectral measure is called a **spectral algebra** (over K). If A is a l.c. algebra over K with unit e, a subalgebra of A which contains e and is a spectral algebra is called a **spectral subalgebra** of A.

(a) A l.c. algebra over R with unit e is a spectral algebra if and only if there exists an order of A such that (i) A is an ordered algebra; (ii) e is an order unit, and $[-e, e]$ is bounded; (iii) every positive sequence of type l^1 is order summable. (Use Exercise 24.)

(b) A l.c. algebra over C with unit e is a spectral algebra if and only if there exists a real subalgebra A_1 containing e such that each $a \in A$ has a unique representation $a = b + ic$ $(b, c \in A_1)$, and such that A_1 is a spectral algebra over R (cf. (8.3)).

(c) Let μ be a spectral measure on X with range A. Define the support of μ to be the complement X_0 of the largest open set $G \subset X$ such that $\mu(f) = 0$ whenever f has its support in G. Then μ induces a spectral measure μ_0 on X_0 with range A which is biunivocal. For μ_0 to be a homeomorphism (equivalently, for μ to be a topological homomorphism), it is necessary and sufficient that $a \to r(a)$ be continuous on A, where $r(a)$ denotes the spectral radius of $a \in A$. (Chapter IV, Exercise 40.)

(d) If A is a l.c. algebra, an element $a \in A$ is called a **spectral element** if a is contained in a spectral subalgebra of A. (If A is an algebra of continuous endomorphisms of a l.c.s. E, the spectral elements of A are called (scalar type) **spectral operators** on E.) If a is a spectral element of A and μ is a spectral measure on X such that $a = \mu(f)$, then $f(X)$ is the spectrum of a, $f(X) = \sigma(a)$. (Cf. Schaefer [9], II, Theorem 3.)

(e) Let A be a spectral algebra over K, and $a \in A$. There exists a spectral measure ν on $\sigma(a)$ such that $a = \nu(\tilde{1})$, where $\tilde{1}$ denotes the identity function on $\sigma(a)$, and such that the range of ν is the smallest spectral subalgebra of A that contains a. (Consider the mapping $g \to \mu(g \circ f)$ of $\mathscr{C}_K(\sigma(a))$ into A.)

26. The following are typical examples of spectral algebras:

(a) The algebra $L^\infty(\tau)$ (where (Z, Σ, τ) is a measure space and multiplication is defined by pointwise multiplication of representatives). This algebra can be viewed as a spectral subalgebra of the Banach algebra of continuous endomorphisms, $\mathscr{L}(E)$, where $E = L^p(\tau)$ $(1 \leqq p \leqq + \infty)$.

(b) Every norm-closed (real) algebra of Hermitian operators on a Hilbert space (in particular, the closure of each such algebra under the topology of simple convergence).

(c) Let E be a l.c. vector lattice on which every positive linear form is continuous (cf. (5.5)), and suppose that the positive cone C is weakly sequentially complete. Endow $\mathscr{L}(E)$ with its canonical order and the topology of simple convergence, and denote by e the identity map of E. The linear hull of $[-e, e]$ in $\mathscr{L}(E)$ is a spectral subalgebra of $\mathscr{L}(E)$. This can be extended to the case where E is a l.c.s. over C such that its underlying real space E_0 is a l.c.v.l. (Schaefer [8], Theorem 7.)

Examples (a) and (b) are special cases of (c). (For (b), see Schaefer [3], (11.3).)

27. (Extension of Spectral Measures). Let X be a compact space, μ a spectral measure on X with range A_1, where A_1 is a spectral subalgebra of the l.c. algebra A. Denote by J the unit interval of A for the finest order on A such that μ is positive (where $\mathscr{C}(X)$ is ordered as usual), and suppose that J is weakly sequentially complete. There exists a continuous

extension $\bar{\mu}$ of μ to the Banach algebra $\mathscr{B}(X)$ of bounded Baire functions on X, such that $\bar{\mu}$ is an algebraic homomorphism of $\mathscr{B}(X)$ into A. Moreover, $\bar{\mu}$ induces a homomorphism of the Boolean algebra of Baire subsets of X onto a σ-complete Boolean algebra of idempotents of A. (Cf. Schaefer [9], II, Theorem 8.)

Appendix

SPECTRAL PROPERTIES OF
POSITIVE OPERATORS

It has been discovered around the turn of the century (Frobenius [1], [2], Perron [1]) that the spectrum of $n \times n$ matrices with real entries ≥ 0 has certain special features; in particular, the spectral radius is an eigenvalue with a positive eigenvector (for the canonical order of R^n). Since that time a slow but steady development has taken place which, in an abstract setting, reached a climax with the advent of the well-known memoir by Krein-Rutman [1]. In fact, it appears that apart from normal operators on Hilbert space (and their generalizations, usually called spectral operators), positive operators on ordered topological vector spaces are the most interesting class from a spectral point of view; in addition, it should be noted that the theory of spectral operators is largely governed by order theory (cf. Chapter V, Exercises 25–27). (In this Appendix, the term "operator" will be used synonymously with "continuous endomorphism".) Hence a good deal of the motivation for the study of ordered topological vector spaces has its origin in spectral theory, and it is the objective of this Appendix to introduce the reader to some spectral theoretic applications of the results of Chapter V. We assume familiarity with the most elementary facts about the algebra of operators on a Banach space; Section 1 enumerates in detail what will be needed in the sequel.

In the investigation of the spectrum of a positive operator defined on an ordered Banach space, various routes of approach can be followed according to the type of problem under consideration; Sections 2 and 3 each are devoted to a certain mode of attack. Section 2, exploiting function theoretic properties of the resolvent, contains most of what seems to be attainable in a very general setting; the generalized Pringsheim theorem (2.1) is the unifying theme of the section. Of course, results on the existence of eigenvectors are bound to involve compactness in one form or another. Section 3 is devoted to a study of the peripheral point spectrum of positive operators under more specific assumptions, both on the operator and its space of definition. A salient feature of these results is that compactness is not invoked. Theorem (3.4) is an example of a rather general result in this area. The reader himself will notice that this direction of research is far from being exhausted and in some respects appears to be quite promising.

Except where other references are given, the results in this appendix can, for their greater part, be found in the author's papers [6] and [9]–[11].

1. ELEMENTARY PROPERTIES OF THE RESOLVENT

Let $(E, \| \ \|)$ be a complex Banach space, and denote by $\mathscr{L}(E)$ the Banach algebra (Chapter IV, Exercise 40) of continuous endomorphisms of E, under the standard norm $u \to \|u\| = \sup\{ \|u(x)\| : \|x\| \leq 1\}$. If $u \in \mathscr{L}(E)$, the **spectrum** $\sigma(u)$ is the complement in C of the largest open set $\rho(u)$ in which $\lambda \to (\lambda e - u)^{-1}$ exists and is locally holomorphic (Chapter IV, Exercise 39); here and in the following, e denotes the unit of $\mathscr{L}(E)$ (that is, the identity map of E). For $\lambda \in \rho(u)$, we set $(\lambda e - u)^{-1} = R(\lambda)$; $\lambda \to R(\lambda)$ is called the **resolvent**, $\rho(u)$ the **resolvent set** of u. Supposing that E is not reduced to $\{0\}$, it is a well-known fact (cf. Hille-Phillips [1] and Chapter IV, Exercise 40) that $\sigma(u)$ is a non-empty compact subset of C; the radius $r(u)$ of the smallest circle of center 0 in C that contains $\sigma(u)$ is called the **spectral radius** of u; the set $\{\lambda \in C: |\lambda| = r(u)\}$ is termed the **spectral circle** of u. Moreover, if $\lambda \in \rho(u)$ and $\mu \in \rho(u)$ one has the resolvent equation

$$R(\lambda) - R(\mu) = -(\lambda - \mu)R(\lambda)R(\mu). \tag{1}$$

Here we denote the composite $u \circ v$ of $u, v \in \mathscr{L}(E)$ by juxtaposition uv; for the simple proof of (1) see, e.g., Hille-Phillips [1].

If, more generally, E is a l.c.s. and u is a continuous endomorphism of E, we can define the spectrum, resolvent set and the resolvent of u as before, considering $\mathscr{L}(E)$ under an $\mathfrak{S}$-topology for which $\mathscr{L}(E)$ is a l.c. algebra (Chapter IV, Exercise 40). This is true for every $\mathfrak{S}$-topology on $\mathscr{L}(E)$ such that $u(\mathfrak{S}) \subset \mathfrak{S}$ for all $u \in \mathscr{L}(E)$; in particular, $\mathscr{L}(E)$ is a l.c. algebra for the topologies of simple and bounded convergence. Most of the properties of $\sigma(u)$ familiar from Banach spaces fail for continuous endomorphisms of more general l.c.s.; however, if E is a semi-complete l.c.s. and u is a bounded endomorphism of E, then by transition to a suitable Banach space $\tilde{E}_U$ (Chapter III, Section 7) the classical results can be shown to hold for u (cf. Schaefer [3], Section 10). In the same manner, most of the results on positive operators in ordered Banach spaces, derived in Sections 2 and 3 below, can be generalized to bounded positive endomorphisms of an ordered l.c.s.; since no essentially new methods are involved, we shall restrict attention to Banach spaces. This does not mean, of course, that the consideration of non-normable topologies is eliminated; as an example we refer to the proof of (2.4) below.

We return to the assumption that E is a complex Banach space. By virtue of Banach's homomorphism theorem (Corollary 1 of (III, 2.1)) and the fact that the set of invertible elements of a Banach algebra is open, for each $u \in \mathscr{L}(E)$ the spectrum $\sigma(u)$ can be characterized as the set of those $\lambda \in C$ for which $\lambda e - u$ fails to be an algebraic automorphism of E. In view of this, we have the following result.

Let $u \in \mathscr{L}(E)$, where E is a complex Banach space, and assume that $\{\lambda_n: n \in N\}$ is a sequence in $\rho(u)$ converging to some $\lambda \in C$. Then $\lambda \in \sigma(u)$ if and only if $\lim_n \|R(\lambda_n)\| = +\infty$.

In fact, the condition is clearly sufficient for $\lambda \in \sigma(u)$. To prove its necessity, suppose there exists a subsequence $\{\mu_n\}$ of $\{\lambda_n\}$ such that $\{R(\mu_n): n \in N\}$ is bounded; by (1) the latter is a Cauchy sequence in $\mathscr{L}(E)$ and hence convergent to some $v \in \mathscr{L}(E)$. This implies $\lim_n R(\mu_n)(\mu_n e - u) = v(\lambda e - u) = e$ and, similarly, $(\lambda e - u)v = e$; hence we obtain $\lambda \in \rho(u)$, which is contradictory.

The subset of $\sigma(u)$ in which $(\lambda e - u)$ fails to be one–to–one is called the **point spectrum** $\pi(u)$ of u. An element $\lambda_0 \in \pi(u)$ is called an **eigenvalue** of u, the null space of $(\lambda_0 e - u)$ the corresponding **eigenspace** $N(\lambda_0)$. The dimension of $N(\lambda_0)$ is called the (geometric) **multiplicity** of λ_0, and the non-zero elements of $N(\lambda_0)$ are termed **eigenvectors** of u for λ_0. (The terms characteristic value, characteristic space, and characteristic vector are also in current use.)

The point spectrum of u contains all poles of the resolvent R. Let λ_0 be a pole of R and let

$$R(\lambda) = \sum_{k=-n}^{\infty} a_k (\lambda - \lambda_0)^k \qquad (a_{-n} \neq 0) \tag{2}$$

be the Laurent expansion of R near λ_0; the integer $n \ (\geq 1)$ is the **order** of the pole λ_0; the partial sum of (2), extending from $k = -n$ to $k = -1$, is the **principal part** of the expansion; a_{-n} the **leading coefficient**, and a_{-1} is the **residue** of R at $\lambda = \lambda_0$. Multiplying (2) by $(\lambda e - u) = (\lambda_0 e - u) + (\lambda - \lambda_0)e$ and comparing coefficients in the resulting identity (which is justified by the identity theorem for analytic functions), we obtain, in particular, $a_{-n}(\lambda_0 e - u)$ $= (\lambda_0 e - u)a_{-n} = 0$ and $a_{-n} = a_{-1}(u - \lambda_0 e)^{n-1}$; clearly, the coefficients a_k commute with u. These relations show that λ_0 is in $\pi(u)$; more precisely, it turns out that a_{-1} is a projection of E onto the null space of $(\lambda_0 e - u)^n$ which contains $N(\lambda_0)$. Let us recall also (cf. Riesz-Nagy [1], Hille-Phillips [1]) that for compact u, the resolvent R is a meromorphic function on the Riemann sphere punctured at 0 (one defines, generally, $R(\infty) = 0$); thus for compact u, $\sigma(u)$ is a countable set, with 0 as its only possible accumulation point, and each non-zero $\lambda \in \sigma(u)$ is an eigenvalue of u of finite multiplicity.

Finally, if $u \in \mathscr{L}(E)$ and $|\lambda| > r(u)$, the resolvent of u is given by

$$R(\lambda) = \sum_{n=0}^{\infty} \lambda^{-(n+1)} u^n \tag{3}$$

$(u^0 = e)$; (3) is the expansion of R at ∞, and is called the C. Neumann's series. It follows from Cauchy's criterion for the convergence of power series that $r(u) = \lim \sup \|u^n\|^{1/n}$; more precisely, it is true that $r(u) = \lim_n \|u^n\|^{1/n}$ (cf. Hille-Phillips [1]). In case $r(u) = 0$, u is called a **topological nilpotent** of the Banach algebra $\mathscr{L}(E)$; clearly, u is a topological nilpotent if and only if $\sigma(u) = \{0\}$ or, equivalently, if and only if the resolvent R (with $R(\infty) = 0$) is an entire function of λ^{-1}.

If E is a Banach space over $\mathbf{R}$ and $u \in \mathscr{L}(E)$, the real spectrum $\sigma_\mathbf{R}(u)$ can be defined as the subset of $\mathbf{R}$ in which $(\lambda e - u)$ fails to be an automorphism of E; analogously, we can define the real resolvent of u as the function $\lambda \rightarrow (\lambda e - u)^{-1}$ with domain $\mathbf{R} \sim \sigma_\mathbf{R}(u)$. (It can happen that $\sigma_\mathbf{R}(u)$ is empty, as the example of a rotation about the origin of the Euclidean plane $\mathbf{R}_0^2$ shows.) We shall not follow this practice, but instead subsume the case of a real Banach space under the preceding by the following standard procedure.

Let $(E, \| \; \|)$ be a Banach space over $\mathbf{R}$; the complexification E_1 (Chapter I, Section 7) of the t.v.s. E is a complete normable space over $\mathbf{C}$. If we desire to have a norm on E_1 such that the imbedding of E into E_1 becomes a real norm isomorphism, the definition

$$\|x + iy\|_1 = \sup_{0 \leq \theta < 2\pi} \|(\cos \theta)x + (\sin \theta)y\|$$

will do; this is a generalization of the definition of the usual absolute value on $\mathbf{C}$, considering $\mathbf{C}$ as the complexification of $\mathbf{R}$. Now every $u \in \mathscr{L}(E)$ has a unique **complex extension** $\bar{u} \in \mathscr{L}(E_1)$, defined by $\bar{u}(x + iy) = u(x) + iu(y)$ for all $x, y \in E$. In case E is a real Banach space and $u \in \mathscr{L}(E)$, we define the spectrum, resolvent, spectral radius of u to be the corresponding objects for $\bar{u}$ as defined above. Sometimes it is even convenient to identify u with its complex extension $\bar{u}$. It is easy to see that for $u \in \mathscr{L}(E)$, we have $\sigma_\mathbf{R}(u) = \sigma(u) \cap \mathbf{R}$, that for $\lambda \in \mathbf{R} \sim \sigma_\mathbf{R}(u)$ the real resolvent of u is the restriction of the resolvent of u to E (considered as a real subspace of E_1), and that the spectral radius $r(u)$ is the smallest real number $\alpha \geq 0$ such that for $|\lambda| > \alpha$, $\lambda \in \mathbf{R}$, the series (3) converges in $\mathscr{L}(E)$.

2. PRINGSHEIM'S THEOREM AND ITS CONSEQUENCES

Perhaps the best-known result on positive operators in ordered Banach spaces is the theorem that whenever the positive cone is total and u is a compact positive endomorphism with spectral radius $r(u) > 0$, then $r(u)$ is an eigenvalue of u with an eigenvector ≥ 0. This theorem, which has a comparatively long history, was first proved in the stated generality by Krein-Rutman [1]. A more general theorem appeared in Bonsall [3] and was extended to locally convex spaces by the author [3]. In this section, we shall derive this and other results (some not dependent on compactness assumptions) in a uniform way from a theorem on vector-valued analytic functions which is an extension of a classical theorem due to Pringsheim. We shall need this theorem for functions taking their values in a locally convex space.

2.1

Theorem. *Let E be an ordered, semi-complete l.c.s. over C such that the positive cone C is weakly normal. If $a_n \in C$ ($n = 0, 1, \ldots$) and if $\sum_0^\infty a_n z^n$ has radius of convergence 1, then the analytic function represented by the power series is singular at $z = 1$. In addition, if this singularity is a pole, it is of maximal order on $|z| = 1$.*

Proof. Let f be the functional element (with values in E) given by $f(z) = \sum_0^\infty a_n z^n$ when $|z| < 1$ and let the radius of convergence of this series be 1. Let x' be any continuous, real linear form on E; the radius $r_{x'}$ of convergence of the series $\sum_0^\infty \langle a_n, x' \rangle t^n$, where t is real, is $\geqq 1$. Further we have $\inf\{r_{x'} : x' \in D\}$ $= 1$, where D denotes the set of all continuous, real linear forms on E that are $\geqq 0$ on C. For if we had $\inf\{r_{x'} : x' \in D\} = \eta > 1$, the series $\sum_0^\infty a_n t^n$ would converge in E for all t, $-\eta < t < \eta$, since by (V, 3.3), Corollary 3, the weak normality of C is equivalent to $E_0' = D - D$, where E_0 is the underlying real space of E (cf. (I, 7.2)). Thus $z \rightarrow f(z)$ would have a holomorphic extension (Chapter IV, Exercise 39) to the open disk $|z| < \eta$, which is contradictory.

Let ρ, $0 < \rho < 1$, be fixed, let $x' \in D$ and define

$$b_k = \sum_{n=k}^\infty \binom{n}{k} \rho^{n-k} a_n$$

for $k = 0, 1, \ldots$. Since for $\rho < t < 1$, all terms in the three series

$$\sum_0^\infty \langle a_n, x' \rangle t^n = \sum_0^\infty \langle a_n, x' \rangle ((t - \rho) + \rho)^n = \sum_0^\infty \langle b_n, x' \rangle (t - \rho)^n$$

are non-negative, it follows that the series

$$\sum_0^\infty \langle b_n, x' \rangle (t - \rho)^n$$

has radius of convergence $r_{x'} - \rho$, and hence that

$$\sum_0^\infty b_n (t - \rho)^n$$

has radius of convergence $1 - \rho$. By a conclusion familiar from the theory of analytic functions (cf. Chapter IV, Exercise 39(d)) this implies that $z = 1$ is singular for f.

Assume now that the singularity of f at $z = 1$ is a pole of order k. If $\zeta = \exp i\theta$ is any complex number of modulus 1, and if $z = t\zeta$, $0 < t < 1$, we have

$$\lim_{t \to 1} f(|z|)|z - \zeta|^p = 0$$

for all $p > k$. Since C is a weakly normal cone, this implies, for any $p > k$, that

$$(1 - t)^p \sum_0^\infty (t^n \cos n\theta)a_n, \quad (1 - t)^p \sum_0^\infty (t^n \sin n\theta)a_n$$

both converge to 0 for $\sigma(E, E')$ as $t \to 1$. Thus if ζ is a pole of f of order m, it follows that $m \leq k$ and the theorem is proved.

In the first two of the following applications of the Pringsheim theorem, E can be any ordered complex Banach space; the remaining results gain in transparency by starting from a real space. The reader will notice that (2.2) applies, in particular, to any ordered Banach space over R whose positive cone C is normal and generates E (for instance, a Banach lattice); just apply (2.2) to the complexification E_1 of E, ordered with positive cone $C + iC$.

2.2

Let E be an ordered complex Banach space, not reduced to $\{0\}$, with positive cone C such that C is normal and $E = C - C$. For any positive (necessarily continuous) endomorphism u of E, the spectral radius $r(u)$ is an element of $\sigma(u)$; if $r(u)$ is a pole of the resolvent, it is of maximal order on the spectral circle of u.

Proof. By (V, 5.5), any positive endomorphism u of E is continuous. In view of the corollary of (V, 3.5), C is a $\mathfrak{B}$-cone and hence (V, 5.2) implies that the cone H of positive endomorphisms of E is normal (hence weakly normal) in $\mathscr{L}(E)$ for the topology of bounded convergence, that is, for the norm topology of $\mathscr{L}(E)$. If $r(u) > 0$, Theorem (2.1) applies to $z \to f(z) = R(r(u)/z) = \sum_0^\infty u^n(z/r(u))^{n+1}$ (Section 1, Formula (3)); if $r(u) = 0$, u is a topological nilpotent, $\sigma(u) = \{0\}$, and the assertion is equally true.

2.3

Let $\dot{E}$ be an ordered complex (B)-space satisfying the hypothesis of (2.2), and let u be a positive endomorphism of E. If $\lambda \in \rho(u)$, then $R(\lambda)$ is positive if and only if λ is real and $\lambda > r(u)$.

Proof. It is clear that $\lambda > r(u)$ is sufficient for $R(\lambda) \geq 0$ (with respect to the canonical order of $\mathscr{L}(E)$), in view of Formula (3) of Section 1. Suppose that $R(\lambda) \geq 0$ for some $\lambda \in \rho(u)$. Choose an $x_0 > 0$ and define recursively $x_n = R(\lambda)x_{n-1}$ $(n \in N)$. Each x_n satisfies the relation

$$\lambda x_n = u(x_n) + x_{n-1} \quad (n \in N). \qquad (*)$$

Clearly, $x_n \in C$ for all n and, in fact $x_n > 0$ (for, $x_n = 0$ for some $n \in N$ would imply $x_0 = 0$). Moreover, by induction on n it is shown from $(*)$ that $\lambda^n x_n \in C$ and $\lambda^{n-1} x_n \in C$ for all $n \in N$, and that

$$\lambda^n x_n \geq \lambda^{n-1} x_{n-1} \geq x_0 \quad (n \in N).$$

This necessitates $\lambda \neq 0$ and we can assume that $|\lambda| = 1$, for if $R(\lambda)$ is positive at $\lambda \neq 0$, then the resolvent of $|\lambda^{-1}|u$ is positive at $\lambda|\lambda^{-1}|$. Let $\lambda = \exp i\theta$, $0 \leq \theta < 2\pi$, and suppose that $\theta > 0$. It is clear that $n\theta \not\equiv \pi \pmod{2\pi}$ for all positive integers n, or else C would not be a proper cone. Hence there exists a smallest integer $n_0 > 0$ such that the triangle with vertices 1, $\exp i(n_0 - 1)\theta$, $\exp in_0\theta$ in the complex plane contains 0 in its interior. Consider the unique real subspace M of E of dimension 2 that contains the points x_{n_0}, $\lambda^{n_0 - 1}x_{n_0}$ and $\lambda^{n_0}x_{n_0}$. It follows that $M \cap C$ contains 0 as an interior point, which conflicts with the fact that C is a proper cone; thus $\theta = 0$, and hence $\lambda > 0$.

Up to this point of the proof we have used C only as a proper cone $\neq \{0\}$. Assume now that C is normal and $E = C - C$; as before, it follows from (V; 3.5), Corollary, and (V, 5.2) that the positive cone $\mathcal{H} \subset \mathcal{L}(E)$ is normal. If it were true that $R(\lambda) \geq 0$ for some λ, $0 < \lambda \leq r(u)$, then the resolvent equation (Section 1, Formula (1)) would imply that $0 \leq R(\mu) \leq R(\lambda)$ for all $\mu > r(u)$ and hence, in view of the normality of $\mathcal{H}$, that $\{R(\mu): \mu > r(u)\}$ is a bounded family in $\mathcal{L}(E)$. This clearly contradicts (2.2) above (cf. Section 1), and hence it follows that $\lambda > r(u)$.

REMARK. The preceding proof shows that whenever E is an ordered Banach space with positive cone $C \neq \{0\}$ and u is a positive operator (continuous endomorphism) of E, then $R(\lambda) \geq 0$ implies $\lambda > 0$.

We now turn to the Krein-Rutman theorem on compact positive operators mentioned at the beginning of this section; for a historical account and a bibliography of earlier work on the subject, the reader should consult the memoir of Krein-Rutman [1]. We derive the Krein-Rutman theorem from the following result which establishes the conclusion of (2.2) for a restricted class of positive operators but with no restriction on the ordering, except that the positive cone be (closed and) total. It is clear that this latter condition is indispensable; the additional condition on the operator cannot be dropped (Bonsall [3]) but probably it can be further relaxed. For continuous endomorphisms u of a real Banach space E, the terms *spectrum*, *resolvent*, etc., refer to the complex extension of u to the complexification of E (cf. end of Section 1).

2.4

Let E be an ordered real Banach space with total positive cone C, and assume that u is a continuous positive endomorphism of E whose resolvent has a pole on the spectral circle $|\lambda| = r(u)$. Then $r(u) \in \sigma(u)$, and if $r(u)$ is a pole of the resolvent it is of maximal order on the spectral circle.

Proof. Since C is a closed, proper, total cone in E, its dual cone C' has the same properties with respect to $\sigma(E', E)$, and hence $G = C' - C'$ is a dense subspace of the weak dual E'_σ. If F denotes the space $(E, \sigma(E, G))$, then C is a normal cone in F by (V, 3.3), Corollary 3. Denote by E_1, F_1 the complexifications (Chapter I, Section 7) of E, F respectively. We consider E_1 as ordered with positive cone C; then the canonical order of $\mathscr{L}(E_1)$ is determined by the positive cone $\mathscr{H} = \{w \in \mathscr{L}(E_1): w(C) \subset C\}$. Moreover, we shall identify $u \in \mathscr{L}(E)$ with its complex extension to E_1.

Let us further denote by $\mathscr{L}_\sigma(E_1, F_1)$ the space of continuous linear maps of E_1 into F_1, provided with the topology of simple convergence on C. Then there is a natural imbedding ψ of $\mathscr{L}(E_1)$ into $\mathscr{L}_\sigma(E_1, F_1)$ which is continuous; for simplicity of notation, we denote the images of elements and subsets of $\mathscr{L}(E_1)$ under ψ by an index zero. We note first that by (V, 5.2) and the normality of C in F_1, the image $\mathscr{H}_0$ of the cone $\mathscr{H}$ is normal in $\mathscr{L}_\sigma(E_1, F_1)$. Now let ζ, $|\zeta| = r(u)$, be a pole of order k (≥ 1) of the resolvent $\lambda \to R(\lambda)$ of u, and let $a \in \mathscr{L}(E_1)$ be the leading coefficient of the principal part at $\lambda = \zeta$; one has $a = \lim_{\lambda \to \zeta} (\lambda - \zeta)^k R(\lambda)$; hence also $a_0 = \lim_{\lambda \to \zeta} (\lambda - \zeta)^k R_0(\lambda)$. Suppose that $r(u) \notin \sigma(u)$; then $\lambda \to R(\lambda)$ and a fortiori $\lambda \to R_0(\lambda)$ would be holomorphic at $\lambda = r(u)$. Since the coefficients of the expansion $R_0(\lambda) = \sum_0^\infty \lambda^{-(n+1)} u_0^n$ of R_0 at infinity are elements of the normal cone $\mathscr{H}_0$, (2.1) implies that R_0 has an extension, with values in the completion of $\mathscr{L}_\sigma(E_1, F_1)$, which is holomorphic for $|\lambda| > \tau$, where $0 \leq \tau < r(u)$; in particular, $\{R_0(\lambda): |\lambda| > r(u)\}$ is a bounded family in $\mathscr{L}_\sigma(E_1, F_1)$. Clearly, this implies $a_0 = 0$ and hence $a = 0$, which is contradictory. Thus $r(u) \in \sigma(u)$.

For the final assertion we note that any pole of $\lambda \to R(\lambda)$ on $|\lambda| = r(u)$ is a pole of the same order for R_0; thus the assertion follows again from Pringsheim's theorem (2.1). The theorem is proved.

COROLLARY (Krein-Rutman). *Let E be an ordered real Banach space with total positive cone C, and let u be a compact positive endomorphism of E. If u has a spectral radius $r(u) > 0$, then $r(u)$ is a pole of the resolvent of maximal order on the spectral circle, with an eigenvector in C. A corresponding result holds for the adjoint u' in E'.*

Proof. Since u is compact the only possible singularities $\neq 0$ of the resolvent are poles, and there is at least one such singularity on $|\lambda| = r(u)$. Hence $\lambda = r(u)$ is a pole of some order k (≥ 1) of the resolvent, and we have $p = \lim_{\lambda \to r(u)} (\lambda - r(u))^k R(\lambda)$ for the leading coefficient of the corresponding principal part. Since $R(\lambda) \geq 0$ (for the canonical order of $\mathscr{L}(E)$) whenever $\lambda > r(u)$, it follows that $p \geq 0$, since the positive cone of $\mathscr{L}(E)$ is closed (cf. (V, 5.1)). Since C is total in E, there exists $y \in C$ such that $p(y) > 0$; in view of $(r(u)e - u)p = 0$ it follows that $p(y)$ is an eigenvector in C pertaining to $r(u)$. Finally, if u' is the adjoint of u in the strong dual E', we have $\sigma(u) = \sigma(u')$ and $\lambda \to R(\lambda)'$ is the resolvent of u' (cf. (IV, 7.9)). In particular, $\lambda \to R(\lambda)'$ has a pole at $\lambda = r(u')$ $= r(u)$, and we obtain the assertion for u' by taking adjoints throughout in the preceding proof; in particular, $p(C) \subset C$ implies $p'(C') \subset C'$, and p' does not vanish on C', since C' is total in E_σ' and p' is continuous.

REMARK. If C is total in E, the preceding proof shows that for any continuous, positive endomorphism u of E whose resolvent has a pole at $\lambda = r(u)$, there exist eigenvectors of u in C and of u' in C' pertaining to $r(u)$.

The theorem of Krein-Rutman has generalizations in various directions; of course, some compactness assumption has to be made. The remaining two results in this section are typical for the sort of generalization that one obtains.

Let E be an ordered real Banach space with positive cone C; a linear map u of E into itself is called C-**compact** if u is continuous on C into C, and if $u(U \cap C)$ is relatively compact, U denoting the unit ball of E. We define the C-**spectral radius** of u to be the number

$$r_C = \lim_{n \to \infty} (\sup \|u^n(x)\| : x \in C, \|x\| \leq 1)^{1/n}$$

(the proof below will show that the limit always exists). In the following theorem, we assume the normality of C merely for the convenience of proof; this assumption is actually dispensable. The reader who wishes to obtain further information is referred to Bonsall [4] and the author's paper [3], Section 10.

2.5

Let E be an ordered real Banach space with normal positive cone C. If u is a C-compact mapping in E such that $r_C > 0$, then r_C is an eigenvalue of u with an eigenvector in C.

Proof. Denote by U the unit ball of E and by W the convex, circled hull of $U \cap C$. Then $\{\varepsilon W : \varepsilon > 0\}$ is a 0-neighborhood base for a normable topology $\mathfrak{T}$ on the subspace $E_0 = C - C$ of E. It is readily seen that $\mathfrak{T}$ is identical with the topology $\mathfrak{T}_1$ introduced in Chapter V, Section 3, Lemma 2. Thus if q is the gauge function of W, (E_0, q) is a Banach space; moreover, on C the norm q agrees with the original norm of E. Thus C is a normal closed cone in (E_0, q), and since r_C is nothing other than the spectral radius $r(v)$ of the restriction v of u to E_0, it follows from (2.2) that $r(v) \in \sigma(v)$. Hence $\{R_v(\lambda_n) : n \in N\}$ is unbounded in $\mathcal{L}((E_0, q))$ for any decreasing real sequence $\{\lambda_n\}$ such that $\lim_n \lambda_n = r(v)$, and by the principle of uniform boundedness there exists $y \in C$ such that $\lim_n q(R_v(\lambda_n)y) = +\infty$ for a given sequence $\{\lambda_n\}$ tending monotonically to $r(v)$. Let $x_n = R_v(\lambda_n)y/q(R_v(\lambda_n)y)$; then $x_n \in C$ and $q(x_n) = \|x_n\| = 1$ for all n. Moreover, $\lim_n q(\lambda_n x_n - v(x_n)) = \lim_n \|\lambda_n x_n - u(x_n)\| = 0$; this implies $\lim_n (r_C e - u)x_n = 0$ in E. Therefore, since the range of the sequence $\{u(x_n)\}$ is relatively compact in E by hypothesis, the sequence $\{x_n\}$ has a cluster point x in E (and hence in C, since C is closed). Clearly, this cluster point satisfies $r_C x = u(x)$ and $\|x\| = 1$, which completes the proof.

The second generalization that we have in mind concerns convex cones with compact base. Let us recall (Chapter II, Exercise 30) that a convex cone C of vertex 0 in a l.c.s. E has a compact base if there exists a (real) affine subspace N of E not containing 0, such that $C \cap N$ is compact and $C = \{\lambda x: \lambda \geqq 0, x \in N \cap C\}$. From the separation theorem (II, 9.2) it is clear that there exists a closed real hyperplane H strictly separating $N \cap C$ from $\{0\}$; clearly, then, $C = \{\lambda x: \lambda \geqq 0, x \in H \cap C\}$.

2.6

Theorem. *Let E be a l.c.s. over $\mathbf{R}$ and let C be a cone in E with compact base. If u is an endomorphism of the subspace $C - C$ of E such that $u(C) \subset C$ and the restriction of u to C is continuous, then u has an eigenvalue $\geqq 0$ with an eigenvector in C.*

Proof. Let $H = \{x: f(x) = 1\}$ be a hyperplane in E such that $H \cap C$ is a compact base of C. Denote by V the convex hull of $\{0\} \cup (H \cap C)$ in E and set $U = V - V$; then $\{\varepsilon U: \varepsilon > 0\}$ is a 0-neighborhood base in $E_0 = C - C$ for a normable topology $\mathfrak{T}$; it is not difficult to verify that the norm

$$z \to \|z\| = \inf\{f(x) + f(y): z = x - y, x, y \in C\}$$

generates the topology $\mathfrak{T}$ on E_0. Moreover, since U is compact and hence complete in E, and since $\mathfrak{T}$ is finer on E_0 than the topology induced by E, it follows from (I, 1.6) that $(E_0, \mathfrak{T})$ is complete, hence $(E_0, \| \; \|)$ is a Banach space. Further, C is closed in this space and clearly normal, and by (V, 5.5) u is a continuous, positive endomorphism of $(E_0, \| \; \|)$ for the order of E_0 whose positive cone is C. Thus from (2.2) above, it follows that the spectral radius $r(u)$ is a number in $\sigma(u)$. (It is quite possible that $r(u) = 0$, even if $u \neq 0$.) As in the proof of (2.5), we construct a sequence $\{x_n\}$ in C such that $\|x_n\| = f(x_n) = 1$ for all n, and such that $\lim_n \|r(u)x_n - u(x_n)\| = 0$. Since $H \cap C$ is compact in E and u is continuous on C by hypothesis, every cluster point $x \in H \cap C$ (for the topology induced by E) of the sequence $\{x_n\}$ satisfies $r(u)x = u(x)$. This completes the proof.

The following corollary is due to Krein-Rutman [1].

COROLLARY. *Let E be an ordered real Banach space whose positive cone C has interior points. If u is any positive (necessarily continuous) endomorphism of E, there exists an eigenvalue $\geqq 0$ of the adjoint u' with an eigenvector in the dual cone C'. If, in addition, C is normal (in particular, if E is an (AM)-space with unit), then the spectral radius $r(u)$ of u is such an eigenvalue of u'.*

Proof. In fact, if x_0 is interior to C, then the hyperplane $H = \{x': \langle x_0, x' \rangle = 1\}$ in E' has a $\sigma(E', E)$-compact intersection with C'; hence C' is a cone with compact base in E'_σ, and u' (satisfying $u'(C') \subset C'$) is $\sigma(E', E)$-continuous so that (2.6) applies. If, in addition, C is a normal cone in E, then $E' = C' - C'$ by (V, 3.3), Corollary 3, and it is readily seen that the topology $\mathfrak{T}$ constructed in the proof of (2.6) is the topology of the strong dual E'_β. Hence the number

$r(u')$, which was proved to be an eigenvalue of u', is the spectral radius of u' in E'_β and therefore equals $r(u)$.

3. THE PERIPHERAL POINT SPECTRUM

Let u be a continuous positive endomorphism of an ordered Banach space E. The subset of $\sigma(u)$ located on the spectral circle $\{\lambda: |\lambda| = r(u)\}$ will be called the **peripheral spectrum**, and its intersection with $\pi(u)$ the **peripheral point spectrum** of u. It has been pointed out earlier that it is apparently unknown whether (2.4) is true for all positive operators on E; on the other hand, it is natural to ask if, for example, under the hypothesis of (2.2) the peripheral spectrum of u is possibly subject to further restrictions. The following example shows that the answer is negative.

Example. Let E be a real Banach space of dimension at least 2; denote by F any closed subspace of codimension 1, and by $G = \{\lambda x_0: \lambda \in \mathbf{R}\}$ a complementary subspace so that $E = G \oplus F$ (cf. (I, 3.5)). For each $x \in E$, let $x = \lambda x_0 + y$ be the unique representation such that $y \in F$. The set $C = \{x \in E: \lambda \geq \|y\|\}$ is readily seen to be a closed normal cone such that $E = C - C$; thus E, with C as its positive cone, is an ordered Banach space such that the positive cone of $\mathscr{L}(E)$ is normal. Now let $v \in \mathscr{L}(F)$ satisfy $\|v\| \leq 1$, and define $u \in \mathscr{L}(E)$ by

$$u(x) = \lambda x_0 + v(y) \qquad (x \in E);$$

that is, $u = p + v \circ (e - p)$, where p denotes the projection $x \to \lambda x_0$ of E onto G. Because of $\|v\| \leq 1$, we have $u(C) \subset C$, and hence u is positive; clearly, $r(u) = 1$ and $\sigma(u) = \{1\} \cup \sigma(v)$, $\sigma(v)$ denoting the spectrum of $v \in \mathscr{L}(F)$. By choosing E so that F is isomorphic with a suitable Banach space and by a suitable choice of $v \in \mathscr{L}(F)$, it can be arranged that $\sigma(u)$ is a preassigned closed subset of $\{\lambda: |\lambda| = 1\}$ containing 1.

Therefore, for any further fruitful investigation of spectral properties of positive operators, it is necessary to consider more restricted types of positive maps and/or of ordered spaces. We present in this section a number of results on the peripheral point spectrum that are valid under reasonably general assumptions.

Several such results have been obtained by Krein-Rutman [1] for what they called strongly positive operators on ordered Banach spaces whose positive cone has non-empty interior; these operators u are such that for each non-zero $x \in C$, there exists $n \in N$ for which $u^n(x)$ is interior to C. This is a severe restriction, since spaces without order units (such as $L^p(\mu)$, $1 \leq p < +\infty$) are excluded from the discussion (cf. Chapter V, Exercise 10); on the other hand, if E is a Banach lattice with order units, then in view of (V, 8.4) and (V, 8.5), E is isomorphic (as an ordered t.v.s.) with $\mathscr{C}_\mathbf{R}(X)$ for a suitable compact space X, and we obtain very strong results (see (3.3) below). We generalize the notion of strong positivity as follows:

Let E be an ordered real Banach space with positive cone $C \neq \{0\}$. A continuous, positive endomorphism u of E is called **irreducible** if for some scalar $\lambda > r(u)$ and each non-zero $x \in C$, the element

$$uR(\lambda)x = \sum_{n=1}^{\infty} \lambda^{-n}u^n(x)$$

is a quasi-interior point of C. Recall that $y \in E$ is quasi-interior to C (Chapter V, Section 7) if the order interval $[0, y]$ is a total subset of E. If E is a normed lattice (Chapter V, Section 7), the quasi-interior points of C are weak order units of E; the following characterization of irreducible endomorphisms of Banach lattices justifies our terminology.

3.1

A continuous positive endomorphism $u \neq 0$ of a Banach lattice is irreducible if and only if no closed solid subspace, distinct from $\{0\}$ and E, is invariant under u.

Proof. Let u be irreducible and let $F \neq \{0\}$ be a closed solid subspace invariant under u. If $0 \neq x \in F \cap C$, then $y = uR(\lambda)x$ is, for suitable $\lambda > r(u)$, a quasi-interior point of C contained in F; hence $F = E$.

Conversely, if u leaves no closed proper solid subspace $\neq \{0\}$ of E invariant, then $u(x) > 0$ for each $x > 0$; for, $u(x_0) = 0$ for some $x_0 > 0$ would imply that u leaves the closed solid subspace G invariant which is generated by x_0, hence $G = E$, and it would follow that $u = 0$, which is contradictory. Therefore, if $x > 0$, then $y = uR(\lambda)x > 0$ ($\lambda > r(u)$ being arbitrary), and since $u(y) \leqq \lambda y$ it follows that the closed solid subspace F generated by y is invariant under u. Hence $F = E$, so that y is a quasi-interior point of E.

In the proof of our first result on irreducible endomorphisms, we shall need this lemma.

LEMMA 1. *Let E be an ordered Banach space with positive cone C and $p \in \mathscr{L}(E)$ a positive projection. If $x \in p(C)$ is quasi-interior to C, then x is quasi-interior to $p(C)$ in $p(E)$.*

Proof. Letting $C_1 = p(C)$ and $E_1 = p(E)$, we observe that $C_1 \cap (x - C_1) = [0, x]_1 = [0, x] \cap E_1 = p([0, x])$, since p is a positive projection. Since the linear hull of $[0, x]$ is dense in E, the linear hull of $[0, x]_1$ is dense in the subspace E_1 by the continuity of p.

A linear form f on an ordered vector space E over $\boldsymbol{R}$ is termed **strictly positive** if $x > 0$ implies $f(x) > 0$. Note also that the existence of an irreducible positive endomorphism on an ordered Banach space E implies that the positive cone C of E is total (for C contains quasi-interior points); hence the dual cone $C' \subset E'$, being a closed proper cone in E'_σ, defines the canonical order of E'.

3.2

Let E be an ordered real Banach space with positive cone C, and suppose that u is an irreducible positive endomorphism whose spectral radius r is a pole of the resolvent. Then:

(i) *$r > 0$ and r is a pole of order 1.*

(ii) *There exist positive eigenvectors, pertaining to r, of u and u'. Each positive eigenvector for r is quasi-interior to C, and each positive eigenvector of u' for r is a strictly positive linear form.*

(iii) *Each of the following assumptions implies that the multiplicity $d(r)$ of r is 1: (a) C has non-empty interior, (b) $d(r)$ is finite, (c) E is a Banach lattice.*

REMARK. (iii) can be replaced by the assertion "$d(r) = 1$" if there exists no ordered Banach space of dimension > 1 where each $x > 0$ is quasi-interior to C. On the other hand, if such a (necessarily infinite-dimensional) space exists, the identity map e is irreducible and such that $d(r) = \dim E$.

Proof of (3.2). Let p denote the leading coefficient of the principal part of the resolvent at $\lambda = r$ and let q be the residue at r. Then p is positive (cf. proof of (2.4), Corollary), and $p = q(u - re)^{k-1}$, where k is the order of the pole; moreover, q and its adjoint q' are projections such that $q(E), q'(E')$ are the null spaces of $(re - u)^k, (re' - u')^k$, respectively.

(ii): Since C and C' are total subsets of E and E'_σ, respectively, there exist eigenvectors (pertaining to r) of u in C, and of u' in C'. Let x_0, x'_0 be any such eigenvectors. From

$$uR(\lambda)x_0 = \sum_{1}^{\infty} \left(\frac{r}{\lambda}\right)^n x_0 \qquad (\lambda > r)$$

it follows that $r > 0$ and that x_0 is quasi-interior to C. Similarly, from

$$\langle x, x'_0 \rangle \sum_{1}^{\infty} \left(\frac{r}{\lambda}\right)^n = \sum_{1}^{\infty} \lambda^{-n} \langle u^n(x), x'_0 \rangle = \langle uR(\lambda)x, x'_0 \rangle \ (\lambda > r)$$

it follows that $\langle x, x'_0 \rangle > 0$ whenever $0 \neq x \in C$, for x'_0 must be > 0 at the quasi-interior point $uR(\lambda)x$ of C.

(i): We have to show that r, which is > 0 by the preceding, is a pole of order 1. In fact, let $x \in C$ be such that $x_0 = p(x)$ is not zero and let $x' \in C'$ be an eigenvector of u' for r. In view of $p = q(u - re)^{k-1}$ and $q'(x'_0) = x'_0$, we obtain

$$0 < \langle x_0, x'_0 \rangle = \langle q(u - re)^{k-1}x, x'_0 \rangle = \langle x, (u' - re')^{k-1}x'_0 \rangle,$$

which implies that $k = 1$.

(iii): Since $\lambda = r$ is a simple pole of the resolvent, we have $p = q$; hence p is a positive projection and, by (ii), every non-zero element of $p(C)$ is quasi-interior to C. Therefore, by Lemma 1 every non-zero $x \in p(C)$ is quasi-interior to $p(C)$ in $p(E)$. In case (a), C has interior points, hence so does $p(C)$ in $p(E)$. Since $p(C)$ (which is contained in C) is a closed proper cone in $p(E)$, it follows that $p(E)$ has dimension 1 and hence that $d(r) = 1$. If (b) $d(r)$ is finite, then every quasi-interior point of $p(C)$ in $p(E)$ is interior to $p(C)$ (cf. the lemma preceding (V, 4.1)) and the conclusion is the same as before.

There remains to show that $d(r) = 1$ if E is a Banach lattice. If x is any eigenvector of u pertaining to r, then from $rx = u(x)$ it follows that $r|x| = |u(x)| \leq u(|x|)$; if $x_0' \in C'$ is an eigenvector of u' for r, we obtain

$$r\langle |x|, x_0' \rangle \leq \langle u(|x|), x_0' \rangle = r\langle |x|, x_0' \rangle,$$

and this implies that $r|x| = u(|x|)$, since $r > 0$ and since x_0' is a strictly positive linear form by (ii). Now $x = x^+ - x^-$ and $|x| = x^+ + x^-$; hence both x^+, x^- are positive elements of the eigenspace of u pertaining to r. Since they are lattice disjoint and the lattice operations are continuous in E, both cannot be quasi-interior points of C; thus either $x^+ = 0$ or else $x^- = 0$. Therefore, if x is an eigenvector of u for r, we have either $x \in C$ or $x \in -C$. Hence this eigenspace is totally ordered; since this order is also Archimedean, it follows that $d(r) = 1$ (Chapter V, Exercise 2).

This completes the proof of (3.2).

Our principal result on irreducible positive maps is concerned with Banach lattices E of type $\mathscr{C}_{\mathbf{R}}(X)$, X being a compact space. To avoid confusion with the unit e of $\mathscr{L}(E)$, we shall denote the constantly-one function $t \to 1$ on X by 1. It will be convenient to employ the following terminology: A positive endomorphism u of $\mathscr{C}_{\mathbf{R}}(X)$ with spectral radius r is said to have a **cyclic** peripheral point spectrum if $r\alpha f = u(f)$, $|\alpha| = 1$, and $f = |f|g \in \mathscr{C}_{\mathbf{C}}(X)$ imply that $r\alpha^n |f| g^n = u(|f| g^n)$ for all integers ($n \in \mathbf{Z}$); here u is identified with its unique extension to $\mathscr{C}_{\mathbf{C}}(X)$ (the complexification of $\mathscr{C}_{\mathbf{R}}(X)$), $|f|$ denotes the usual absolute value of $f \in \mathscr{C}_{\mathbf{C}}(X)$, and fg denotes the function $t \to f(t)g(t)$ (pointwise multiplication). We shall need the following lemma.

LEMMA 2. *Let v be a positive endomorphism of $\mathscr{C}_{\mathbf{R}}(X)$ such that $v(1) = 1$. If $\alpha g = v(g)$, where $|\alpha| = 1$ and $|g| = 1$, $g \in \mathscr{C}_{\mathbf{C}}(X)$, then $\alpha^n g^n = v(g^n)$ for all $n \in \mathbf{Z}$.*

Proof. For each $s \in X$, the mapping $f \to v(f)(s)$ is a continuous positive linear form on $\mathscr{C}_{\mathbf{R}}(X)$, hence a positive Radon measure μ_s on X; from $v(1) = 1$ we conclude that $\|\mu_s\| = \mu_s(1) = 1$. Hence $\alpha g(s) = \int_X g(t)\, d\mu_s(t)$ for each $s \in X$, and from $|g| = 1$ it follows that $g(t)$ is constant on the support of μ_s, namely equal to $\alpha g(s)$. Therefore, $\alpha^n g^n(s) = \int_X g^n(t)\, d\mu_s(t)$ ($s \in X$, $n \in \mathbf{Z}$), which is the assertion.

In view of the representation theorem (V, 8.5), the following result is valid for irreducible positive maps of an arbitrary (AM)-space with unit.

3.3

Theorem. *Let X be a compact space and suppose that u is an irreducible positive endomorphism of $\mathscr{C}_R(X)$. Then the following assertions hold:*

(i) *The spectral radius r of u is > 0, and the condition $\|u\| = r$ is equivalent with $u(1) = r1$.*

(ii) *The peripheral point spectrum of u is cyclic.*

(iii) *Each eigenvalue $r\alpha$, $|\alpha| = 1$, of u has multiplicity 1 and the corresponding eigenfunction is $\neq 0$ throughout; moreover, $\sigma(u)$ is invariant under the rotation through θ, where $\alpha = \exp i\theta$.*

(iv) *If the peripheral point spectrum contains an isolated point, then it is of the form rH, where H is the group of nth roots of unity for some $n \geqq 1$.*

(v) *If the peripheral point spectrum contains a pole of the resolvent of u, then all its points are poles of order 1 of the resolvent.*

(vi) *r is the only possible eigenvalue of u with an eigenfunction $\geqq 0$. If X is connected, the peripheral point spectrum cannot contain points $r\alpha$ such that α is a root of unity distinct from 1.*

REMARK. It can happen that the peripheral point spectrum of u is empty; on the other hand, even if X is connected, the peripheral point spectrum can be dense in the spectral circle (see examples below). In the latter case, it is still of the form rG where G is a subgroup of the circle group.

Proof of (3.3). By the corollary of (2.6), there exists a positive Radon measure μ_0 on X such that $r\mu_0 = u'(\mu_0)$. Since $\{f : \mu_0(|f|) = 0\}$ is a solid subspace of $\mathscr{C}_R(X)$ invariant under u, μ_0 is strictly positive which is equivalent to the assertion that the support S_0 of μ_0 equals X.

We proceed to prove the statements of the theorem in the order of enumeration.

(i): The assumption $r = 0$ implies $u'(\mu_0) = 0$, hence $\int u(1) d\mu_0 = 0$; since $S_0 = X$, it follows that $u(1) = 0$, hence $u = 0$, which is contradictory. Thus $r > 0$. Clearly, $u(1) = r1$ implies $\|u\| = r$; conversely, if $\|u\| = r$, then $u(1) \leqq r1$, and $\langle r1 - u(1), \mu_0 \rangle = r\mu_0(1) - \langle u(1), \mu_0 \rangle = 0$; from $S_0 = X$ we conclude that $u(1) = r1$.

(ii): Suppose that $r\alpha f = u(f)$, where $f \neq 0$ and $|\alpha| = 1$. We obtain $r|f| = r|\alpha f| = |u(f)| \leqq u(|f|)$; since $u'(\mu_0) = r\mu_0$, it follows that $\langle u(|f|) - r|f|, \mu_0 \rangle = 0$, and we conclude, as before, that $r|f| = u(|f|)$. This implies $f(s) \neq 0$ for all $s \in X$. In the opposite case, the closed subspace of $\mathscr{C}_R(X)$ generated by the order interval $[0, |f|]$ would be a closed solid sublattice, neither $\{0\}$ nor the whole space, and clearly invariant under u; this is impossible by (3.1).

Now let $f = |f|g$ and define a positive endomorphism v of $\mathscr{C}_R(X)$ by

$$v(h)(s) = r^{-1}|f(s)|^{-1}u(|f|h)(s) \qquad (s \in X).$$

We have $v(1) = 1$ and the complex extension of v (again denoted by v)

satisfies $\alpha g = v(g)$. From Lemma 2 it follows that $\alpha^n g^n = v(g^n)$ for all $n \in \mathbf{Z}$ which clearly implies the assertion.

(iii): Let $r\alpha$, where $\alpha = \exp i\theta$, be an eigenvalue of u. It has been shown under (ii) that any eigenfunction f pertaining to $r\alpha$ is $\neq 0$ throughout; moreover, $f_0 = |f|$ is an eigenfunction for r. If h is any other eigenfunction for r, we can assume that h is real valued; set $c = \sup\{h(t)/f_0(t): t \in X\}$. The function $cf_0 - h$ belongs to the eigenspace $N(r)$ and vanishes for at least one $t \in X$, since the supremum is assumed in X. Hence $cf_0 - h = 0$ by the preceding, that is, the multiplicity of r is 1.

Denote by w the endomorphism of $\mathscr{C}_c(X)$ defined by $h \to gh$, where g (with $|g| = 1$) is the function $f/|f|$ introduced above. We define v by $v = \alpha^{-1}w^{-1}uw$; as a continuous endomorphism of $\mathscr{C}_c(X)$, v is given by

$$v(h)(s) = \int h(t)\, dv_s(t) \qquad (s \in X),$$

where each v_s is a uniquely determined complex Radon measure on X. Similarly, let $u(h)(s) = \int h(t)\, d\mu_s(t)(s \in X)$; in view of $|g| = 1$, we obtain

$$\left| \int h(t)\, dv_s(t) \right| = |g(s)^{-1}u(gh)(s)| \leq u(|h|)(s) = \int |h(t)|\, d\mu_s(t)$$

for all $s \in X$. Thus if $v_s = \rho_s + i\tau_s$ is the decomposition of v_s by means of real Radon measures (Chapter I, Section 7), it follows that $\rho_s \leq \mu_s$ for all s. On the other hand, $v(f_0) = rf_0$ and $u(f_0) = rf_0$, whence $\int f_0 d(\mu_s - \rho_s) = 0$ for all $s \in X$; since $f_0(t) > 0$ throughout, it follows that $\rho_s = \mu_s$. Therefore, $\tau_s = 0$ for all s and hence $v = u$; in other words,

$$u = \alpha^{-1}w^{-1}uw. \qquad (*)$$

From this it is clear that $\sigma(u) = \sigma(\alpha u)$, hence $\sigma(u)$ is invariant under rotation through θ ($\alpha = \exp i\theta$). Formula (*) also shows that if there are any elements in the peripheral point spectrum, then their common multiplicity is 1 (namely, equal to that of r).

(iv): Suppose that $r\alpha$, where $\alpha = \exp i\theta \neq 1$, is an isolated element of the peripheral point spectrum of u; it follows from (*) that r is such an element. Consequently, there exists an eigenvalue $r \exp i\theta_1$ ($0 < \theta_1 < 2\pi$) for which θ_1 is minimal; for, since r is isolated, (ii) implies that the peripheral point spectrum consists of a finite number of roots of unity. In particular, $\theta_1 = 2\pi/n$ for some $n > 1$ and (again by (ii)) the numbers $r \exp im\theta_1$ ($m = 0, 1, \ldots, n - 1$) are all eigenvalues. Now let $r \exp i\theta$ be any eigenvalue on $|\lambda| = r$ and denote by k the smallest integer > 0 such that $\theta + k\theta_1 \geq 2\pi$. Let $\chi = \theta + k\theta_1$; since $r \exp ik\theta_1$ is an eigenvalue of u, it follows from (*), applied to $\alpha = \exp i\theta$, that $r \exp i\chi$ is an eigenvalue of u. Now if we had $\chi > 2\pi$, we would also have $\chi < 2\pi + \theta_1$, which contradicts the definition of θ_1. Thus $\theta + k\theta_1 = 2\pi$ and hence $\theta = (n - k)\theta_1$; this shows that the peripheral point spectrum of u is exactly the set rH, where H denotes the group of nth roots of unity.

(v): If the peripheral point spectrum contains a pole of the resolvent of u, then, clearly, the preceding applies; it follows from (*) that each element of rH is a pole of the same order, this common order being 1 by (3.2).

(vi): Suppose that $\lambda f = u(f)$, where $0 \neq f \geq 0$; from $\lambda \langle f, \mu_0 \rangle = \langle u(f), \mu_0 \rangle = r \langle f, \mu_0 \rangle > 0$ it follows that $\lambda = r$.

Let $r\alpha$, where α is a primitive nth root of unity, be an eigenvalue of u with eigenfunction $f = |f|g$; if v is the mapping $h \to r^{-1}|f|^{-1}u(|f|h)$, then $\alpha g = v(g)$, and v satisfies the hypothesis of Lemma 2. We define $M_k = g^{-1}(\alpha^k)$ ($k = 0$, 1, ...), and without loss in generality we can assume that $M_0 \neq \varnothing$, that is, $g(t) = 1$ for some $t \in X$. As in the proof of Lemma 2, we write $v(h)(s) = \int h(t) d\mu_s(t)$ ($s \in X$) and conclude from $\alpha g = v(g)$, $|g| = 1$, that whenever $s \in M_k$ then the support of μ_s is contained in M_{k+1}. Since $k \equiv k' \pmod n$ implies $M_k = M_{k'}$ (sets with indices incongruent mod n being disjoint), the map v induces a cyclic permutation $M_k \to M_{k-1}$ ($k \bmod n$). From this it follows that the closed solid sublattice $F \subset \mathscr{C}_{\mathbf{R}}(X)$ of functions vanishing on $M = \bigcup_0^{n-1} M_k$ is invariant under v and hence under u; hence by (3.1), $F = \{0\}$ or, equivalently, $X = M$, since M is closed. Therefore, $g(X)$ is the cyclic group generated by α; on the other hand, $g(X)$ is connected, since X is connected and g is continuous, and this implies that $\alpha = 1$.

This completes the proof of (3.3).

An example of an irreducible positive operator with empty point spectrum is furnished by the endomorphism u of $\mathscr{C}_{\mathbf{R}}[0, 1]$ defined by

$$u(f)(s) = sf(s) + \int_0^s f(t)\, dt + \int_0^1 (1-t)^2 f(t)\, dt$$

for $s \in [0, 1]$. It is immediate that u is irreducible, and it is not difficult to verify that u has no eigenvalues.

For our second example, let X be the unit circle $\{z: |z| = 1\}$, let α be a fixed element of X which is not a root of unity, and let u be the endomorphism of $\mathscr{C}_{\mathbf{R}}(X)$ defined by $u(f)(z) = f(\alpha z)$; then u is positive, and irreducible since the set $\{\alpha^n: n \in \mathbf{N}\}$ is dense in X. It is easy to see that the group $H = \{\alpha^n: n \in \mathbf{Z}\}$ belongs to the point spectrum of u; in fact the peripheral point spectrum of u is exactly H.

Our final theorem is concerned with a condition under which a positive endomorphism of a general Banach lattice has cyclic peripheral point spectrum (in a slightly weakened sense). If E is a Banach lattice, E_1 the Banach space which is the complexification of E, we shall say that the absolute value $x \to |x|$ of E can be extended to E_1 if for each pair $(x, y) \in E \times E$, $\sup\{|x \cos \theta + y \sin \theta|: 0 \leq \theta < 2\pi\}$ exists in E; in these circumstances, the supremum serves to define $|x + iy|$. The reader will observe that the absolute value of E can be extended to E_1 whenever E is order complete; however, this condition is not a necessary one as the example of the Banach lattices $\mathscr{C}_{\mathbf{R}}(X)$ (X compact) shows. The extension of $x \to |x|$ to E_1, if it exists, satisfies the relations

$|z_1 + z_2| \leqq |z_1| + |z_2|$, $|\rho z| = |\rho| \, |z|$ $(z, z_1, z_2 \in E_1; \ \rho \in C)$, and $|u(z)| \leqq u(|z|)$ whenever u is (the extension to E_1 of) a positive endomorphism of E.

3.4

Theorem. *Let E be a Banach lattice whose absolute value can be extended to the complexification E_1 of E, and let u be a positive endomorphism of E, with spectral radius r, for which there exists a strictly positive linear form x_0' satisfying $u'(x_0') \leqq rx_0'$. If the peripheral point spectrum of u contains $r\alpha$, $|\alpha| = 1$, then it contains rH, where H is the cyclic group generated by α.*

Proof. We can suppose that $r > 0$, and hence for convenience we assume $r = 1$. Let $\alpha x = u(x)$, where $0 \neq x = x_1 + ix_2 \in E_1$ $(x_1, x_2 \in E)$ and $|\alpha| = 1$; it follows that $|x| = |\alpha x| = |u(x)| \leqq u(|x|)$. Now $\langle |x|, x_0' \rangle \leqq \langle u(|x|), x_0' \rangle \leqq \langle |x|, x_0' \rangle$ by virtue of the hypothesis $u'(x_0') \leqq x_0'$; we obtain $\langle u(|x|) - |x|, x_0' \rangle = 0$ and hence $|x| = u(|x|)$, since x_0' is strictly positive.

Define $x_0 = |x|$, and consider the solid sublattice $F = \bigcup_1^\infty n[-x_0, x_0]$ of E. Since E is complete and the order interval $[-x_0, x_0]$ is closed, every positive sequence of type l^1 in F is order summable to an element of F, and x_0 is an order unit of F; hence by (V, 6.2), F is complete under its order topology (in general, F is not a closed subspace of E). Moreover, by (V, 8.4), (F, p) is an (AM)-space with unit x_0 where p stands for the gauge function of $[-x_0, x_0]$. Hence by (V, 8.5) there exists an isomorphism ψ of (F, p) onto $\mathscr{C}_R(X)$ (X compact), which extends to an isomorphism of the complexification F_1 of F onto $\mathscr{C}_C(X)$; in addition, F_1 can be identified with a subspace of E_1. Clearly, F and F_1 are invariant under u, and the restriction u_0 of u to F induces a positive endomorphism v of $\mathscr{C}_R(X)$ (precisely, $v = \psi \circ u_0 \circ \psi^{-1}$). Since $\psi(x_0) = 1$ (the constantly-one function on X), we have $v(1) = 1$ from $u(x_0) = x_0$; moreover, we have $x \in F_1$ and hence $g = \psi(x)$ for some $g \in \mathscr{C}_C(X)$ satisfying $\alpha g = v(g)$, and $|g| = 1$, since ψ preserves absolute values. The assertion follows now from Lemma 2, and the proof is complete.

The preceding theorem applies to all Banach lattices $L^p(\mu)$, (Y, Σ, μ) being an arbitrary measure space (Chapter II, Section 2, Example 2). If, in this particular case, u satisfies the hypothesis of (3.4) and (assuming $r(u) = 1$) $\alpha f = u(f)$, then the Banach lattice (F, p) constructed in the preceding proof can be identified with the set of all classes (mod μ-null functions) $|f|g$, where g contains a function bounded on Y_0 and vanishing on $Y \sim Y_0$, Y_0 being the set on which some fixed representative of the class f is $\neq 0$. The space (F, p) is therefore essentially $L^\infty(\mu, Y_0)$, and the isomorphism ψ can be chosen to be an isomorphism of the Banach algebra $L^\infty(\mu, Y_0)$ onto the Banach algebra $\mathscr{C}_R(X)$ (cf. Chapter V, Exercise 24). It follows that if we write $f = |f|g$ with $g \in L^\infty(\mu, Y_0)$, then $\alpha^n |f| g^n = u(|f| g^n)$ for all $n \in Z$; it is thus reasonable to extend the meaning of the term "cyclic peripheral point spectrum" to the present case, where $E = L^p(\mu)$ $(1 \leq p \leq \infty)$. In particular, we obtain the following corollary.

COROLLARY (G.-C. Rota [1]). *Every positive endomorphism u of $L^1(\mu)$, satisfying $r(u) = \|u\|$, has cyclic peripheral point spectrum.*

Proof. In fact, $f \to \int f \, d\mu$ is a strictly positive linear form h on $L^1(\mu)$, and $\|u\| = r(u)$ implies that $\langle u(f), h \rangle = \int u(f) \, d\mu \leq \|u\| \int f \, d\mu = r(u) \langle f, h \rangle$ whenever $f \geq 0$; it follows that $u'(h) \leq r(u)h$, completing the proof.

INDEX OF SYMBOLS

PREREQUISITES

Section A, 1: $x \in X,\ x \notin X$

$X \subset Y,\ X = Y$

$\{x \in X : (p)x\},\ \{x : (p)x\},\ \{x\}$

$Y \sim X,\ \varnothing$

$\Rightarrow,\ \Leftrightarrow$

$\mathfrak{P}(X)$

2: $f : X \to Y,\ x \to f(x)$

$f(A),\ f^{-1}(B)$

$f \circ g$

f_A

$\{x_\alpha : \alpha \in A\},\ \{x_\alpha\},\ \{x_n\}$

N

4: $\cup,\ \cap,\ \prod$

X^A

X/R

5: $\leqq,\ <,\ \geqq,\ >$

$\sup A,\ \inf A,\ \sup(x, y),\ \inf(x, y)$

B, 1: $(X, \mathfrak{T})$

$\mathring{A},\ \overline{A}$

6: $(X, \mathfrak{W})$

$\tilde{X}$

C, 2: $\sum$

$R,\ C$

3: L/M

$L(L_1, L_2)$

4: K_0

L^*

5: R_+

BIBLIOGRAPHY

ANDO, T.

[1] Positive linear operators in semi-ordered linear spaces. *J. Fac. Sci. Hokkaido Univ.*, Ser. I., *13* (1957), 214–228.

[2] On fundamental properties of a Banach space with a cone. *Pacific J. Math.*, *12* (1962), 1163–1169.

ARENS, R. F.

[1] Duality in linear spaces. *Duke Math J.*, *14* (1947), 787–794.

ARENS, R. F. AND J. L. KELLEY

[1] Characterizations of the space of continuous functions over a compact Hausdorff space. *Trans. Amer. Math. Soc.*, *62* (1947), 499–508.

BAER, R.

[1] Linear algebra and projective geometry. New York 1952.

BANACH, S.

[1] Théorie des opérations linéaires. Warsaw 1932.

BANACH, S. ET H. STEINHAUS

[1] Sur le principe de la condensation de singularités. *Fund. Math.*, *9* (1927), 50–61.

BARTLE, R. G.

[1] On compactness in functional analysis. *Trans. Amer. Math. Soc.*, *79* (1955), 35–57.

BAUER, H.

[1] Sur le prolongement des formes linéaires positives dans un espace vectoriel ordonné. *C. R. Acad. Sci. Paris*, *244* (1957), 289–292.

[2] Über die Fortsetzung positiver Linearformen. *Bayer. Akad. Wiss. Math.-Nat. Kl. S.-B.*, *1957* (1958), 177–190.

BERGE, C.

[1] Topological spaces. New York 1963.

BIRKHOFF, G.

[1] Lattice theory. 3rd ed. New York 1961.

BIRKHOFF, G. AND S. MAC LANE
[1] A survey of modern algebra. 3rd ed. New York 1965.

BONSALL, F. F.
[1] Endomorphisms of partially ordered vector spaces. *J. London Math. Soc.*, *30* (1955), 133–144.
[2] Endomorphisms of a partially ordered vector space without order unit. *J. London Math. Soc.*, *30* (1955), 144–153.
[3] Linear operators in complete positive cones. *Proc. London Math. Soc.* (3), *8* (1958), 53–75.
[4] Positive operators compact in an auxiliary topology. *Pacific J. Math.*, *10* (1960), 1131–1138.

BOURBAKI, N.   ÉLÉMENTS DE MATHÉMATIQUE.
[1] Théorie des ensembles. Fascicule de résultats. 3rd ed. Paris 1958.
[2] Algèbre, chap. 2. 3rd ed. Paris 1962.
[3] Algèbre, chap. 3. 2nd ed. Paris 1958.
[4] Topologie générale, chap. 1 et 2. 3rd ed. Paris 1961.
[5] Topologie générale, chap. 9. 2nd ed. Paris 1958.
[6] Topologie générale, chap. 10. 2nd ed. Paris 1961.
[7] Espaces vectoriels topologiques, chap. 1 et 2. Paris 1953.
[8] Espaces vectoriels topologiques, chap. 3–5. Paris 1955.
[9] Intégration, chap. 1–4. Paris 1952.
[10] Intégration, chap. 5. Paris 1956.
[11] Intégration, chap. 6. Paris 1959.

BOURGIN, D. G.
[1] Linear topological spaces. *Amer. J. Math.*, *65* (1943), 637–659.

BRACE, J. W.
[1] The topology of almost uniform convergence. *Pacific J. Math.*, *9* (1959), 643–652.
[2] Approximating compact and weakly compact operators. *Proc. Amer. Math. Soc.*, *12* (1961), 392–393.

BUSHAW, D.
[1] Elements of general topology. New York 1963.

CIVIN, P. AND B. YOOD
[1] Quasi-reflexive spaces. *Proc. Amer. Math. Soc.*, *8* (1957), 906–911.

COLLINS, H. S.
[1] Completeness and compactness in linear topological spaces. *Trans. Amer. Math. Soc.*, *79* (1955), 256–280.

DAY, M. M.
[1] The spaces L^p with $0 < p < 1$. *Bull. Amer. Math. Soc.*, *46* (1940), 816–823.
[2] Normed linear spaces. 2nd ed. Berlin-Göttingen-Heidelberg 1962.
[3] On the base problem in normed spaces. *Proc. Amer. Math. Soc.*, *13* (1962), 655–658.

DE BRANGES, L.
[1] The Stone-Weierstrass theorem. *Proc. Amer. Math. Soc.*, *10* (1959), 822–824.

DIEUDONNÉ, J.
[1] La dualité dans les espaces vectoriels topologiques. *Ann. Sci. École Norm. Sup.* (*3*), *59* (1942), 107–139.
[2] Natural homomorphisms in Banach spaces. *Proc. Amer. Math. Soc.*, *1* (1950), 54–59.
[3] Sur les espaces de Köthe. *J. Analyse Math.*, *1* (1951), 81–115.
[4] Sur un théorème de Šmulian. *Arch. Math.*, *3* (1952), 436–440.
[5] Complex structures on real Banach spaces. *Proc. Amer. Math. Soc.*, *3* (1952), 162–164.
[6] Sur les propriétés de permanence de certains espaces vectoriels topologiques. *Ann. Soc. Polon. Math.*, *25* (1952), 50–55 (1953).
[7] Sur les espaces de Montel métrisables. *C. R. Acad. Sci. Paris*, *238* (1954), 194–195.
[8] On biorthogonal systems. *Michigan Math. J.*, *2* (1954), 7–20.
[9] Denumerability conditions in locally convex vector spaces. *Proc. Amer. Math. Soc.*, *8* (1957), 367–372.

DIEUDONNÉ, J. ET L. SCHWARTZ
[1] La dualité dans les espaces (F) et (LF). *Ann. Inst. Fourier* (*Grenoble*), *1* (1949), 61–101 (1950).

DUNFORD, N. AND J. T. SCHWARTZ
[1] Linear operators. Part I: General theory. New York 1958.
[2] Linear operators. Part II: Spectral theory. New York 1963.

DVORETZKY, A. AND C. A. ROGERS
[1] Absolute and unconditional convergence in normed linear spaces. *Proc. Nat. Acad. Sci. U.S.A.*, *36* (1950), 192–197.

EBERLEIN, W. F.
[1] Weak compactness in Banach spaces. I. *Proc. Nat. Acad. Sci. U.S.A.*, *33* (1947), 51–53.

FREUDENTHAL, H.
[1] Teilweise geordnete Moduln. *Nederl. Akad. Wetensch. Proc.*, *39* (1936), 641–651.

FROBENIUS, G.
[1] Über Matrizen aus positiven Elementen. *S.-B. Preuss. Akad. Wiss. Berlin* 1908, 471–476; 1909, 514–518.
[2] Über Matrizen aus nicht negativen Elementen. *S.-B. Preuss. Akad. Wiss. Berlin* 1912, 456–477.

GILLMAN, L. AND M. JERISON
[1] Rings of continuous functions. Princeton 1962.

GORDON, H.
[1] Topologies and projections on Riesz spaces. *Trans. Amer. Math. Soc.*, *94* (1960), 529–551.
[2] Relative Uniform Convergence. *Math. Ann.*, *153* (1964), 418–427.

GROSBERG, J. ET M. KREIN
[1] Sur la décomposition des fonctionnelles en composantes positives. *C. R. (Doklady) Acad. Sci. URSS* (*N.S.*), *25* (1939), 723–726.

GROTHENDIECK, A.
[1] Sur la complétion du dual d'un espace vectoriel localement convexe. *C. R. Acad. Sci. Paris, 230* (1950), 605–606.
[2] Quelques résultats relatifs à la dualité dans les espaces (F). *C. R. Acad. Sci. Paris, 230* (1950), 1561–1563.
[3] Critères généraux de compacité dans les espaces vectoriels localement convexes. Pathologie des espaces (LF). *C. R. Acad. Sci. Paris, 231* (1950), 940–941.
[4] Quelques résultats sur les espaces vectoriels topologiques. *C. R. Acad. Sci. Paris, 233* (1951), 839–841.
[5] Sur une notion de produit tensoriel topologique d'espaces vectoriels topologiques, et une classe remarquable d'espaces vectoriels liée à cette notion. *C. R. Acad. Sci. Paris, 233* (1951), 1556–1558.
[6] Critères de compacité dans les espaces fonctionnels généraux. *Amer. J. Math., 74* (1952), 168–186.
[7] Sur les applications linéaires faiblement compactes d'espaces du type C(K). *Canadian J. Math., 5* (1953), 129–173.
[8] Sur certains espaces de fonctions holomorphes, I, II. *J. reine angew. Math., 192* (1953), 35–64, 77–95.
[9] Sur les espaces de solutions d'une classe générale d'équations aux dérivées partielles. *J. Analyse Math., 2* (1953), 243–280.
[10] Sur les espaces (F) et (DF). *Summa Brasil. Math., 3* (1954), 57–123.
[11] Leçons sur les espaces vectoriels topologiques. Instituto de Matemática Pura e Aplicada Universidade de São Paulo. 2nd ed. São Paulo 1958.
[12] Résumé des résultats essentiels dans la théorie des produits tensoriels topologiques et des espaces nucléaires. *Ann. Inst. Fourier (Grenoble), 4* (1952), 73–112 (1954).
[13] Produits tensoriels topologiques et espaces nucléaires. *Mem. Amer. Math. Soc. no. 16* (1955).
[14] Une caractérisation vectorielle-métrique des espaces L^1. *Canadian J. Math., 7* (1955), 552–561.

HALMOS, P. R.
[1] Measure theory. 5th ed. Princeton 1958.
[2] Introduction to Hilbert space and the theory of spectral multiplicity. 2nd ed. New York 1957.
[3] Naive set theory. Princeton 1960.

HILLE, E. AND R. S. PHILLIPS
[1] Functional analysis and semi-groups. 2nd ed. Providence, R.I. 1957.

HOFFMAN, K.
[1] Banach spaces of analytic functions. Englewood Cliffs, N.J., 1962.

HYERS, D. H.
[1] Locally bounded linear topological spaces. *Rev. Ci. (Lima), 41* (1939), 555–574.
[2] Linear topological spaces. *Bull. Amer. Math. Soc., 51* (1945), 1–24.

JAMES, R. C.
[1] Bases and reflexivity of Banach spaces. *Ann. of Math. (2), 52* (1950), 518–527.
[2] A non-reflexive Banach space isometric with its second conjugate space.

Proc. Nat. Acad. Sci. U.S.A., *37* (1951), 174–177.

[3] Weak compactness and reflexivity. *Israel J. of Math.*, *2* (1964), 101–119.

JERISON, M. (See Gillman, L., and M. Jerison.)

KADISON, R. V.

[1] A representation theory for commutative topological algebras. *Mem. Amer. Math. Soc. no. 7* (1951).

KAKUTANI, S.

[1] Concrete representation of abstract (L)-spaces and the mean ergodic theorem. *Ann. of Math.* (2), *42* (1941), 523–537.

[2] Concrete representation of abstract (M)-spaces. *Ann. of Math.* (2), *42* (1941), 994–1024.

KARLIN, S.

[1] Unconditional convergence in Banach spaces. *Bull. Amer. Math. Soc.*, *54* (1948), 148–152.

[2] Bases in Banach spaces. *Duke Math. J.*, *15* (1948), 971–985.

[3] Positive operators. *J. Math. Mech.*, *8* (1959), 907–937.

KELLEY, J. L. (See also R. F. Arens)

[1] General topology. New York 1955.

KELLEY, J. L., I. NAMIOKA AND CO-AUTHORS

[1] Linear topological spaces. Princeton 1963.

KIST, J.

[1] Locally *o*-convex spaces. *Duke Math. J.*, *25* (1958), 569–582.

KLEE, V. L.

[1] Invariant metrics in groups (solution of a problem of Banach). *Proc. Amer. Math. Soc.*, *3* (1952), 484–487.

[2] Boundedness and continuity of linear functionals. *Duke Math. J.*, *22* (1955), 263–270.

[3] Extremal structure of convex sets. *Arch. Math.*, *8* (1957), 234–240.

[4] Extremal structure of convex sets. II. *Math. Z.*, *69* (1958), 90–104.

[5] Convexity. Princeton (to appear).

KOLMOGOROFF, A.

[1] Zur Normierbarkeit eines allgemeinen topologischen linearen Raumes. *Studia Math.*, *5* (1934), 29–33.

KÖTHE, G.

[1] Die Stufenräume, eine einfache Klasse linearer vollkommener Räume. *Math. Z.*, *51* (1948), 317–345.

[2] Über die Vollständigkeit einer Klasse lokalkonvexer Räume. *Math. Z.*, *52* (1950), 627–630.

[3] Über zwei Sätze von Banach. *Math. Z.*, *53* (1950), 203–209.

[4] Neubegründung der Theorie der vollkommenen Räume. *Math. Nachr.*, *4* (1951), 70–80.

[5] Topologische lineare Räume. I. Berlin-Göttingen-Heidelberg 1960.

KOMURA, Y.

[1] Some examples on linear topological spaces. *Math. Ann.*, *153* (1964), 150–162.

KREIN, M. (See also Grosberg, J. and M. Krein.)

[1] Sur quelques questions de la géometrie des ensembles convexes situés dans un espace linéaire normé et complet. *C. R. (Doklady) Acad. Sci. URSS (N.S.), 14* (1937), 5–8.

[2] Propriétés fondamentales des ensembles coniques normaux dans l'espace de Banach. *C. R. (Doklady) Acad. Sci. URSS (N.S.), 28* (1940), 13-17.

KREIN, M. AND D. MILMAN

[1] On extreme points of regular convex sets. *Studia Math., 9* (1940), 133–138.

KREIN, M. G. AND M. A. RUTMAN

[1] Linear operators leaving invariant a cone in a Banach space. *Uspehi Mat. Nauk (N.S.) 3, no. 1 (23)* (1948), 3–95. (Russian). Also *Amer. Math. Soc. Transl. no. 26* (1950).

KREIN, M. AND V. ŠMULIAN

[1] On regularly convex sets in the space conjugate to a Banach space. *Ann. of Math. (2) 41* (1940), 556–583.

LANDSBERG, M.

[1] Pseudonormen in der Theorie der linearen topologischen Räume. *Math. Nachr., 14* (1955), 29–38.

[2] Lineare topologische Räume, die nicht lokalkonvex sind. *Math. Z., 65* (1956), 104–112.

MACKEY, G. W.

[1] On infinite dimensional linear spaces. *Proc. Nat. Acad. Sci. U.S.A., 29* (1943), 216–221.

[2] On convex topological linear spaces. *Proc. Nat. Acad. Sci. U.S.A., 29* (1943), 315–319.

[3] Equivalence of a problem in measure theory to a problem in the theory of vector lattices. *Bull. Amer. Math. Soc., 50* (1944), 719–722.

[4] On infinite-dimensional linear spaces. *Trans. Amer. Math. Soc., 57* (1945), 155–207.

[5] On convex topological linear spaces. *Trans. Amer. Math. Soc., 60* (1946), 519–537.

MAC LANE, S. (See Birkhoff, G., and S. Mac Lane.)

MAHOWALD, M.

[1] Barrelled spaces and the closed graph theorem. *J. London Math. Soc., 36* (1961), 108–110.

MARTINEAU, A.

[1] Sur une propriété caractéristique d'un produit de droites. *Arch. Math., 11* (1960), 423–426.

MAURIN, K.

[1] Abbildungen vom Hilbert-Schmidtschen Typus und ihre Anwendungen. *Math. Scand., 9* (1961), 359–371.

MILMAN, D. P. (See also Krein, M., and D. P. Milman.)

[1] Characteristics of extreme points of regularly convex sets. *Dokl. Akad. Nauk. SSSR (N.S.), 57* (1947), 119–122.

NACHBIN, L.

[1] Topological vector spaces of continuous functions. *Proc. Nat. Acad. Sci. U.S.A.*, *40* (1954), 471–474.

NAMIOKA, I. (See also Kelley, J. L., I. Namioka, and co-authors.)

[1] Partially ordered linear topological spaces. *Mem. Amer. Math. Soc. no. 24* (1957).

[2] A substitute for Lebesgue's bounded convergence theorem. *Proc. Amer. Math. Soc. 12* (1961), 713–716.

V. NEUMANN, J.

[1] On complete topological linear spaces. *Trans. Amer. Math. Soc.*, *37* (1935), 1–20.

PERESSINI, A. L.

[1] On topologies in ordered vector spaces. *Math. Ann.*, *144* (1961), 199–223.

[2] Concerning the order structure of Köthe sequence spaces. *Michigan Math. J. 10* (1963), 409–415.

[3] A note on abstract (M)-spaces. *Illinois J. Math.*, *7* (1963), 118–120.

PERRON, O.

[1] Zur Theorie der Matrices. *Math. Ann.*, *64* (1907), 248–263.

PHILLIPS, R. S. (See Hille, E. and R. S. Phillips.)

PIETSCH, A.

[1] Unbedingte und absolute Summierbarkeit in F-Räumen. *Math. Nachr.*, *23* (1961), 215–222.

[2] Verallgemeinerte vollkommene Folgenräume. Berlin 1962.

[3] Zur Theorie der topologischen Tensorprodukte. *Math. Nachr. 25* (1963), 19–30.

[4] Eine neue Charakterisierung der nuklearen lokalkonvexen Räume I. *Math. Nachr.*, *25* (1963), 31–36.

[5] Eine neue Charakterisierung der nuklearen lokalkonvexen Räume II. *Math. Nachr.*, *25* (1963), 49–58.

[6] Absolut summierende Abbildungen in lokalkonvexen Räumen. *Math. Nachr.*, *27* (1963), 77–103.

[7] Zur Fredholmschen Theorie in lokalkonvexen Räumen. *Studia Math.*, *22* (1963), 161–179.

[8] Nukleare lokalkonvexe Räume. Berlin 1965.

POULSEN, E. T.

[1] Convex sets with dense extreme points. *Amer. Math. Monthly*, *66* (1959), 577–578.

PTAK, V.

[1] On complete topological linear spaces. *Czechoslovak. Math. J.*, *3 (78)* (1953), 301–364.

[2] Compact subsets of convex topological linear spaces. *Czechoslovak. Math. J.*, *4 (79)* (1954), 51–74.

[3] Weak compactness in convex topological linear spaces. *Czechoslovak. Math. J.*, *4 (79)* (1954), 175–186.

[4] On a theorem of W. F. Eberlein. *Studia Math.*, *14* (1954), 276–284.

[6] Two remarks on weak compactness. *Czechoslovak. Math. J.*, 5 (80) (1955), 532–545.

[6] Completeness and the open mapping theorem. *Bull. Soc. Math. France, 86* (1958), 41–74.

[7] A combinatorial lemma on the existence of convex means and its application to weak compactness. *Proc. Sympos. Pure Math., Vol. VII.* Convexity, 437–450. Providence, R. I. 1963.

RIESZ, F.

[1] Sur quelques notions fondamentales dans la théorie générale des opérations linéaires. *Ann. of Math.*, (2) 41 (1941), 174–206.

ROBERTSON, A. AND W. ROBERTSON

[1] On the closed graph theorem. *Proc. Glasgow Math. Assoc.*, 3 (1956), 9–12.

[2] Topological vector spaces. Cambridge 1964.

ROBERTSON, W.

[1] Contributions to the general theory of linear topological spaces. Thesis, Cambridge 1954.

ROGERS, C. A. (See Dvoretzky, A. and C. A. Rogers.)

ROTA, G.-C.

[1] On the eigenvalues of positive operators. *Bull. Amer. Math. Soc.*, 67 (1961), 556–558.

RUTMAN, M. A. (See Krein, M. G. and M. A. Rutman.)

SCHAEFER, H. H.

[1] Positive Transformationen in lokalkonvexen halbgeordneten Vektorräumen. *Math. Ann.*, 129 (1955), 323–329.

[2] Halbgeordnete lokalkonvexe Vektorräume. *Math. Ann.*, 135 (1958), 115–141.

[3] Halbgeordnete lokalkonvexe Vektorräume. II. *Math. Ann.*, 138 (1959), 259–286.

[4] Halbgeordnete lokalkonvexe Vektorräume. III. *Math. Ann.*, 141 (1960), 113–142.

[5] On the completeness of topological vector lattices. *Michigan Math. J.*, 7 (1960), 303–309.

[6] Some spectral properties of positive linear operators. *Pacific J. Math., 10* (1960), 1009–1019.

[7] On the singularities of an analytic function with values in a Banach space. *Arch. Math. 11* (1960), 40–43.

[8] Spectral measures in locally convex algebras. *Acta Math., 107* (1962), 125–173.

[9] Convex cones and spectral theory. *Proc. Sympos. Pure Math., Vol. VII.* Convexity, 451–471. Providence, R. I. 1963.

[10] Spektraleigenschaften positiver Operatoren. *Math. Z.*, 82 (1963), 303–313.

[11] On the point spectrum of positive operators. *Proc. Amer. Math. Soc., 15* (1964), 56–60.

SCHAEFER, H. H. AND B. J. WALSH

[1] Spectral operators in spaces of distributions. *Bull. Amer. Math. Soc., 68* (1962), 509–511.

SCHATTEN, R.
[1] A theory of cross spaces. *Ann. Math. Studies no. 26* (1960).
[2] Norm ideals of completely continuous operators. Berlin-Göttingen-Heidelberg 1960.

SCHAUDER, J.
[1] Zur Theorie stetiger Abbildungen in Funktionalräumen. *Math. Z.*, *26* (1927), 47–65 und 417–431.
[2] Über lineare stetige Funktionaloperationen. *Studia Math.*, *2* (1930), 183–196.

SCHWARTZ, J. T. (See Dunford, N. and J. T. Schwartz.)

SCHWARTZ, L. (See also Dieudonné, J. et L. Schwartz.)
[1] Théorie des distributions. Tome I. 2nd ed. Paris 1957.
[2] Théorie des distributions. Tome II. 2nd ed. Paris 1959.

SHIROTA, T.
[1] On locally convex vector spaces of continuous functions. *Proc. Jap. Acad.*, *30* (1954), 294–298.

ŠMULIAN, V. L. (See also Krein, M. and V. Šmulian.)
[1] Sur les ensembles faiblement compacts dans les espaces linéaires normés. *Comm. Inst. Sci. Mat. Mec. Univ. Charkov (4)*, *14* (1937), 239–242.
[2] Sur les ensembles régulièrement fermés et faiblement compacts dans les espaces du type (B). *C. R. (Doklady) Acad. Sci. URSS (N.S.) 18*, (1938), 405–407
[3] Über lineare topologische Räume. *Mat. Sbornik N.S.*, *7 (49)* (1940), 425–448.

STEINHAUS, H. (See Banach, S. et H. Steinhaus.)

STONE, M. H.
[1] The generalized Weierstrass approximation theorem. *Math. Mag. 21* (1948), 167–183 and 237–254.

SZ.-NAGY, B.
[1] Spektraldarstellung linearer Transformationen des Hilbertschen Räumes. Berlin-Göttingen-Heidelberg 1942.

TAYLOR, A. E.
[1] Introduction to functional analysis. New York 1958.

TYCHONOFF, A.
[1] Ein Fixpunktsatz. *Math. Ann.*, *111* (1935), 767–776.

WALSH, B. J. (See Schaefer, H. H. and B. J. Walsh.)

WEHAUSEN, J. V.
[1] Transformations in linear topological spaces. *Duke Math. J.*, *4* (1938), 157–169.

YOOD, B. (See Civin, P. and B. Yood.)

Index

Graduate Texts in Mathematics

For information

A student approaching mathematical research is often discouraged by the sheer volume of the literature and the long history of the subject, even when the actual problems are readily understandable. The new series, Graduate Texts in Mathematics, is intended to bridge the gap between passive study and creative understanding; it offers introductions on a suitably advanced level to areas of current research. These introductions are neither complete surveys, nor brief accounts of the latest results only. They are textbooks carefully designed as teaching aids; the purpose of the authors is, in every case, to highlight the characteristic features of the theory.

Graduate Texts in Mathematics can serve as the basis for advanced courses. They can be either the main or subsidiary sources for seminars, and they can be used for private study. Their guiding principle is to convince the student that mathematics is a living science.